Practical Guide
to Industrial
Boiler Systems

MECHANICAL ENGINEERING
A Series of Textbooks and Reference Books

Founding Editor

L. L. Faulkner

*Columbus Division, Battelle Memorial Institute
and Department of Mechanical Engineering
The Ohio State University
Columbus, Ohio*

Additional Volumes in Preparation

Mechanical Engineering Software

Practical Guide to Industrial Boiler Systems

Ralph L. Vandagriff
Consultant—Steam and Power
North Little Rock, Arkansas

CRC Press
Taylor & Francis Group
Boca Raton London New York

CRC Press is an imprint of the
Taylor & Francis Group, an **informa** business

CRC Press
Taylor & Francis Group
6000 Broken Sound Parkway NW, Suite 300
Boca Raton, FL 33487-2742

© 2001 by Taylor & Francis Group, LLC
CRC Press is an imprint of Taylor & Francis Group, an Informa business

No claim to original U.S. Government works

**Visit the Taylor & Francis Web site at
http://www.taylorandfrancis.com**

**and the CRC Press Web site at
http://www.crcpress.com**

To

My mother, Martha Louise Lunday Vandagriff,
Native American of the Delaware Nation

My father, Ralph B. Vandagriff,
a true gentleman

My wife, Sue Chapman Vandagriff,
who has put up with me for over 45 years

Thank you for your love, help, and guidance.
Thank you for teaching me about God and how to trust in Him.

Preface

Much time was spent in researching data in the 35-plus years of my involvement in boiler house work. This text is a compilation of most of that data and information. The purpose of this book is to make the day-to-day boiler house work easier for the power engineer, the operators, and the maintenance people, by supplying a single source for hard-to-find information.

Nontechnical people with an interest in boiler house operation include plant management personnel, safety personnel, and supervisory personnel in government and industry. The technical material in this book, including the spreadsheet calculations and formulas, should be of interest to the boiler engineer, boiler designer, boiler operator, and the power engineering student.

Ralph L. Vandagriff
North Little Rock, Arkansas

Contents

Requirements of a Perfect Steam Boiler

1. Proper workmanship and simple construction, using materials which experience has shown to be the best, thus avoiding the necessity of early repairs.
2. A mud drum to receive all impurities deposited from the water, and so placed as to be removed from the action of the fire.
3. A steam and water capacity sufficient to prevent any fluctuation in steam pressure or water level.
4. A water surface for the disengagement of the steam from the water, of sufficient extent to prevent foaming.
5. A constant and thorough circulation of water throughout the boiler, so as to maintain all parts at the same temperature.
6. The water space divided into sections so arranged that, should any section fail, no general explosion can occur and the destructive effects will be confined to the escape of the contents. Large and free passages between the different sections to equalize the water line and pressure in all.
7. A great excess of strength over any legitimate strain, the boiler being so constructed as to be free from strains due to unequal expansion, and, if possible, to avoid joints exposed to the direct action of the fire.
8. A combustion chamber so arranged that the combustion of the gases started in the furnace may be completed before the gases escape to the chimney.
9. The heating surface as nearly as possible at right angles to the currents of heated gases, so as to break up the currents and extract the entire available heat from the gases.
10. All parts readily accessible for cleaning and repairs. This is a point of the greatest importance as regards safety and economy.
11. Proportioned for the work to be done, and capable of working to its full rated capacity with the highest economy.
12. Equipped with the very best gauges, safety valves, and other fixtures.

Source: List prepared by George H. Babcock and Stephen Wilcox, in 1875 [31].

Tables and Spreadsheets

Practical Guide to Industrial Boiler Systems

1

Experience

Design Notes; Boiler Operation and Maintenance; Experience.

I. DESIGN NOTES

A. Industrial Power Plant Design*

It is not the intent to go into the matter of steam power plant design in any detail, but merely to indicate a few points that come up during the course of the study, to give a little flavor of the kinds of practical considerations that must be taken into account.

1. Steam Piping

High process steam pressures are costly in terms of by-product power generation. Failure to increase steam pipe sizes as loads increase results in greater pressure drops, which can lead to demands for higher pressures than are really needed. This reduces the economy of power generation and can introduce serious temperature-control problems as well.

2. Plant Location

If a new steam and power installation is being put in, careful consideration should be given to its location in relation to the largest steam loads. Long steam lines are very expensive and can result in pressure and temperature losses that penalize power production.

---

* Extract from Seminar Presentation, 1982. Courtesy of W. B. Butler, retired Chief Power Plant Superintendent and Chief Power Engineer for Dow Chemical Co., Midland, Michigan. (Deceased)

3. Boiler Steam Drum

Although many field-erected boilers are custom designed, considerable engineering is required, and experienced personnel are scarce. A known and proven design can be offered for much less than a corresponding special design. A boiler-maker might be asked, for example, for a 200,000-lb/hr boiler of 600 psi working steam pressure. He may have a proven design for a 300,000-lb/hr boiler of 700 psi working steam pressure that would fill the requirements, so he might build according to that design and stamp the drum according to the customer's order. If so, the customer is losing an opportunity for additional economical power generation, so he should explore this possibility before the drum is stamped and the data sheets submitted to the national board. Also, the proper size safety valve nozzles must be installed before the drum is stress relieved.

4. Steam Turbine Sizing

The ratio of steam pressure entering the turbine to that leaving should be at least 4:1 for reasonable turbine efficiency, and as much higher as feasible on other grounds. For example, assume our usual boiler conditions of 900 psi and 825°F, and a process steam requirement of 400,000 lb/hr. If the process steam pressure is 150 psi, about 21.2 MW of gross by-product power generation is possible. If the process steam pressure is 300 psi, this drops to near 14.4 MW.

5. Turbine Manufacturers

Turbine manufacturers may use the same frame for several sizes and capacities, especially in the smaller sizes, which will be sufficiently designed to withstand the highest pressure for which it will be used. Many turbine frames have extraction nozzles for feedwater heating, which are merely blanked off if not required. Knowing the practices of the selected turbine manufacturer, here, can help obtain the most for the money.

6. Stand-Alone Generation

If self-generation is installed in an industrial plant with the idea of becoming independent of the local utility, some thought should be given to auxiliary drives in event of a power failure, momentary or longer. If the auxiliaries are electrically driven, they should have mechanically "latched in" or permanent magnet starters to prevent many false trip-outs.

7. Auxiliary Steam Turbine Drives

Steam turbine drives for auxiliaries have a number of advantages besides alleviating some problems during shutdowns and start-ups. They do require special maintenance, however. The advantages of turbine drives elsewhere throughout the plant should also be explored once it is planned to have higher-pressure steam available.

8. Deaerating Feedwater Heater

Many small steam plants have become extinct owing to boiler and condensate system corrosion problems that could have been prevented with a good deaerating heater.

9. Synchronous Generators and Motors

Synchronous generators and synchronous motors have the capability of feeding as much as ten times their rated maximum currents into a fault or short circuit. The impact is capable of breaking foundation bolts, shearing generator shaft coupling keys, tearing out windings, and exploding oil circuit breakers. Precautions include installing breakers of adequate interrupting capacity, installing current-limiting reactors in the armature circuit, using a transformer to change the generator voltage and limit short-circuit fault currents with its impedance, and using separate breakers and external circuits for the separate windings of the generator.

10. Unbalanced Loads

Electric loads leading to unbalanced circuits should be avoided or, at most, be a small fraction of the total load. As much as 10% unbalance between phases can be troublesome. A large unbalanced load on a small generator will usually cause serious damage to the field coil insulation by pounding it from one side of the slot to the other. A small industrial power plant should never attempt to serve such a load as a large single-phase arc furnace, no matter how economically attractive it might appear.

11. Cogeneration Problem Areas

Many of the problems that will need to be considered will be specific to the individual case, and only some of the more general ones will be mentioned. The listing is illustrative rather than comprehensive.

a. Management Philosophy.   The attitude and policies of the management of the industrial concern involved can be a key factor. Those with policies and experience favoring backward integration into raw materials would not have much trouble with the idea of generating their own power. On the other hand, a management (perhaps even in the same industry) whose policy has been not to make anything they can buy, short of their finished products for sale, might well say, "We're not in the power business and we're not going into the power business as long as we can buy from the utility." In such a case, return on investment is of little consequence. Examples can be found in the automotive industry, the chemical industry, and doubtless others.

The influence of management philosophy can also extend into the operation and maintenance of the steam and power plant, which has its own characteristics and needs. The steam and power plant should be considered a key and integral part of the manufacturing system and not just a necessary evil. Failure to do this can lead to injudicious decisions or demands that accommodate manufacturing at the price of serious or even disastrous trouble later on.

b. Return on Investment. Standards for acceptable return on investment (ROI) will differ, and the 20% ROI used in this study is intended only as a typical average figure. A rapidly growing company having trouble raising capital for expanding its primary business, for example, could well set its sights higher.

c. Difference in Useful Plant Life. A difference in time scales needs to be realized and reconciled. Many manufacturing processes or major equipment installations become obsolete and are replaced or changed after perhaps 10 or 12 years. The useful life of a power plant is probably closer to 30 years, and this must be considered in making the investment commitment. Along the same vein, any substantial shift toward coal as a boiler fuel (which seems almost inevitable at this time) will require opening new mines, as it is quite evident that this will necessitate commitment to long-term purchase contracts. Many products have shorter lifetimes than the periods just mentioned.

d. Outage. A workable, economic solution to many total-energy problems may seem easy until the question is asked, "What do we do when this generator is out of service?" Two weeks of outage in a year is a reasonable estimate for a well-maintained steam-powered system. Under favorable conditions, this maintenance period can be scheduled; many industries also require such periodic maintenance. Some industries can easily be shut down as needed, but others, however, would sustain significant losses if forced to shut down. Stand-by power can be very expensive, whether generated in spare equipment or contracted for from the local utility.

Consideration should also be given to a similar problem on a shorter time scale. Small power plants using gas, oil, or pulverized coal firing are subject to codes such as National Fire Protection Association (NFPA) and others to prevent explosive fuel–air mixtures in boiler furnaces. One measure usually required is a prolonged purge cycle through which the draft fans must be operated before any fuel can be introduced into the furnace. A 5-min purge can be tolerated in a heating or process steam boiler. A flameout, and the required purge in a power boiler serving a loaded turbine–generator, will usually result in a loss of the electrical load. Whether or not this can be tolerated for the type of manufacturing involved should be studied before undertaking power generation.

e. Selling Power to the Utility. If power is to be generated for sale, the attitude of the utility's management also becomes an important factor. Most utilities have strongly discouraged the private generation of power in the past, and old habits and policies sometimes die hard in any industrial organization. Wheeling of power through utility transmission lines has been acceptable to some, although usually only on behalf of another investor-owned utility, and unacceptable to others. Where policies have discouraged these practices in the past, there will have been little experience to shape relationships in the future, and it would be natural for many utilities to begin with a tighter control over industrial power generation than might be necessary in the long run. Each industrial concern must consider the effect on and compatibility with their own patterns of operation, production schedules, load curves, and similar items.

B. Wood-Fired Cogeneration*

1. Fuel Preparation and Handling

Initially, remove all tramp iron from the fuel material before entering the hammer mill or pulverizer by use of a properly placed electromagnet. This is considerably more expensive than use of a metal detector to trip the feed conveyor system; however, a detector alone requires an operator to search for the piece of metal and to restart the conveyor system.

In general, design the conveyor system for free-flowing drop chutes and storage bins. Almost any necked-down storage bins or silos are certain to bridge or hang-up. Wood chips and bark, when left in place, will generate heat (owing to moisture content) and will set up to an almost immovable solid mass.

2. Boiler Unit

Make sure that the furnace and boiler heat-exchange surfaces are designed for the fuel being fired and in accordance with standard boiler design criteria. Provide excess capacity so that the boiler does not have to operate at a wide-open condition.

3. I.D. Fan and Boiler Feedwater Pump

These two items are the heart of any boiler plant. Alone, they can amount to 70% of the power requirement for the total plant. Select equipment that has the best efficiency. Design ductwork and breeching for minimum resistance to flow to reduce the I.D. fan static pressure requirements.

Check the boiler feedwater pump-operating curve pressure at a low or cutoff flow point. This pressure will be higher than at the normal operating condition (could be considerably higher depending on pump selection or flatness of curve). Make sure that all piping components will handle the increased pressure.

---

* Grady L. Martin, P.E. General Considerations for Design of Waste Fuel Power Plants.

4. Boiler Feedwater Controls

As boiler drum pressure swings with steam consumption or load swings, the drum water level swells (at reduced pressure) owing to gases in the boiler water volume. Make sure that the feedwater level control is capable of overriding these swells.

5. Safety Relief Valves

Make sure that all safety valves are securely anchored for reaction jet forces. The pipe stub to which a valve is mounted can bend and cause damage or injury if not externally supported.

6. Stack Emissions Monitoring

There are strict Environmental Protection Agency (EPA) requirements for emissions monitoring. This is a major cost item involving expensive specialized instrumentation (in the 75,000–100,000-dollar range). Carefully check all EPA requirements at the project beginning.

7. Dust Collection Equipment '

This is the same situation as in Section I.B.6. Carefully check the EPA requirements at the beginning of the project.

8. Ash Handling

Select equipment and design the system to control and to contain all dust.

9. Water Treatment

Water treatment is a specialty that is usually done by a water treatment chemical company. They will provide a turnkey installation if desired. Provide equipment and storage tanks for handling large amounts of hydrochloric acid and sodium hydroxide for use in regeneration of demineralizers in the water treatment plant. This usage involves truckload quantities.

10. Control System

Provide flowmeters, pressure indicators, and temperature indicators with recorders for same at all separate flow points in the total boiler system. There will be upset conditions and tripouts during operation. The complete recorded information will help determine the source and cause of a problem.

11. Cooling Tower

Provide adequate bleed-off drainage point and fresh water makeup source. Drainage must be to an EPA-permitted location. Cooling tower water will cloud-up owing to concentration of solids. Drift water from the tower can be a major nuisance if allowed to settle on car windows or other surfaces.

C. Problems Corrected

1. Packaged Boilers [Experience of Gene Doyle, Chief Field Service Engineer, Erie City Energy Div., Zurn Industries.]

Unit: 160,000 lb/hr, 850 psig, 825°F, natural gas and No. 6 fuel oil, continuous operation.

Problem: Superheat temperature was erratic or was low.

Solution: After numerous trips to plant site and rigorous inspection of the boiler in operation, it was found that the contractor erecting the boiler had piped the fuel oil steam atomizing line to the superheater header instead of the plant steam system of 160 psig saturated. Consequently, the flame length of the unit when firing No. 6 oil, was only half as long as it should be. After repiping the atomizing steam line to the plant saturated steam system, the superheat temperature went up and stabilized, the problem was corrected and the boiler performed as it was supposed to.

2. Field Erected Boilers [Experience of Ralph L. Vandagriff, Consultant]

Unit: 14,000-lb/hr hybrid boiler, underfeed stoker, 315 psig saturated, 6% moisture content furniture plant waste, continuous operation.

Problem: After completion of unit and during acceptance testing, unit would not meet steaming capacity, pressure, and emissions all at the same time. Especially not for the 8 hr required in the acceptance test section of the purchase contract.

Solution: The ash from the cyclones was tested and found to contain 76% pure carbon. It became obvious that the furnace section of the boiler was not large enough. Calculations were made that determined that the firebox had less than 1 sec retention time and needed to be increased in height by 42 in. This was done and the unit performed satisfactorily. *Note*: The hybrid boiler is a unit consisting of a waterwall enclosed furnace area with refractory inside the walls, part of the way up the waterwalls. Then the hot combustion gases go through a horizontal tube section and to the dust collectors. The heated water from the waterwalls feeds the horizontal tube section which has a steaming area in the top of its drum. This particular unit fed steam to a backpressure turbine generator system, when the dry kilns were running, and to a condensing turbine generator system when the dry kilns did not need the steam. Maximum of 535 KW generated.

II. BOILER OPERATION AND MAINTENANCE

A. Boiler Operator Training Notes and Experience: Instructors Guide [Courtesy of Lee King, Field Services, RENTECH Boiler Services, Abilene, Texas]

The following guide is for instruction of operators and maintenance personnel in safety, preventive maintenance, operation of the boiler(s) and equipment, troubleshooting, and calibration of their specific boiler equipment.

Instruction is given for day-to-day operation and procedural checks and inspection of the equipment. The hope is that the operators will acquire information to equip themselves with the tools to keep the equipment and the facility in which they work in good operating condition.

B. Training Program

- I. Safety
  - A. General
    1. Boiler equipment room
    2. Pump equipment room
  - B. Chemical
    1. Boiler equipment room
    2. Pump equipment room
  - C. Electrical
    1. Boiler equipment room
    2. Pump equipment room
  - D. Gas, oil, and air
    1. Boiler equipment room
- II. Preventive Maintenance
  - A. Boiler
    1. Internal
    2. External
  - B. Controls
    1. Electrical
    2. Mechanical
  - C. Steam appliances
    1. Safety relief valves
    2. Blowdown valves
    3. Isolation valves

III. Boiler Operation
  A. Prestart check
    1. Valve line up
      a. Steam
      b. Fuel (gas and oil)
      c. Fuel oil levels
    2. Electrical
      a. Main
      b. Control
    3. Safety resets
      a. Fuel
      b. Limits
      c. Electrical
    4. Water
      a. Levels pumps
      b. Chemicals
  B. Start-up
    1. Ignition
      a. Pilot check (gas and oil)
      b. Main flame check (gas and oil)
    2. Run cycle
      a. Flame condition
      b. Controls levels
  C. Normal operation
    1. Temperature: stack
    2. Pressure: steam
    3. Water level(s)
    4. Fuel: levels and pressure
    5. Blowdown
    6. Stories of mishaps
  D. Shutdown
    1. Normal
      a. Secure valves
      b. Secure fuel(s)
      c. Secure electrical
    2. Emergency
      a. Secure valves
      b. Secure fuel(s)
      c. Secure electrical
    3. Long term
IV. Troubleshooting
  A. Electrical

1. Preignition interlocks
2. Running interlocks
3. Level control(s)
   B. Mechanical
   1. Linkage rods
   2. Valves
   3. Louvers
   4. Filters
   5. Orifices: pilot and gas
   6. Oil nozzle
V. Calibration
   A. Gauges
   1. Steam
   2. Temperature
   3. Gas
   4. Oil
   B. Controls
   1. Operating
   2. Limit
   3. Level
   C. Burner
   1. Gas
   2. Oil
   D. Pumps
   1. Water supply
   2. Fuel supply
VI. Daily, Monthly, and Yearly Inspections
   A. Daily inspections
   1. Operating controls
   2. Water levels
   3. Boiler firing
   B. Weekly inspection
   1. Controls
   2. Levels (water, oil, etc.)
   3. O_2 and CO settings
   4. Filters
   C. Monthly inspection
   1. Safety relief valves (pop-offs)
   2. Blow-down operations
   3. Fireside gaskets
   4. Waterside gaskets

D. Yearly inspections
   1. Open, clean and close fireside
   2. Open waterside
      a. Manways
      b. Handholes
      c. Plugs
   3. Open burner
      a. Filters
      b. Louvers
      c. Valves
      d. Ignitor(s)
      e. Wiring
      f. Forced draft fan

VII. Summary
   A. What and when to replace
      1. Bi-annually
      2. Yearly

1. Safety

a. General Safety. As we are all aware, being operators and maintainers of equipment, it is to everyone's benefit to be safety conscious. Your company should have a safety policy, or safety guidelines to follow. Some of the things that we want to be aware of are the common things we may forget from time to time.

We should make a habit of wearing safety glasses or safety goggles where required; ear plugs where required (OSHA guidelines and/or decibel testing); safety shoes, boots, or safety rubber boots; long-sleeved shirts and long pants; also rubber gloves when required. Kidney belts are also required by OSHA or company guidelines when lifting by hand. There may also be a weight limit for lifting objects by hand. Check with you safety engineer or supervisor if you are not sure. Hard hats or bump hats may also be required headgear.

When entering the boiler room or mechanical area, pay attention to all safety warning signs. These may include "Hearing Protection Required," "Hard Hat Area," "Safety Glasses Required," or others. Be on the lookout for safety or warning signs that say "No Smoking in this Area," "High Voltage," "Chemicals," "Flammable Liquids," "Gases," or others.

You should be aware of your surroundings in the mechanical room. Use your senses. You want to look, hear, and smell. A steam leak can be a cause of severe burn or even death. You never know when water, oil, or a chemical has either been spilled or has leaked out of a container. Gas leaks are not always easy to find. Natural gas leaks can cause explosions and fires, which can cause serious injury or death.

b. Chemical Safety. Chemicals in the mechanical or boiler room areas are necessary because of the need for water treatment, descaling, solvents for oils, and so on. One of the first things you should know about chemicals is the labeling of the chemical and what the labeling means. Become familiar with and read all labeled chemicals and materials for "Warnings." All chemicals are required to have information (minimum) listing the following: ingredients, hazards, first aid and disposal procedures. Material safety data sheet (MSDS) information should also be posted in an area accessible to personnel for their review. If you are unsure of a chemical, do not use or open it until you know what you are dealing with. You should have protective equipment such as goggles, face shield, rubber gloves, rubber apron, rubber shoes, and mask. Some chemicals may not be toxic but may be CORROSIVE. If you do not know what a chemical or liquid is, do not mess with it. (Use common sense) until you can determine what it is and take the necessary precautions for use, removal, clean up, or disposal. Keep all empty containers stored in their designated places. Keep all containers tightly closed and covered and properly labeled. Do not change containers without proper labeling.

If chemicals and chemical equipment are supplied and maintained by a "Chemical Company," make sure they supply all required information on the equipment and chemicals even though they may be maintaining the equipment and chemical for you. (See discussion in Section VI). When using spray cleaners and chemicals, do not use around electrical equipment. Do not discard chemicals down drains. Always follow EPA guidelines for removal and disposal of chemicals. (Ask trainees for questions on chemical safety before continuing.)

c. Electrical Safety. Electrical safety in the boiler and mechanical areas is essential. Caution and common sense around electricity should always be observed. Untrained personnel should be oriented and trained before any introduction to electrical components. We as professional maintainers and operators should be constantly aware of the dangers and possible hazards of electrical equipment. Wiring that has been wet can cause short circuits, major malfunctions, explosions, severe injury, and even death. (Illustrate lax electrical safety, use a story about electrical hazards to drive home the point or near miss of injury). Any person can become lax about electrical safety. Most people are aware that high voltage is very dangerous, but forget about everyday electrical current, such as 110/120-V electricity. Even 24 V electricity can be deadly.

When working on electrical appliances or trouble-shooting electrical controls always use proper tools and properly insulated tools and protective clothing, such as rubber-soled footwear and gloves. Make sure all equipment is shutdown and all circuits are disconnected (or fuses pulled) before working on the equipment. Lock and tag-out all equipment. If using a team or buddy system, do not assume anything or take your team member for granted. Any one can make an error and a small error can be deadly. The main thing is to "work and be safe."

(Relate another story about buddy or team safety.)

Ask for questions before continuing.

d. Gas, Oil, and Air. When we talk about gas such as "natural gas," we do not pay too much attention because it is in our everyday lives and hardly ever dealt with. It remains inside piping and well hidden from exposure to us. The fact is, gas (natural gas) is colorless and odorless and *very deadly*. Natural gas will not ignite normally unless it is introduced to air or oxygen and ignition or a spark from a source. This is where we get the term "combustion." For our discussions on combustion and gas, combustion can be very dangerous unless properly controlled. We are concerned with *uncontrolled combustion*.

When operating, maintaining, or trouble-shooting a boiler with gas or oil fuel, always look for leaking valves and fittings, and for proper boiler firing. Check for proper pressures, and if leaks are found in gas, oil, or air lines, properly locate and mark the leaks. If necessary or required, shutdown the equipment as soon as possible or practical and make repairs or notify the proper personnel to make the repair(s). Oil such as No. 2 fuel oil can also be hazardous, even lying on the floor. Clean all fuel oil and oil spills, repair the source of the leak as soon as possible or practical. Use absorbent for clean up and removal of spilled oil and discard according to EPA and OSHA requirements. Large fuel oil spills should be dealt with immediately, as fuel oil is highly volatile. Compressed air can be dangerous also. Introduced to a fuel source helps complete the combustion. It would only need a spark to cause ignition of some kind. One of the most common dangers of compressed air is using it to blow out or clean equipment. Your eyes are the most likely target of a propelled particle. Always use proper safety equipment when using compressed air and approved air too. Make a practice of not using modified air tools.

(Use story or personal experience with any gas, oil or compressed air hazard for an example of safety)

Ask question of class before continuing.

Movie: Safety in the work place.

2.   Preventive Maintenance

a. Boiler. When discussing preventative maintenance on the boiler and mechanical room equipment, we want to do our best to keep the equipment running and avoid nuisance shutdowns or even major breakdowns. Boiler owners and operators have been striving for years to reduce costs of major rebuilding, replacement, and equipment repairs. We will discuss measures to help assure long-term operation with minimum cost.

Waterside. On the boiler "waterside," after shutdown and isolation of the boiler, let the unit cool from steaming temperature to below 200°F before draining of the boiler. This needs to be done naturally and not by means of induced air or cold water. Use of either induced air or cold water to reduce boiler temperature, can cause boiler and refractory damage and lead to major repairs for tube shrinkage or undue metal stress. Let the boiler cool to 140°F or lower before removing hand-hole or manway covers. OSHA standards and requirements for "personnel protection" are greater than 140°F. Remove all plugs on the water column(s) and low-water cutout piping and tees. Flush out with water to remove debris. Also remove the low-water cutout control head and flush with water to remove debris from the bowl or cavity. If any sludge or scale buildup is evident, scrape and flush out. Make sure to flush the drain piping on the waterside of the boiler. Flush, using high-pressure water. Remove all debris by scraping and flushing. If feedwater chemical treatment is working well, you should have soft sludge in the bottom of the mud drum on waterwall boilers and the bottom of the steam drum on firetube boilers. Make sure that the bottom blowdown opening is flushed and clear of debris. Special areas of attention on firetube boilers are the rear tubesheet and tube-to-tubesheet connections, tubesheet-to-fire tube area, fire tubes, boiler shell, and shell bottom. Also the water feed inlet baffle. *Note*: during this procedure the chemical representative should be there to observe and gather samples of the sludge, or other debris. This will give the representative a hands-on look at the boiler internals and will be important in future water treatment recommendations to you.

Fireside. After completing the internal waterside of the boiler, attention is turned to the fireside of the boiler. Open inspection doors (for firetube boilers, front and rear doors) for visual inspection and debris removal. You may encounter soot, red dust, scale, or dry chemical residue. If any of these residues are present, your boiler service representative should be called in to see the problem and fix it.

Example. Firetube boiler. If soot is present (if the firetubes have turbulators, remove them), brush out the firetubes and tubesheets (fireside) removing the soot. The burner then needs to be adjusted before returning to full service. If red dust is present, this means there may be a problem with fireside condensation. If scale or chemical residue is present, you may have leaking tube joints. In all these cases, your boiler service professional should be called in to identify and fix the problem. Complete the fireside inspection by visually inspecting the boiler tubes, tubesheets, furnace tube (Morrison tube) for damage or leaking areas and make any repairs needed. The burner cone refractory and refractory on the front and rear doors (refractory in the furnace) should be inspected and patch coated or replaced as needed. The jurisdiction inspector will note any repairs or replacement necessary to return the boiler back to good condition and return to service.

b. Controls. On the controls, remove waterside probes (such as LWCO Warrick probes) and inspect, clean, and reinstall or replace if necessary. Inspect all electromechanical controls for ruptured bellows (seals) and bare or frayed wiring, repair as necessary and replace their covers. Check all linkages, oil levels, and switches, where practical, for excessive wear or loose fittings and repair or replace as required. Remove and clean flame scanner or rectifier and reinstall. Check the packing on all valve stems and repair or replace as needed.

c. Appendages. Check all appendages such as safety relief valves (pull levers to check for frozen seats, and if valve seat is frozen, replace the valve). Check all blowdown valves, check shaft packing and replace if required. A large amount of the foregoing section should be taught by "hands-on." Use spare valves or illustrations, cut-aways or diagrams. Using an actual boiler, while out of service, is the best.

3. Boiler Operation

To begin, you need a standardized start-up, operation and shutdown check list available for each boiler and its related equipment.

Sample

Piping
——— Check that all valves are oriented in the proper flow direction.
——— Check linkages on all regulating devices, valves, and dampers.
——— Check that all metering devices have been replaced in accordance with recommendations.
——— Check all piping for leakage during the field hydrostatic test.
——— Check with owner's water treatment consultant to assure that feedwater and chemical feed piping arrangements are satisfactory.
——— Check that all flange bolting has been torqued to proper levels.
Vent and drain piping
——— Check all the drain and vent lines for obstructions or debris.
——— Check that all drain and vent lines terminate away from platforms and walkways.
Water columns
——— Check all connecting piping joints for leakage.
——— Check all safety and alarm system wiring.
——— Check isolation valves to be sure they are locked open.
Safety valves
——— Check for blockage on the outlet.
——— Check that all vent pipe supports have been installed in accordance with recommendations.
——— Verify all valves for manufacturer's settings (set pressures are shown on valve tag).
——— Verify that gags have been removed from all valves.

Others
——— Check proper alignment on all ducting and expansion joints.
——— Check all sliding pad installations to ensure proper movement.
——— Check that all normal service gaskets have been installed and have been properly torqued.

Summary of Valve Positions					
Valve	Shutdown	Hydro	Boil-out	Start-Up	Operating
——— Steam shutoff	Close	Close	Close	Open	Open
——— Steam stop/check	Close	Close	Close	Open	Open
——— Drum vent	Close	Close	Close	Open	Open
——— Feedwater control	Close	Close	Close	Close	Open
——— Feedwater control isolation valve	Close	Close	Close	Close	Open
——— Intermittent blow down	Close	Close	Intermittent	Close	Intermittent
——— Chemical feed	Close	Close	Close/open	Close	Open
——— Water column drain	Close	Close	Close	Close	Close
——— Water gauge drain	Close	Close	Close	Close	Close
——— Safety valve	Free	Gag	Free	Free	Free
——— Steam gauge shutoff	Open	Close	Open	Open	Open

The summary of valve positions are basic and standard for most boilers. The concern here is to have a *checklist* for start-up and operation.

The summary of valve positions includes positions for shutdown and lock-out when boiler is to be shut down for scheduled or nonscheduled work.

Line up your valves per the "summary of valve position." The drum vent is left open until you achieve approximately 5 psig steam. This drives out the oxygen from the boiler and water and helps prevent oxygen corrosion. Make sure all water makeup valves at the boiler, return and deaerator system are in open position. Also make sure you have water to the boiler feed pumps before starting the pumps. If you run the boiler feed pumps dry, it will more than likely mean expensive pump repairs. *Do not dry run the boiler feed water pumps.*

Line up the gas valves or oil valves to the burner. Check fuel oil level supply before starting. Check the fuel oil pump and make sure this pump does not dry start as it may cause expensive repairs. Check the air and oil filters and clean or replace them as needed. Check all electrical resets (i.e., BMS Control, High Limit, Air Switch, GP Switches, etc.).

Before starting the boiler, let us make one more trip around the unit to make sure everything is in place and we did not forget something. Check the boiler water level, water level gauge glass cocks, fireside door or furnace access bolts and nuts, and fire chamber sight glass. If any of these items are in need of repair, or glass is cracked, repair or replace before starting the unit.

b. Start-up. Push the RESET button on the boiler management system (BMS), set the firing rate control to manual, and set the rate on "0" or "minimum" position, turn the boiler control switch to ON. Switch the BMS RUN/CHECK switch to CHECK when the pilot/ignition starts. This allows the BMS to stay in ignition mode until you can check the pilot flame and scanner signal (or if initial start-up, perform pilot turndown test). Visually check the pilot to see if the flame is steady or separating from the pilot assembly. No separation should be seen. *Note.* The pilot flame should rotate approximately one-third the way around the burner face, although it is permissible to be as short as 6–8 in. The pilot pressure should be set per "factory recommendations." Now move the RUN/CHECK switch to run to start the main flame.

On dual-fuel burners make sure, if gas is the primary fuel, that calibration of the burner on gas is performed first, then set and calibrate on fuel oil. We will assume at this time that all calibrations of fuel and air ratio are correct. This will be discussed under "calibration."

Now that we have established the main flame and we have noted that the flame is stable, the boiler needs to warm up. Leave the boiler on low or minimum fire until *all refractory is dried out and hot.* On steam boilers, warm the boiler until you have reached approximately 5-psig steam pressure. Close the steam drum vent valve. Most of the oxygen will have been removed from the boiler water by this time. This will help assure that no oxygen corrosion takes place. Recheck all pressure and temperature gauges, boiler water level, and makeup or return tank levels. Now the boiler can be manually fired or ramped up to about 50% firing rate. Take a moderate amount of time to accomplish the manual ramp up. This will allow moisture and condensation to be removed from the fire chamber and stack. This process can take as little as 4 hr or as much as 24 hr, depending on type of boiler and amount of refractory.

c. Normal Operation. While the unit(s) are operating under normal conditions, we want to maintain operational checks. These should include (but not be limited to)

1.	Steam pressure and temperature	Is boiler maintaining designed steam pressure and temperature under all load conditions?
2.	Modulating control of boiler	Are boiler controls following steam demand promptly and accurately? Are set points correct?
3.	Is boiler going to "low fire" properly?	Under low-load conditions, is boiler cycling (or shutting down) at proper pressure/setpoint?

4. Do any of the setpoints change slightly each time the boiler cycles?

This is information that is vital to correct boiler operation, and should be monitored regularly. Also, check the water level in the boiler sight glass for stability. Check for rapid fluctuation in the steam drum water level. The sight glass should remain clean. Another check point during normal operation is the condensate return tank level and temperature. If the level is very high and the temperature is high, it could mean you have a serious malfunction in the steam traps. This high temperature can cause vapor-locking of the condensate transfer pumps and possibly the boiler feed pumps. The result is expensive system shutdown and pump repair. Check the steam traps, isolate the bad traps, and repair or replace them. One of the best ways to check for a malfunctioning steam trap is with an infrared temperature-reading device. You can also check a steam trap with a temperature gauge. Place it on both the inlet and outlet of the trap piping. You should see a moderate temperature difference. Trap maintenance can save on fuel costs, pumping electricity, and such. Check the fuel oil levels and fuel pressures on a regular basis. Blowdown is necessary and is one of the most neglected operations of boiler operators and owners. This one operational check can save a boiler and avoid thousands of dollars in downtime and repairs. Proper blowdown procedure along with proper boiler water chemicals, can keep the boiler in a good operating condition. Blowing down a boiler is a procedure for removal of total dissolved solids (TDS), such as rust, sludge, and sediment, that are carried in with the boiler feedwater. The sludge and sediment mainly come from the groundwater chemicals in the boiler feedwater, such as calcium. Usually blowdown is performed during light boiler load periods or at the start of each shift.

d. Shutdowns

Short-Term Shutdown.   Short-term may be defined as 1 week or less.

Normal short-term shutdown may be performed in this order. Secure header valve, close, tag, and lock out. Allow boiler to cycle off normally. Secure electrical, tag, and lock out. *Note*: Do not blowdown while, and if, the water feed valve is closed. It is better, however, to drain and dry out the boiler to avoid condensation and prevent rust from forming on the waterside of the boiler. If the idle boiler must be left full of water for over a week, the recommendations for dry set up should be followed. *Note*: If the boiler is left tied into a multiboiler system, make sure the water feed is kept lined up.

Emergency Shutdown.   In "emergency" situations, keep a level head and maintain common sense. An "emergency" shutdown may include some of the following situations.

1. Relief valve: "popping off"

Action. Shut off burner control and allow the boiler to reduce pressure. Isolate the boiler (in multiboiler situations) from the common steam header. Leave water feed system on. Isolate the fuel and electrical systems. After the boiler cools down to 140°F, or below—OSHA required temperature—remove the safety relief valve (SRV) and replace it, or have it repaired, reset, and stamped by an "SR" code stamp shop. If this emergency takes place, this indicates a problem with the boiler control system, you will have to locate the problem and fix it. The SRV is the last safety device for the boiler and operates only when the control system fails to shut down the boiler on excessive pressure or temperature. The repairs to the control system may require the services of a licensed and reputable boiler service organization. You need to consult your boiler insurance carrier.

2. Boiler firing with *no* visible water in glass.

Action. Isolate the boiler electrical control system. Isolate the boiler feedwater (*turn it off*). As soon as is practical, isolate the header and the fuel valves. Let the boiler cool down "naturally." Notify your boiler insurance carrier. *Note*: *Do not* add cold water to a very hot boiler under any circumstances. This can cause severe damage to the boiler or cause an explosion and possibly serious injury or death to operators.

3. Furnace explosion

Action. Furnace explosions can come in varying degrees of intensity, from unnoticeable to a major explosion. If one occurs, shut the boiler down completely, isolate all utilities. Notify your service repair organization and your boiler insurance carrier. *Note*: In all major mishaps or emergencies, notify your insurance carrier and jurisdiction authority. The cause of major mishaps must be determined and repaired before returning the boiler to operating status.

4. Fuel gas leak

Action. Immediately shut down all electrical circuits in the boiler room and isolate the leak area. Clear the room of all combustibles. Determine the cause of the gas leak and fix it. Before trying to start up the boiler, purge the gas chambers and exhaust stacks of any combustibles.

Long-Term Shutdown. This means a shutdown for an extended period. This will vary with the individual plant needs. I suggest longer than 1 month duration.

Action. Isolate main outlet valve and drain the boiler after the unit has cooled down (below 200°F, minimum). Isolate the fuel, electrical, and water supplies. Drain the unit completely and remove all access opening closures (i.e., plugs, hand-holes, and manway covers). Wash the unit down removing all scale, sludge, and foreign material from the waterside of the boiler. Remove all rust and carbon buildup from the fireside. Coat the fireside surface with a thin coat of very light oil to keep the metal surfaces from rusting. Dry out the waterside and install desiccant to keep the moisture in the air from attacking the metal. Close all access openings to the waterside. Make sure all utilities are tagged and locked out.

4. Calibration

a. Gauges.   The calibration of equipment gauges, and especially the boiler gauges, should be performed yearly. Calibration should be performed by a qualified (licensed) testing laboratory. This ensures that you obtain a "Certificate of Calibration." Calibrate the steam and temperature gauges. Calibrate the system signal inputs and outputs on 4–20 mA type controls. You should keep a spare set of calibrated gauges on hand for boiler hydro in case a hydrostatic test is requested by the authorized inspector. Most of the other gauges such as air pressure, fuel pressure, and others, should be replaced when found in bad working order.

b. Controls.   Calibration or resetting of controls, such as limit and operating controls, should be done in conjunction with a calibrated gauge and set for the desired temperature or pressure. Steam pressure High Limit should be set at least 10 psi below the safety relief valve pressure setting. Confirm this with your insurance carrier.

On units with float-type level controls and pump controls, the level control should be set to allow the water supply to engage at least 1 in. before low-water shutdown occurs. There may be some variation to this owing to pump size and steam usage. Adjustments in the MM-150 controller are accomplished by adjusting set screws. *Note*: This should only be performed by qualified personnel.

c. Burner Combustion Analysis and Calibration.   For dual-fuel units, using natural gas as the primary fuel and fuel oil as the standby or secondary fuel, calibration of the burner should be performed on gas first and then calibrate on oil. None of the air linkages should be modified while setting on oil, just the fuel oil setting. To perform this you should have a combustion analyzer, the original factory fast-fire report (or data), calibrated gauges (gas and oil), and assorted hand tools. The calibration should be performed by factory-trained personnel. You may also have to make regular scheduled combustion analysis on your equipment, and for that, you will require a portable analyzer to check O_2,CO, excess air, efficiency, and so on. Testing should be attempted when a load or demand is on the system.

Remember to check all linkages for slippage. After all settings are complete, make sure all settings are marked with paint or drilled and pinned. If you have problems with the boiler being out of adjustment on a regular basis, pinning is the best way to ensure that settings do not get changed. Pinning is usually performed on larger, water tube boilers.

d. Calibration of Pump Equipment. Calibrating of water supply pressure, temperature, and pump motor balance, will require calibration instruments. Check for the correct rpm on motors, correct voltage, and temperature of the deaerator system proportioning valve. Check the manufacturers recommendations on the pressure rating and temperature ratings. Check the "dead head" (maximum pressure the pump is capable of) on the boiler feedwater supply. This needs to be performed especially after pump rebuilding is done or motors are changed. Make sure when motors are changed that they are changed kind for kind (rpm, HP, voltage, enclosure, etc.). *Note*: Do not oversize the amp breakers or heaters on a pump motor.

5. Troubleshooting

Troubleshooting boiler problems should usually be left to the boiler service professionals. In plant operations and maintenance there are troubleshooting methods and techniques that can be used to minimize "service call outs." First of all, the preventive maintenance procedures you develop can greatly reduce the need for troubleshooting. Second, when the need arises for troubleshooting it may be a potentially dangerous situation. Never bypass any safety control. There are many other less dangerous ways to troubleshoot safety controls.

Example. Let us say you go into the boiler room for operational checks and you smell something like hot metal or insulation smoking. You notice the stack is extremely hot. You see the stack temperature gauge is way up, 800–1000°F. What should you do? Shut the boiler off immediately. What if you notice after a short time the temperature is still very high? You may have a *soot* fire problem in the boiler or breeching. What then do you do? Call the fire department or the in-plant emergency response team. They are better equipped to put out the soot fire.

Example. Now, let us say you are on callout for maintenance, and operations calls you. "The number 1 unit is down." You arrive at the boiler room and what should be your first move?

1. Talk to the operator on duty. Find out what he knows about the shutdown. Did he reset the unit or adjust anything? Remember you are not looking for someone to blame, you just want information so you can make a decision.
2. Take a look around the boiler and look at the overall situation. Shut off the manual firing fuel valve. Go around the boiler and check the water level, electrical supply, controller, stack temperature, and manual resets on limits (gas, steam, oil, air, water, or other). Note if you find any manual reset thrown. Now that you have checked everything, let us say you found the Hi-Limit switch thrown. What do you do next?

You might be tempted to just reset it and go, but there is usually a reason that the Hi-Limit switch is thrown. This may mean the operating limit has malfunctioned and needs to be replaced. Just replace it. These parts are reasonably priced. Lives and property are not. *Note*: Make sure all pressure is off the unit before replacement.

a. Electrical. When troubleshooting electrical controls, such as limits, interlocks,and level controls, check for continuity between contacts. Most controls are electrical over mechanical. Check the mechanical conditions of the control (for instance a control with a set screw may have moved owing to a boiler burner vibration on a gas pressure switch, causing a shutdown).

b. Mechanical. Troubleshooting mechanical parts is mostly common-sense. Look for worn parts such as worn linkage rods, loose nuts, worn surfaces, leaking valve stems, dirty filters, or clogged orifices and nozzles. Replace if needed.

When cleaning clogged fuel oil-firing nozzles, use a degreaser or a very soft copper brush, or both. This will keep from distorting the nozzle holes. Also make sure you assemble the nozzle back together properly before reinstalling.

6.   Inspections on a Daily, Monthly, or Yearly Basis

a. Daily. Daily inspections should include, but not be limited to, checking operating controls, water levels, and boiler firing. You should use the operational checklist provided or a list devised by your organization. Lists have been published and given by companies, such as Hartford Insurance Company (52), and others.

b. Monthly. Monthly inspections should include, but not be limited to, making the same check as the daily checks and checking the low water cutoff, lifting the SRV seat to ensure it is not galled or wire drawn, checking combustion, and opening the waterside of the boiler to remove sludge and scale. Change the SRV if the seat is galled, wire drawn, stuck, or will not reseat.

c. Yearly. Yearly inspections should include opening of all interior openings (i.e., hand-holes, manways, water column plugs in cross tees, feedwater line, and blowdown lines). Remove all foreign matter and scrape out and flush with water or high-pressure wash. Check all internals in the boiler for corrosion or malfunction. Replace as necessary. Replace all plugs and use an antiseize compound. Replace all hand-hole and manway opening gaskets with new gaskets, making sure the gaskets are correct for the pressure and temperature of the boiler. Remove and clean the Low-Water Cutoff control, which may be float or probe type, clean the inside of the controls, and replace worn or damaged parts. Replace with new gaskets and tighten to manufacturers recommendations.

Open the fireside of the boiler (furnace area). Remove all foreign materials, such as soot, ash, fallen refractory, or other. Repair or replace broken refractory and inspect the burner head for cracks or plugging. On oil-firing equipment, check the fuel oil nozzle for plugging or damage and replace if necessary. Make sure you replace the oil nozzle in the same position as when you removed it. Make sure that you mark any linkages before removal so it can be replaced in the same position. Check all other linkages for worn or damaged parts and replace or repair. Remove and replace all air and oil filters (fuel and motor oil filters). Remove and check the ignitor assembly and replace if it is worn or damaged. Remove and replace all bad-order gauges or have them recalibrated. Check for worn or bare wiring and replace if necessary. On a mercury switch-type control that is found to be bad or broken, do not replace just the mercury bulb part of the switch, replace the complete switch. The switches are calibrated and set by the manufacturer. They now make nonmercury switches for the same control purpose. Inspect all valves, motors, and valve cocks and repair or replace if excessively worn.

At least once a year, remove the forced draft (FD) fan shroud, louver, and linkages, and clean all foreign debris from the fan blades and fan body. You can accomplish this by using a wire brush and scraper. The fan may have to be removed from the motor shaft for complete cleaning. Solvent works well for a cleaning tool especially if the air entering the fan is greasy. Be sure not to move or remove any of the fan balancing weights. If there has been some vibration noticed during operation, rebalance the fan, blow dust and debris out from the fan housing. Replace the shroud, louver, and linkages in the same order and original position. Before boiler operation, check the RPM of the FD fan motor. If the motor is not running at the correct rpm, check bearings, shaft alignment, amperage, and voltage to determine the cause of low (or high) rpm. If needed, buy, rebuild or replace the motor.

7. Summary

If possible, take pictures of all repairs and inspected conditions. Document and record all repairs and inspections for your future use. In other words, start a boiler history file. This should contain records from start-up of the new boiler to present. Any time a valve is replaced, tubes are replaced, burner is recalibrated, or any part is changed, record this in the unit history file. Document all boiler failures, pump failures, and other related equipment failures in this master record file. This "Record File—Boiler No. X," will be immensely helpful in future troubleshooting the particular boiler. It will also help train new personnel. Include a copy of the daily, monthly, and yearly check and maintenance logs.

Do not try to repair something you are not qualified to repair. You will not save any time. Call in your service professional, for you and your company's safety are of the highest priority. Make sure the organization that performs repairs on you boiler is licensed and qualified. This usually means that they carry a *current* state license and have a *current* "R" stamp certification. It helps to be cooperative with the service technician as he trys to gather information about your problem. Also good communication between in-house boiler operations personnel and maintenance personnel is vitally important.

In shutdowns, refer to the manufacturer's recommendations. You should be able to find most information in the M&O Manual (maintenance and operations). Keep good through records of all operational and maintenance checks—this cannot be emphasized enough. Keep your boiler room clean and orderly and clean up spills as soon as they occur. Work safely around steam, hot water, electricity, gas, oil, and chemicals, as each of these, separately or together, can be dangerous. Remember your boiler room safety, it could save your life and the life of others.

III. EXPERIENCE

Unusual experiences of Lee King in his 30-plus years of working with steam boilers follow.

1.   The company I worked for in the 1960s was called out to look at a 500-hp Scotch Marine firetube boiler in western New Mexico. The owner was always adjusting the flame. The boiler suffered a furnace explosion while lighting off. At that time the owner was looking through the rear door view port of the boiler. Normally, on Scotch Marine firetube boilers, the rear door has swing davits and large lug bolts holding it in place. This furnace explosion blew the rear door off completely and it came to rest against a wall about 20 ft away. Needless to say, the owner was killed immediately.

2.   While I was training in boiler work under my father, who was a Master Boilermaker, we were working on a firebox boiler. My father was inside the firebox in the process of rebricking the furnace. For some reason, a boiler operator turned on the electrical boiler controller. In those days, this type of boiler had a 24-V slow-opening gas valve. My father, on hearing the click of the pilot igniter, dove for the access opening where I was standing watching him work. He made it out the access opening just as the gas flames started to burn the soles of his boots. That was an extremely close call. From then on, we made very sure that all utilities were locked out and the operating handles removed.

3.   During the 1970s, again in western New Mexico, I was working with another crew member and rolling tubes in a mud drum of a water tube boiler. This boiler was a 60,000-lb/hr unit producing saturated steam and was connected to a common blowdown line with an adjacent boiler. From past experience, we had tagged and locked out the utilities and chained and locked the blowdown line valves. It was standard practice at this particular plant that after the shift change, the operators would blow down the system. Somehow the chain and lock were removed from the blowdown valve by someone. We were still in the mud

drum rolling tubes when suddenly we heard a loud noise like a freight train. We then saw steam roaring down the drum straight toward us. We both dove out the drum access door and made it out without a scratch. This was really too close for comfort. We then proceeded to totally disconnect the common blowdown line and valves.

4.   I was called out to look at a boiler in West Texas because the boiler would not start up. The owner/operator said "he had reset the boiler but it wouldn't run." This was about a 700-mile round trip and a service call that far away is very expensive. I began to look at the resets on the boiler, and traced the problem to a burned out 10-amp fuse. Once I had replaced the fuse and checked the circuitry for irregularities, I put the boiler on line. It seems he must have had an electrical surge/spike to his system. This service call cost them quite a bit of money. The owner/operator bought some extra fuses.

5.   I received a call from a boiler owner who said, "my boiler won't make steam." When I arrived, the first thing I checked was the boiler steam drum water level sight glass. There did not appear to be any water in it. I opened the boiler blowdown line and no water came out. I then shut the boiler down, shut off all the utilities, let the boiler cool off, and opened the boiler waterside. I found that the boiler was full of scale and mud. This failure to operate and maintain his boiler properly cost the owner a lot of money to remove all of the scale buildup and the mud and then completely retube the boiler. It was plain that the owner did not know what the boiler blowdown system was for. He does now.

Refer to the Basic Powerplant Checklist (Fig. 1.1) for a summary of the procedures in this chapter.

BASIC POWER PLANT CHECK LIST	E.F.W., INC.
R.L.Vandagriff	1420 Starfield Rd., North Little Rock, Arkansas USA 72116
Program: BlrHsCkList	Phone: 501-753-9986
Date:12/31/93	Fax: 501-753-7165
Customer: SCMS Engineering, Houston, Texas	
Plant Location: Guyana, SA	
Phone Number:	
Fax Number:	
Plant Water System	**Boiler House, Continued:**
Raw Water Analysis:	Air compressor #1: op.psig: hp:
Raw Water: Main Size:	Air compressor #2: op.psig: HP:
Raw Water Gal/Day Required:	Air receiver size:
Raw Water Cost: $ / 1000 gal:	Air dryer:
Storage Tank Size: Gallons:	
Strainers:	Steam Main: Size:Op.Psig:
Filters:	Steam Flow:lb/hr:
Main Pump:	Pressure drop/100 feet; velocity FPS:
Standby Pump:	Insulation: Heat Loss / Btu/Hr/Ft. of run
Water Treatment # 1:	Length of run: feet:
Water Treatment # 2:	Flowmeter/Temp/Psig/Recorder
Flowmeter/pressure meter/recorder	
	Steam Line #1: Size:Op.Psig:
	Steam Flow:lb/hr:
Boiler Feedwater System	Pressure drop/100 feet; velocity FPS:
Deaerator, pressure & type:	Insulation: Heat Loss / Btu/Hr/Ft. of run
Deaerator steam requirement:#/Hr, psig:	Length of run: feet:
Deaerator vent: #/hr:	Flowmeter/Temp./Psig/Recorder
Boiler feed pump No. 1: gpm: psig out:	
Boiler Feed pump No. 2: turbine drive: gpm:psig out:	Steam Line #2: Size:Op.Psig:
Steam req'd:#2 pump steam turbine: lb/hr, psig	Steam Flow:lb/hr:
Flowmeter/pressure meter	Pressure drop/100 feet; velocity FPS:
Flow recorder	Insulation: Heat Loss / Btu/Hr/Ft. of run
No flow alarm & boiler shutdown	Length of run: feet:
	Flowmeter/Temp./Psig/Recorder
Boiler feedwater req'd. for desuperheater: gpm	
Flowmeter/Pressure meter	Steam Line #3: Size:Op.Psig:
Flow recorder	Steam Flow:lb/hr:
	Pressure drop/100 feet; velocity FPS:
Boiler House	Insulation: Heat Loss / Btu/Hr/Ft. of run
Boiler No. One:	Length of run: feet:
Lb. per hour steam:	Flowmeter/Temp./Psig/Recorder
Degree superheat:	
Operating pressure:	Steam Line #4: Size:Op.Psig:
Lb. per hour blowdown:	Steam Flow:lb/hr:
Non return valve: size:	Pressure drop/100 feet; velocity FPS:
Economizer: BFW outlet temp.:	Insulation: Heat Loss / Btu/Hr/Ft. of run
Stack Height: feet:	Length of run: feet:
	Flowmeter/Temp./Psig/Recorder
Boiler No. Two:	
Lb. per hour steam:	Steam Line #5: Size:Op.Psig:
Degree superheat:	Steam Flow:lb/hr:
Operating pressure:	Pressure drop/100 feet; velocity FPS:
Lb. per hour blowdown:	Insulation: Heat Loss / Btu/Hr/Ft. of run
Non return valve: size:	Length of run: feet:
Economizer: BFW outlet temp.:	Flowmeter/Temp./Psig/Recorder
Stack Height: feet:	
Boiler No. Three:	Condensate return system: gpm:
Lb. per hour steam:	condensate return system insulation:
Degree superheat:	condensate return system temp.,F.:
Operating pressure:	steam traps:
Lb. per hour blowdown:	Condensate Receiver: size:gallons:
Non return valve: size:	Condensate pump to deaerator:
Economizer: BFW outlet temp.:	Flowmeter/Temperature/recorder
Stack Height: feet:	
	Condensate polisher, if required.
Blowdown heat recovery unit: gpm:	
Ash handling system: Tons per day:	

FIGURE 1.1

Check List, continued:	
Plant Electrical:	**Plant Waste Flow**
Local utility name:	Paper: pounds per day:
Electricity delivered: voltage,phase, hertz:	Waxed boxes: pound per day:
Plant electricity: voltage, phase, hertz:	Polyethylene: pound per day:
Transformer #1:	Other plastic: pound per day:
Transformer#2:	Waste pallets: pounds per day:
Standby rate: per kwh:	Wood waste: pound per day:
Regular rate: per kwh:	Rice Hulls: pound per day:
Rate # 2: per kwh:	
Summer max Kw demand:	Land fill charges: $ per ton:
Summer min. Kw demand:	
Winter max Kw demand:	
Winter min. Kw demand:	**Plant Wastewater Flow:**
	Flow: GPD to sewer:
Plant Generating System:	Sewer charge: $ / 1000 gal.:
T/G #1: lb/hr steam;KwH;	BOD allowed
T/G #2: lb/hr steam;KwH;	COD allowed:
Synchronous generator:	
Alternating generator:	**Wastewater Treatment Required**
Condenser, cooling tower system:	Settling pond
Switchgear, etc.:	Clarifier
	Other:
Fuels:	
Natural gas: $/mcf; mcf/day req'd.:Mcf/Month:	Disposal of Ash from Boilers:
Gas line size; gas line pressure: water:	
#2 Fuel oil: $/gal; GPD req'd.:Gal/Month:	Landfill for waste:
#2 Fuel Oil Storage and Pumps	
#6 Fuel Oil: $/gal; GPD req'd:Gal/Month:	
Coal: $/ton: ton/day:ton/month:	**Plant Air Emissions:**
Lignite: $/ton: ton/day: ton/month:	Lb. per day CO_2:
Wood: $/ton:MC(Wb); ton/day; ton/month:	Lb. per day CO_2 allowed:
Rice Hulls: $/ton; MC(Wb); ton/day:ton/month:	Lb. per day CO:
Other: $/ton; MC(Wb); ton/day; ton/month:	Lb. per day CO allowed:
	Lb. per day NOx:
Solid Fuel System	Lb. per day NOx allowed:
Receiving and Storage	Lb. per day particulate:
Conveyors: Storage to Dryer	Lb. per day particulate allowed:
Fuel Dryer	Other: lb/day
Conveyors: Dryer to Day Bin	Other: lb/day allowed:
Day Bin	
Conveyors: Day Bin to Feeder Bin	**Plant Operating Schedule: Hours**
	Shifts: ONE TWO THREE
Plant Refrigeration:	Sunday
Total tons per hour required:	Monday
TPH above 38 degree F:	Tuesday
TPH below 38 degree F:	Wednesday
Chilled water # 1: gpm, deg. F:	Thursday
Chilled water # 2: gpm, deg. F:	Friday
Chilled water # 3: gpm, deg. F:	Saturday
Ice: lb per day required:	
CO_2: lb. per day required:	**Plant Steam Load: Lb./Hour:**
	Shifts: ONE TWO THREE
Air Emission Equipment Required:	Sunday
Multicyclones	Monday
Baghouse	Tuesday
SO2 scrubber	Wednesday
Other	Thursday
	Friday
Energy Recovery Auxiliary Equipment	Saturday
Blowdown Flash Steam System	
Blowdown Liquid Heat Exchanger/Water to Deaerator	Notes:
	Ralph L. Vandagriff 12/31/1993

2

General Data

Personnel Safety; Operating Safety Precautions; Abnormal Boiler Operation; Common Boiler House Terms; Boiler Design; Conversion Factors and Unit Equivalents; British Thermal Unit and Flame Temperature; Fuel Combustion; Heating Value of Fuel; Boiler Tubing; Refractory; Corrosion; ph Values; Screening; Electric Motor Selection; Emmisivity and Emittance.

I. PERSONNEL SAFETY

Operating instructions usually deal primarily with the protection of equipment. Rules and devices for personnel protection are also essential. The items listed here are based on actual operating experience and point out some personnel safety considerations [1].

1. When viewing flames or furnace conditions, always wear tinted goggles or a tinted shield to protect the eyes from harmful light intensity and flying ash or slag particles.
2. Do not stand directly in front of open ports or doors, especially when they are being opened. Furnace pulsations caused by firing conditions, sootblower operation, or tube failure can blow hot furnace gases out of open doors, even on suction-fired units. Aspirating air is used on inspection doors and ports of pressure-fired units to prevent the escape of hot furnace gases. The aspirating jets can become blocked, or the aspirating air supply can fail. Occasionally, the entire observation port or door can be covered with slag, causing the aspirating air to blast slag and ash out into the boiler room.
3. Do not use open-ended pipes for rodding observation ports or slag on furnace walls. Hot gases can be discharged through the open-ended pipe directly onto its handler. The pipe can also become excessively hot.
4. When handling any type of rod or probe in the furnace—especially in coal-fired furnaces—be prepared for falling slag striking the rod or probe. The fulcrum action can inflict severe injuries.

5. Be prepared for slag leaks. Iron oxides in coal can be reduced to molten iron or iron sulfides in a reducing atmosphere in the furnace resulting from combustion with insufficient air. This molten iron can wash away refractory, seals, and tubes, and may leak out onto equipment or personnel.

6. Never enter a vessel, especially a boiler drum, until all steam and water valves, including drain and blowdown valves, have been closed and locked or tagged. It is possible for steam and hot water to back up through drain and blowdown piping, especially when more than one boiler or vessel is connected to the same drain or blowdown tank.

7. Be prepared for hot water in drums and headers when removing manhole plates and handhole covers.

8. Do not enter a confined space until it has been cooled, purged of combustible and dangerous gases, and properly ventilated with precautions taken to keep the entrance open. Station a worker at the entrance, notify a responsible person, or run an extension cord through the entrance.

9. Be prepared for falling slag and dust when entering the boiler setting or ash pit.

10. Use low-voltage extension cords, or cords with ground fault interrupters. Bulbs on extension cords and flashlights should be explosion proof.

11. Never step into flyash. It can be cold on the surface yet remain hot and smoldering underneath for weeks.

12. Never use toxic or volatile fluids in confined spaces.

13. Never open or enter rotating equipment until it has come to a complete stop and its circuit breaker is locked open. Some types of rotating equipment can be set into motion with very little force. This type should be locked with a brake or other suitable device to prevent rotation.

14. Always secure the drive mechanism of dampers, gates, and doors before passing through them.

II. OPERATING SAFETY PRECAUTIONS

A. Water Level

*The most important rule in the safe operation of boilers is to keep water in the boiler at proper level. **Never** depend entirely on automatic alarms, feedwater regulators, or water level controls.* When going on duty, determine the level of water in the boiler. The gage glass, gage cocks, and connecting lines should be blown several times daily to make sure that all connections are clear and in proper working order. The gage glass must be kept clean because it is of extreme importance that the water level be accurately indicated at all times. *If there is any question on the accuracy of the water level indicated, and the true level cannot be determined immediately, the boiler should be removed from service and all water level-indicating attachments should be checked.*

B. Low Water

In case of low water, stop the supply of air and fuel immediately. For hand-fired boilers, cover the fuel bed with ashes, fine coal, or earth. Close the ash pit doors and leave the fire doors open. *Do not change the feedwater supply. Do not open the safety valves or tamper with them in any way.* After the fire is banked or out, close the feedwater valve. After the boiler is cool, determine the cause of low water and correct it. Carefully check the boiler for the effects of possible over-heating before placing it in service again.

C. Automatic Controls and Instructions

Automatic control devices should be kept in good operating condition at all times. A regular schedule for testing, adjustment, and repair of the controls should be adopted and rigidly followed. *Low-water fuel supply cutoffs and water level controls should be tested at least twice daily in accordance with the manufacturer's instructions and overhauled at least once each year.* All indicating and recording devices and instruments, such as pressure or draft gages, steam or feedwater flow meters, thermometers, and combustion meters should be checked frequently for accuracy and to determine that they are in good working order.

D. Safety Valves

Each safety valve should be made to operate by steam pressure with sufficient frequency to make certain that it opens at the allowable pressure. The plant log should be signed by the operator to indicate the date and operating pressure of each test. If the pressure shown on the steam gage exceeds the pressure at which the safety valve is supposed to blow, or if there is any other evidence of inaccuracy, no attempt should be made to readjust the safety valve until the correctness of the pressure gage has been determined.

E. Leakage and Repairs Under Pressure

Any small leaks should be located and repaired when the boiler is removed from service. If a serious leak occurs, the boiler should be removed from service immediately for inspection and repair. No repairs of any kind should be made to a boiler or piping while the parts on which the work is to be done are under pressure. Neglect of this precaution has resulted in many cases of personal injury.

F. Avoid Scalding Men

Attach a sign, **"DO NOT OPEN—MAN IN BOILER"** to each valve in the steam lines, feedwater lines, and blowoff pipes connected to a boiler that is ready for cleaning and repair. Do not remove the signs or open a valve until the boiler is closed and ready for filling. It is well to lock the main steam stop valves and blowoff valves in the closed position when the boiler being cleaned or repaired is in the same battery with other boilers under pressure. Padlocks and chains may be used for this purpose [52].

[Personal note: Check the boiler very carefully before closing it up. Someone may still be inside. It has happened, and you cannot hear someone inside a drum screaming, as the boiler is being filled with water.]

III. ABNORMAL BOILER OPERATION [1]

A. Low Water

If water level in the drum drops below the minimum required (as determined by the manufacturer), fuel firing should be stopped. Because of the potential of temperature shock from the relatively cooler water coming in contact with hot drum metal, caution should be exercised when adding water to restore the drum level. Thermocouples on the top and bottom of the drum will indicate if the bottom of the drum is being rapidly cooled by feedwater addition, which would result in unacceptable top-to-bottom temperature differentials. If water level indicators show there is still some water remaining in the drum, then feedwater may be slowly added using the thermocouples as a guide. If the drum is completely empty, then water may be added only periodically with soak times provided to allow drum temperature to equalize.

B. Tube Failures

Operating a boiler with a known tube leak is not recommended. Steam or water escaping from a small leak can cut other tubes by impingement and set up a chain reaction of tube failures. By the loss of water or steam, a tube failure can alter boiler circulation or flow and result in other circuits being overheated. This is one reason why furnace risers on once-through type boilers should be continuously monitored. A tube failure can also cause loss of ignition and a furnace explosion if reignition occurs.

Any unusual increase in furnace riser temperature on the once–through-type boiler is an indication of furnace tube leakage. Small leaks can sometimes be detected by the loss of water from the system, the loss of chemicals from a drum-type boiler, or by the noise made by the leak. If a leak is suspected, the boiler should be shut down as soon as normal-operating procedures permit. After the leak is then located by hydrostatic testing, it should be repaired.

Several items must be considered when a tube failure occurs. In some cases where the steam drum water level cannot be maintained, shut off all fuel flow and completely shut off any output of steam from the boiler. When the fuel has been turned off, purge the furnace of any combustible gases and stop the feedwater flow to the boiler. Reduce the airflow to a minimum as soon as the furnace purge is completed. This procedure reduces the loss of boiler pressure and the corresponding drop in water temperature within the boiler.

The firing rate or the flow of hot gases cannot be stopped immediately on some waste heat boilers and some types of stoker-fired boilers. Several factors are involved in the decision to continue the flow of feedwater, even though the steam drum water level cannot be maintained. In general, as long as the temperature of the combustion gases is hot enough to damage the unit, the feedwater flow should be continued. The thermal shock resulting from feeding relatively cold feedwater into an empty steam drum should also be considered. Thermal shock is minimized if the feedwater is hot, the unit has an economizer, and the feedwater mixes with the existing boiler water.

After the unit has been cooled, personnel should make a complete inspection for evidence of overheating and for incipient cracks, especially to headers and drums and welded attachments.

An investigation of the tube failure is very important so that the condition(s) causing the tube failure can be eliminated and future failures prevented. This investigation should include a careful visual inspection of the failed tube. Occasionally, a laboratory analysis or consideration of background information leading up to the tube failure is required. This information should include the location of the failure, the length of time the unit has been in operation, load conditions, start-up and shutdown conditions, feedwater treatment, and internal deposits.

IV. COMMON BOILER HOUSE TERMS

1. *Heat*: A form of energy that causes physical changes in the substance heated. Solids, such as metal, when first heated, expand, and at high temperatures, liquefy. Liquids, when heated, vaporize, and the vapor coming in contact with a cooler surface condenses, giving to the surface the heat that caused vaporization. For example, the addition of heat to ice (a solid) will change it to water (a liquid) and the further addition of heat will change the water to steam (a vapor).

2. *Btu*: British thermal unit is the unit measurement of heat. It is that amount of heat required to raise 1 lb (approximately 1 pint) of water 1°F or 1/180 of the amount of heat required to raise 1 lb of water from 32° to 212°F. To raise 10 lb of water 50° will require 10 × 50 or 500 Btu. One Btu will raise approximately 55 ft³ of air 1°F. To raise 300 ft³ of air 20°F will require 300/55 × 20, or 109 Btu.

3. *Latent heat*: The amount of heat required to change the form of a substance without change in temperature. Water at 212°F, in changing to steam at the same temperature, requires the addition of 970.3 Btu/lb.
4. *Specific heat*: The amount of heat required to raise 1 lb of any substance 1°F compared with 1 lb of water which has a specific heat of 1.00.
5. *Transfer of heat*: Heat flows from a body of higher temperature to one at a lower temperature and may be transferred by three following methods:
   a. *Conduction*: The transfer of heat between two bodies or parts of a body having different temperatures. *Example*: the flow of heat from the inside surface of a radiator to the outside surface.
   b. *Convection*: Transmission of heat conveyed by currents of air, water, or other substances passing over a surface having a higher temperature than the currents flowing over it. *Example*: air currents passing over radiator surfaces are heated by convection.
   c. *Radiation*: The transfer of heat from one body to another by heat waves (light waves) that radiate from the body with higher temperature to one at a lower temperature, without heating the air between the two bodies. *Example*: the noticeable difference in temperature between a piece of metal in bright sunlight and similar piece in the shade—the metal exposed to the sun absorbs radiant heat; the metal in the shade does not receive radiant heat and is at the temperature of the surrounding air.
6. *Condensation*: The change from a vapor into a liquid with a transfer of heat from vapor to condensing surface. *Example*: 1 lb of steam at 212°F in condensing gives up to the condensing surface (radiator surface) 970.3 Btu, which is the output of the radiating surface.
7. *Atmospheric pressure*: The weight of the blanket of air (approximately 50 miles thick) that surrounds the earth. At sea level this air exerts a pressure of 14.696 lb/in.^{2} above absolute zero. Under this pressure at sea level water will boil and vaporize at 212°F.
8. *Gauge pressure*: Pressure measured with atmospheric pressure as the starting or zero point. *Example*: 0 on pressure gauge is 14.7 lb absolute pressure, whereas 2 lb on the pressure gauge is 16.7 lb absolute.
9. *Vacuum*: The pressure in a closed chamber below atmospheric pressure caused by complete or partial removal of air. Vacuum is usually measured in inches of mercury with atmospheric pressure as zero. *Example*: 9.6 in. of vacuum is 10 lb/in.^{2} absolute or 4.7 lb/in.^{2} below atmospheric pressure. To indicate the effect of atmospheric pressure on the boiling or vaporizing point, attention is called to an example of

a boiling point of 188°F at an elevation of 12,934 ft. and a vaporizing temperature of 188°F. at 11.7 in. of vacuum at sea level. Therefore, the effect on vaporization is the same for both conditions.

10. *Pressure drop*: The difference in pressure between two points, primarily caused by frictional resistance and condensation in the pipe line. *Example*: With a boiler pressure of 100 lb and a pressure at the end of the steam line of 90 lb, the total pressure drop equals 10 lb. If the line is 100 ft long, the pressure drop per foot of pipe is 0.1 lb.

11. *Velocity of flow*: The rate of flow passing a given point in a unit of time, such as feet per second (ft/sec).

12. Equivalents

1 ft³ of water	= 7.48 gal
weight	= 62.37 lb at 60°F
	= 59.83 lb at 212°F
1 gal of water	= 8.34 lb at 60°F.
	= 7.99 lb at 212°F
1 ton of refrigeration	= 286,600 Btu/day

13. Heating

CFM	= cubic feet per minute
TR	= temperature rise in degrees Fahrenheit.

Heating air: $\dfrac{\text{CFM} \times 60 \times \text{TR}}{55} = \text{Btu/hr required}$

Heating water: Pounds of water per hour $\times$ TR = Btu/hr required
Gallons of water per hour $\times$ 8.34 $\times$ TR = Btu/hr required

14. Miscellaneous

One pound of water, introduced with the fuel into the boiler furnace chamber, occupies 26.8 ft³ at 212°F, and approximately 90 ft³ at 1750°F furnace temperature.

One pound of water at 60°F requires 152 Btu to reach 212°F, and an additional 970.3 Btu to turn from a liquid to a vapor (steam), at atmospheric pressure at 212°F. Then an additional 821.6 Btu to raise steam to 1750°F furnace temperature.

Example: Fuel = Southern pine bark at 35% moisture content.

(Southern pine bark = 9000 Btu/lb, zero moisture [FPRS 1982]).

One pound of fuel has 65% with zero moisture; 35% water

Btu per pound of fuel as fired =	9000 × 0.65 = 5850 Btu
Minus Btu to flash the water =	1122.3 × 0.35 = 393 Btu
Minus Btu to raise steam to 1750°F	821.6 × 0.35 = 288 Btu
Btu/lb of fuel available for steam production:	5169 Btu

V. BOILER DESIGN

A. Definitions

1. *Effective projected radiant surface* (EPRS): Effective projected radiant surface is the total projected area of the planes that pass through the centers of all wall tubes, plus the area of a plane that passes perpendicular to the gas flow where the furnace gases reach the first convection superheater of reheater surface. In calculating the EPRS, the surfaces of both sides of the superheater and reheater platens extending into the furnace may be included.

2. *Furnace volume*: The cubage of the furnace within the walls and planes defined under EPRS.

3. *Volumetric heat release rate*: The total quantity of thermal energy above fixed datum introduced into a furnace by the fuel, considered to be the product of the hourly fuel rate and its high heat value, and expressed in Btu per hour per cubic foot (But/hr ft^{-3}) of furnace volume. This value, does not include the heat added by preheated air nor the heat unavailable through the evaporation of moisture in fuel and that from the combustion of hydrogen.

4. *Heat available on net heat input*: The thermal energy above a fixed datum that is capable of being absorbed for useful work. In boiler practice, the heat available in the furnace is usually taken to be the higher-heating value of the fuel corrected by subtracting radiation losses, unburned combustible, latent heat of the water in the fuel or formed by the burning of hydrogen, and adding sensible heat in the air (and recirculated gas if used) for combustion, all above an ambient or reference temperature.

5. *Furnace release rate*: Furnace release rate is the heat available per square foot of heat-absorbing surface in the furnace (the EPRS).

6. *Furnace plan heat-release rate*: Furnace plan heat-release rate is usually based on the net heat input at a horizontal cross-sectional plane of the furnace through the burner zone, expressed in million Btu/hr ft^{-2}. The area of the plan is calculated from the horizontal length and width of the furnace taken from the centerline of the waterwall tubes.

B. Furnaces for Oil and Natural Gas Fuels [13]

Oil does not require, at what has been normal excess-air requirements, as large a furnace volume as coal to achieve complete combustion. However, the rapid burning of, and high radiation rate from, oil results in high-heat–absorption rates in the active burning zone of the furnace. The furnace size must, therefore, be increased above the minimum required for complete combustion, to reduce heat absorption rates and avoid excessive furnace wall metal temperatures.

Natural gas firing permits the selection of smaller furnaces than for oil firing, primarily because a more uniform heat absorption pattern is obtained.

This brief discussion of furnace sizing relates primarily to tangential firing. Some of the statements made do not necessarily apply to a furnace using parallel or other firing methods.

VI. CONVERSION FACTORS AND UNIT EQUIVALENTS

A. British Thermal Units (Btu)

1. Energy, Heat, and Work
    3413 Btu = 1 kWh
    1 Btu = 0.2931 Wh   = 0.0002931 kWh
           = 252.0 cal   = 0.252 kcal
           = 778 ft lb
           = 1055 joules (J) = 0.001055 MJ   = 1.055 GJ

2. Heat Content and Specific Heat
    1 Btu/lb   = 0.55556 cal/g   = 2326 J/kg
    1 Btu/ft^3 = 0.00890 cal/cm^3 = 8.899 kcal/m^3 = 0.0373 MJ/m^3
    1 Btu/US gal = 0.666 kcal/L
    1 Btu/lb°F   = 1 cal/g·°C   = 4187 J/kg·K   = 4.1868 kJ · kg · K

3. Heat Flow, Power
    1 Btu/hr = 0.252 kcal/hr   = 0.0003931
    hp   = 0.2931 J/sec

4. Heat Flux and Heat Transfer Coefficient
    1 Btu/ft^2 sec^{-1} = 0.2713 cal/cm^2 sec^{-1}
    1 Btu/ft^2 hr^{-1} = 0.0003153 kW/m^2 = 2.713 kcal/m^2 hr^{-1}

5. Thermal Conductivity
    1 Btu · ft/ft^2 · hr · °F = 1.488 kcal/m · hr · K   = 1.730 Wm·K
    1 Btu · in/ft^2 · hr · °F = 0.1442 W/m · K
    1 Btu · ft/ft^2 · hr · °F = 0.004139 cal · cm/cm^2 · sec · °C
    1 Btu · in./ft^2 · hr · °F = 0.0003445 cal · cm/cm^2 · sec · °C

VII. BRITISH THERMAL UNIT AND FLAME TEMPERATURE

1. Most of the British thermal unit (Btu) values shown were derived using Dulong's Formula.

    a. For a slightly modified version, see Ref 31, p 193.

    Heat units in Btu per pound of dry fuel =
    $$14{,}600\ C + 62{,}000\ (H\text{-}O/8) + 4000\ S$$

    where C, H, O, and S are the proportionate parts by weight of carbon, hydrogen, oxygen, and sulfur.

    b. The gross calorific value may also be approximated by Dulong's formula, as follows [16; p 27]:

    $$Btu/lb = 14{,}544\ C + 62{,}028\ (H - O/8) + 4{,}050\ S$$

    c. A more recent application of Dulong's formula, including chlorine can be found in Hazardous Waste Incineration Calculations, Problems and Software. John Wiley & Sons, 1991; p 50.

    Using molecular weight (m):
    $$14{,}000\ m_C + 45{,}000\ (m_H - 1/8\ m_O) - 760\ m_{Cl} + 4{,}5000\ m_S$$

2. Net heating value of fuel (4)

    The gross heating value minus the moisture loss per unit of fuel is equal to the net heating value per unit of fuel.

    As an alternate, the following approximate formula may be used;

    Moisture loss, in Btu/hr =
    $$lb\ H_2O/hr \times (1{,}088 + 0.46 \times [t_2 - 60])$$

    where t_2 is the furnace exit temperature (°F) and 60 is the base temperature (°F) used to evaluate the gross heating value of the fuel.

3. Theoretical adiabatic flame temperature can be calculated several different ways.

    a. *North American Combustion Handbook*, [4,9] says;

    A simplified formula for theoretical adiabatic flame temperature is:

    $$\frac{\text{Net heating value of the fuel} - \text{effect of dissociation}}{(\text{Weight of combustion products}) \times (\text{specific heat of combustion products}).}$$

    Under certain conditions, particularly high temperatures, a phenomenon known as dissociation occurs. Dissociation is simply reverse combustion; that is, it is the breaking down of the combus-

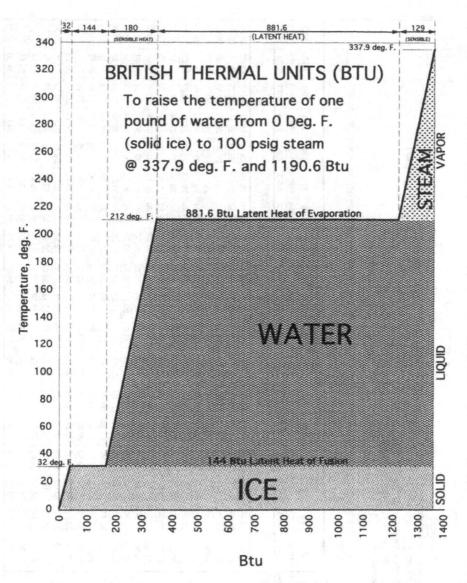

FIGURE 2.1 Relation between British thermal units (Btu), heat, and water conversion.

TABLE 2.1

BOILER HOUSE NOTES:		Page No:

Ralph L. Vandagriff   Date: 8/12/98

Boiler Horsepower

"Boiler Horsepower" is an old boiler rating term that is occasionally used in rating scotch marine firetube boilers.

This old rating unit of capacity = evaporating 34.5 lb. of water "from and at" 212 deg. F. (970.3 x 34.5 = 33,475 Btu per hour).

This means that the steam production is at zero (0) psig pressure with a feedwater temperature of 212 degrees F.

To adjust this rating to actual operating conditions of the boiler, the Factor of Evaporation is used, as follows;

Formula for "Factor of Evaporation" = 34.5 x (1 / (H - h) / 970.3)   (H = Btu/lb. of steam) (h = Btu/lb. of feedwater)

Example: The steam produced per hour, by a 200 horsepower boiler, at 100 psig saturated, with feedwater temperature of 130 deg. F.,
is calculated: 200 x 30.64 = 6,128 lb per hour

Steam Gauge Pressure in Psig

Pounds of Dry Saturated Steam per Boiler Horsepower

Factor of Evaporation

Feedwater Temperature degrees F.	0	10	15	20	40	50	60	80	100	120	140	150	160	180	200	220	240	260
40	29.29	29.04	28.95	28.67	28.65	28.57	28.50	28.39	28.31	28.24	28.19	28.16	28.14	28.10	28.07	28.05	28.02	28.01
50	29.55	29.29	29.20	29.12	28.90	28.82	28.75	28.63	28.55	28.48	28.42	28.40	28.38	28.34	28.31	28.29	28.26	28.24
60	29.81	29.55	29.46	29.37	29.15	29.07	29.00	28.88	28.79	28.73	28.67	28.65	28.62	28.58	28.55	28.53	28.50	28.48
70	30.08	29.81	29.72	29.63	29.40	29.32	29.25	29.13	29.04	28.98	28.92	28.89	28.87	28.83	28.80	28.77	28.75	28.73
80	30.35	30.08	29.99	29.90	29.66	29.58	29.51	29.38	29.30	29.23	29.17	29.14	29.12	29.08	29.05	29.02	29.00	28.98
90	30.63	30.35	30.26	30.17	29.93	29.85	29.77	29.65	29.56	29.49	29.42	29.40	29.38	29.34	29.30	29.27	29.25	29.23
100	30.92	30.63	30.53	30.44	30.20	30.11	30.04	29.91	29.82	29.75	29.68	29.66	29.64	29.60	29.56	29.53	29.51	29.49
110	31.20	30.92	30.81	30.72	30.47	30.39	30.31	30.18	30.09	30.01	29.95	29.93	29.90	29.86	29.82	29.80	29.77	29.75
120	31.50	31.20	31.10	31.01	30.75	30.67	30.59	30.45	30.36	30.29	30.22	30.20	30.17	30.13	30.09	30.06	30.04	30.02
130	31.80	31.50	31.39	31.30	31.04	30.95	30.87	30.73	30.64	30.56	30.50	30.47	30.45	30.40	30.37	30.34	30.31	30.29
140	32.10	31.80	31.69	31.59	31.33	31.24	31.16	31.02	30.92	30.84	30.78	30.75	30.73	30.68	30.64	30.61	30.59	30.57
150	32.41	32.10	31.99	31.89	31.63	31.53	31.45	31.31	31.21	31.13	31.06	31.04	31.01	30.96	30.93	30.90	30.87	30.85
160	32.73	32.41	32.30	32.20	31.93	31.83	31.75	31.60	31.50	31.42	31.35	31.33	31.30	31.25	31.22	31.18	31.15	31.13
170	33.05	32.73	32.61	32.51	32.23	32.14	32.05	31.91	31.80	31.72	31.65	31.62	31.60	31.55	31.51	31.48	31.45	31.43
180	33.38	33.05	32.94	32.83	32.55	32.45	32.36	32.21	32.11	32.02	31.95	31.92	31.90	31.85	31.81	31.78	31.75	31.72
190	33.72	33.38	33.26	33.16	32.87	32.77	32.68	32.53	32.42	32.33	32.26	32.23	32.20	32.15	32.11	32.08	32.05	32.03
200	34.06	33.72	33.60	33.49	33.19	33.09	33.00	32.84	32.74	32.65	32.57	32.54	32.52	32.47	32.42	32.39	32.36	32.34
212	34.5	34.13	34.01	33.90	33.59	33.49	33.40	33.24	33.12	33.04	32.96	32.93	32.90	32.85	32.81	32.77	32.74	32.72
227.4	35.04	34.68	34.55	34.43	34.12	34.01	33.92	33.75	33.64	33.55	33.47	33.44	33.41	33.35	33.31	33.27	33.24	33.22
232.4	35.22	34.86	34.73	34.61	34.30	34.19	34.09	33.92	33.81	33.71	33.63	33.60	33.57	33.52	33.48	33.44	33.41	33.38
249.8	35.88	35.50	35.36	35.24	34.92	34.80	34.70	34.53	34.41	34.32	34.23	34.20	34.17	34.11	34.07	34.03	34.00	33.97
Btu/lb of steam	1150.8	1160.8	1164.4	1167.6	1176.5	1178.6	1182.4	1187.2	1190.6	1193.3	1195.7	1196.6	1197.5	1199.1	1200.4	1201.5	1202.5	1203.2
Latent Btu/lb.@ 0 psig	970.3	970.3	970.3	970.3	970.3	970.3	970.3	970.3	970.3	970.3	970.3	970.3	970.3	970.3	970.3	970.3	970.3	970.3

Boiler House Notes:			Page No:	

Horizontal Return Tube Boilers: Ratings

Horsepower	Diameter of Boiler, Inches	Shell Length Feet	Tube O.D. Inches	No. of Tubes
42	48.0	12.0	4.0	28
44	48.0	12.0	3.5	34
49	48.0	12.0	3.0	46
49	48.0	14.0	4.0	28
51	48.0	14.0	3.5	34
57	48.0	14.0	3.0	46
61	54.0	14.0	4.0	36
65	54.0	14.0	3.5	44
72	54.0	14.0	3.0	60
70	54.0	16.0	4.0	36
73	54.0	16.0	3.5	44
83	54.0	16.0	3.0	60
84	60.0	16.0	4.0	44
88	60.0	16.0	3.5	54
103	60.0	16.0	3.0	76
94	60.0	18.0	4.0	44
100	60.0	18.0	3.5	54
116	60.0	18.0	3.0	76
101	66.0	16.0	4.0	54
112	66.0	16.0	3.5	70
131	66.0	16.0	3.0	98
113	66.0	18.0	4.0	54
126	66.0	18.0	3.5	70
147	66.0	18.0	3.0	98
128	72.0	16.0	4.0	70
150	72.0	16.0	3.5	96
163	72.0	16.0	3.0	124
154	72.0	18.0	4.0	70
168	72.0	18.0	3.5	96
183	72.0	18.0	3.0	124
159	72.0	20.0	4.0	70
186	72.0	20.0	3.5	96
202	72.0	20.0	3.0	124
177	78.0	18.0	4.0	88
210	78.0	18.0	3.5	122
219	78.0	18.0	3.0	150
196	78.0	20.0	4.0	88
232	78.0	20.0	3.5	122
243	78.0	20.0	3.0	150
218	84.0	18.0	4.0	110
245	84.0	18.0	3.5	144
278	84.0	18.0	3.0	194
241	84.0	20.0	4.0	110
272	84.0	20.0	3.5	144
309	84.0	20.0	3.0	194

Boiler Horsepower: This old rating unit of capacity = evaporating 34.5 lb. of water "from and at" 212 deg. F. (970.3 x 34.5 = 33,475 Btu per hour),steam @ 0 psig.

Note: The above boiler horsepower ratings are for coal or solid fuel. These ratings should be doubled for automatic gas or oil firing.

re: Boiler Tube Data Book, American Fuel Economy, Inc.

tion products into combustibles and oxygen again. The higher the temperature, the greater is this tendency to dissociate. So, the hotter the flame, the greater is the amount of heat reabsorbed by this reversing process, and the rising flame temperature comes to a halt at some equilibrium temperature in the range of 3400°F (1870–2090°C) for most fuels.

b. For a complete formula explanation and description of adiabatic flame temperature see Ref. 1; p. 9–13.

For more details see Figure 2.1 and Table 2.1.

VIII. FUEL COMBUSTION [3,4]

A. What Is Combustion?

Combustion, or burning, is a very rapid combination of oxygen (oxidation) with a fuel, resulting in release of heat (rust is the slowest form of oxidation). The oxygen comes from the air, which (by volume) is, 20.99% oxygen, 78.03% nitrogen, 0.94% argon, and 0.04% other gases.

Because most fuels now used consist almost entirely of carbon and hydrogen, burning involves the rapid oxidation of carbon to carbon dioxide, or carbon monoxide, and of hydrogen to water vapor. Perfect combustion is obtained by mixing and burning just exactly the right proportions of fuel and oxygen so that nothing is left over. Perfect combustion happens only theoretically.

If too much oxygen (excess air) is supplied, we say that the mixture is *lean* and that the fire is *oxidizing*. This results in a flame that tends to be shorter and clearer. The excess oxygen plays no part in the process. It only absorbs heat and cools the surrounding gases and then passes out the stack. If too much fuel (or not enough oxygen) is supplied, we say that the mixture is *rich* and that the fire is *reducing*. This results in a flame that tends to be longer and sometimes smoky. This is usually called incomplete combustion; that is, all of the fuel particles combine with some oxygen, but they cannot obtain enough oxygen to burn completely, so carbon monoxide is formed, which will burn later if given more oxygen.

The oxygen supply for combustion usually comes from the air. Because air contains a large proportion of nitrogen, the required volume of air is much larger than the required volume of pure oxygen. The nitrogen in the air is inert and does not take part in the combustion reaction. It does, however, absorb some of the heat, with the result that the heat energy is spread thinly throughout a large quantity of nitrogen and the combustion products. This means that a much lower flame temperature results from using air instead of pure oxygen.

Primary air is that air which is mixed with the fuel at or in the burner.
Secondary air is usually that air brought in around the burner or above the
grate.
Tertiary air is usually that air brought in downstream of the secondary air
through other openings in the furnace.

B. The Three Ts of Combustion:

Time   Enough time for the oxygen to combine with the fuel.
Temperature   High enough temperature for continuous ignition.
Turbulence   Enough mixing of air and fuel for the fuel to find all the
              oxygen it requires.

IX. HEATING VALUE OF FUEL [1]

A. Measurement of Heat of Combustion

In boiler practice, a fuel's heat of combustion is the amount of energy, expressed
in Btu, generated by the complete combustion, or oxidation, of a unit weight of
fuel. Calorific value, fuel Btu value, and heating value are terms also used.

The amount of heat generated by complete combustion is a constant for a
given combination of combustible elements and compounds. It is not affected
by the manner in which the combustion takes place, provided it is complete.

A fuel's heat of combustion is usually determined by direct calorimeter
measurement of the heat evolved. Gas chromatography is also commonly used
to determine the composition of gaseous fuels. For an accurate heating value of
solid and liquid fuels, a laboratory heating value analysis is required. Numerous
empirical methods have been published for estimating the heating value of coal
based on the proximate or ultimate analyses. One of the most frequently used
correlations is Dulong's formula, which gives reasonable accurate results for bituminous coals (within 2–3%). It is often used as a routine check for calorimeter-
determined values. (It also can be used to estimate the calorific value for other
solid fuels.)

Dulong's formula: HHV $= 14,544$ C $+ 62,028$ [H$_2$ $-$ (O$_2$/8)] $+ 4050$ S

where:

HHV $=$ higher heating value, Btu/lb
C $=$ mass fraction carbon
H$_2$ $=$ mass fraction hydrogene
O$_2$ $=$ mass fraction oxygen
S $=$ mass fraction sulfur

A far superior method for checking whether the heating value is reasonable
in relation to the ultimate analysis is to determine the theoretical air on a mass

TABLE 2.2 Theoretical Air Required for Various Fuels

Fuel[a]	Theoretical air (lb/lb fuel)	HHV (Btu/lb)	Theoretical air	
			Typical (lb/10⁴Btu)	Range (lb/10⁴Btu)
Bituminous coal (VM > 30%)	9.07	12,000	7.56	7.35–7.75
Subbituminous coal (VM > 30%)	6.05	8,000	7.56	7.35–7.75
Oil	13.69	18,400	7.46	7.35–7.55
Natural gas	15.74	21,800	7.22	7.15–7.35
Wood	3.94	5,831	6.75	6.60–6.90
MSW and RDF	4.13	5,500	7.50	7.20–7.80
Carbon	11.50	14,093	8.16	—
Hydrogen	34.28	61,100	5.61	

[a] VM, volatile matter, moisture and ash-free basis; MSW, municipal solid waste; RDF, refuse-derived fuel.

per Btu basis. The Table 2.2 indicates the range of theoretical air values. The equation for theoretical air can be rearranged to calculate the higher heating value, HHV, where the median range for theoretical air for the fuel from the table, m_{air}, is used.

$$HHV = 100 \times \frac{11.51C + 34.29\ H_2 + 4.32\ S - 4.32\ O_2}{m_{air}}$$

where:

HHV = higher heating value (Btu/lb)
C   = mass percent carbon (%)
H_2 = mass percent hydrogen (%)
S   = mass percent sulfur (%)
O_2 = mass percent oxygen (%)
m_{air} = theoretical air (lb/10,000 Btu)

B. Higher and Lower Heating Values

Water vapor is a product of combustion for all fuels that contain hydrogen. The heat content of a fuel depends on whether this vapor remains in the vapor state or is condensed to liquid.

For the lower heating value (LHV) or net calorific value (net heat of combustion at constant pressure), all products of combustion including water are as-

COST OF ENERGY

By: Ralph L. Vandagriff
Program: EnergyCost
Date: 5/5/93
Run date: 5/1/00

Boiler House Notes:

Source	Unit	BTU/Unit	Units per MMBTU	Price Unit
Electricity	kilowatt	292,997,953		$/kilowatt
Natural Gas	1,000 cu.ft	1,000,000	1 (1 MCF)	$/1000 cu ft
Butane	Gallon	102,600	9.74656869	$/Gallon
Propane	Gallon	92,000	10.8695652	$/Gallon
#2 Fuel Oil	Gallon	137,080	7.29501021	$/Gallon
#6 Fuel Oil	Gallon	153,120	6.5306265	$/42Gal Bbl
Coal	Pound	12,500	80	$/Ton Delivered
Biomass	Pound	7,100	140.84507	$/Ton Delivered
Cord	128 cu.ft / Rack 64 cu.ft			$/Ton Delivered
Tire Chips	Pound	14,000	71.42857	$/Ton Delivered

Elec $/kW	Elec $/MMBTU	Nat Gas $/1000cf	NG $/MMBTU	Butane $/Gal	Butane $/MMBTU	Propane $/Gal	Propane $/MMBTU	#2 Oil $/Gal	#2 $/MMBTU	#6 Oil $/Bbl	#6 $/MMBTU	Coal $/Ton	Coal $/MMBTU	Biomass $/Ton	Bio $/MMBTU	Cord $/Ton	Cord $/MMBTU	Tire $/Ton	Tire $/MMBTU
0.0200	5.8599	1.90	1.9000	0.50	4.8733	0.50	5.4348	0.45	3.2828	17.50	2.7212	25.00	1.0000	7.00	0.4930	25.00	0.6929	25.00	0.8929
0.0220	6.4459	2.00	2.0000	0.51	4.9708	0.51	5.5435	0.47	3.4287	17.75	2.7601	26.00	1.0400	7.25	0.5106	26.00	0.7206	26.00	0.9286
0.0240	7.0319	2.10	2.1000	0.52	5.0682	0.52	5.6522	0.49	3.5746	18.00	2.7989	27.00	1.0800	7.50	0.5282	27.00	0.7483	27.00	0.9643
0.0260	7.6179	2.20	2.2000	0.53	5.1657	0.53	5.7609	0.51	3.7205	18.25	2.8378	28.00	1.1200	7.75	0.5458	28.00	0.7760	28.00	1.0000
0.0280	8.2039	2.30	2.3000	0.54	5.2632	0.54	5.8696	0.53	3.8664	18.50	2.8767	29.00	1.1600	8.00	0.5634	29.00	0.8038	29.00	1.0357
0.0300	8.7899	2.40	2.4000	0.55	5.3606	0.55	5.9783	0.55	4.0123	18.75	2.9155	30.00	1.2000	8.25	0.5810	30.00	0.8315	30.00	1.0714
0.0320	9.3759	2.50	2.5000	0.56	5.4581	0.56	6.0870	0.57	4.1582	19.00	2.9544	31.00	1.2400	8.50	0.5986	31.00	0.8592	31.00	1.1071
0.0340	9.9619	2.60	2.6000	0.57	5.5556	0.57	6.1957	0.59	4.3041	19.25	2.9933	32.00	1.2800	8.75	0.6162	32.00	0.8869	32.00	1.1429
0.0360	10.5479	2.70	2.7000	0.58	5.6530	0.58	6.3043	0.61	4.4500	19.50	3.0322	33.00	1.3200	9.00	0.6338	33.00	0.9147	33.00	1.1786
0.0380	11.1339	2.80	2.8000	0.59	5.7505	0.59	6.4130	0.63	4.5959	19.75	3.0710	34.00	1.3600	9.25	0.6514	34.00	0.9424	34.00	1.2143
0.0400	11.7199	2.90	2.9000	0.60	5.8480	0.60	6.5217	0.65	4.7418	20.00	3.1099	35.00	1.4000	9.50	0.6690	35.00	0.9701	35.00	1.2500
0.0420	12.3059	3.00	3.0000	0.61	5.9454	0.61	6.6304	0.67	4.8877	20.25	3.1488	36.00	1.4400	9.75	0.6866	36.00	0.9978	36.00	1.2857
0.0440	12.8919	3.10	3.1000	0.62	6.0429	0.62	6.7391	0.69	5.0336	20.50	3.1877	37.00	1.4800	10.00	0.7042	37.00	1.0255	37.00	1.3214
0.0460	13.4779	3.20	3.2000	0.63	6.1404	0.63	6.8478	0.71	5.1796	20.75	3.2265	38.00	1.5200	10.25	0.7218	38.00	1.0532	38.00	1.3571
0.0480	14.0639	3.30	3.3000	0.64	6.2378	0.64	6.9565	0.73	5.3254	21.00	3.2654	39.00	1.5600	10.50	0.7394	39.00	1.0809	39.00	1.3929
0.0500	14.6499	3.40	3.4000	0.65	6.3353	0.65	7.0652	0.75	5.4713	21.25	3.3043	40.00	1.6000	10.75	0.7570	40.00	1.1086	40.00	1.4286
0.0520	15.2359	3.50	3.5000	0.66	6.4327	0.66	7.1739	0.77	5.6172	21.50	3.3432	41.00	1.6400	11.00	0.7746	41.00	1.1364	41.00	1.4643
0.0540	15.8219	3.60	3.6000	0.67	6.5302	0.67	7.2826	0.79	5.7631	21.75	3.3820	42.00	1.6800	11.25	0.7923	42.00	1.1641	42.00	1.5000
0.0560	16.4079	3.70	3.7000	0.68	6.6277	0.68	7.3913	0.81	5.9090	22.00	3.4209	43.00	1.7200	11.50	0.8099	43.00	1.1918	43.00	1.5357
0.0580	16.9938	3.80	3.8000	0.69	6.7251	0.69	7.5000	0.83	6.0549	22.25	3.4598	44.00	1.7600	11.75	0.8275	44.00	1.2195	44.00	1.5714
0.0600	17.5798	3.90	3.9000	0.70	6.8226	0.70	7.6087	0.85	6.2008	22.50	3.4987	45.00	1.8000	12.00	0.8451	45.00	1.2472	45.00	1.6071
0.0620	18.1658	4.00	4.0000	0.71	6.9201	0.71	7.7174	0.87	6.3467	22.75	3.5375	46.00	1.8400	12.25	0.8627	46.00	1.2749	46.00	1.6429
0.0640	18.7518	4.10	4.1000	0.72	7.0175	0.72	7.8261	0.89	6.4926	23.00	3.5764	47.00	1.8800	12.50	0.8803	47.00	1.3027	47.00	1.6786
0.0660	19.3378	4.20	4.2000	0.73	7.1150	0.73	7.9348	0.91	6.6386	23.25	3.6153	48.00	1.9200	12.75	0.8979	48.00	1.3304	48.00	1.7143
0.0680	19.9238	4.30	4.3000	0.74	7.2125	0.74	8.0435	0.93	6.7844	23.50	3.6542	49.00	1.9600	13.00	0.9155	49.00	1.3581	49.00	1.7500
0.0700	20.5098	4.40	4.4000	0.75	7.3099	0.75	8.1522	0.95	6.9303	23.75	3.6930	50.00	2.0000	13.25	0.9331	50.00	1.3858	50.00	1.7857
0.0720	21.0958	4.50	4.5000	0.76	7.4074	0.76	8.2609	0.97	7.0762	24.00	3.7319	51.00	2.0400	13.50	0.9507	51.00	1.4135	51.00	1.8214
0.0740	21.6818	4.60	4.6000	0.77	7.5049	0.77	8.3696	0.99	7.2221	24.25	3.7708	52.00	2.0800	13.75	0.9683	52.00	1.4413	52.00	1.8571
0.0760	22.2678	4.70	4.7000	0.78	7.6023	0.78	8.4783	1.01	7.3680	24.50	3.8096	53.00	2.1200	14.00	0.9859	53.00	1.4690	53.00	1.8929
0.0780	22.8538	4.80	4.8000	0.79	7.6998	0.79	8.5870	1.03	7.5139	24.75	3.8485	54.00	2.1600	14.25	1.0035	54.00	1.4967	54.00	1.9286
0.0800	23.4396	4.90	4.9000	0.80	7.7973	0.80	8.6957	1.05	7.6598	25.00	3.8874	55.00	2.2000	14.50	1.0211	55.00	1.5244	55.00	1.9643
0.0820	24.0256	5.00	5.0000	0.81	7.8947	0.81	8.8043	1.07	7.8057	25.25	3.9263	56.00	2.2400	14.75	1.0387	56.00	1.5521	56.00	2.0000
0.0840	24.6116	5.10	5.1000	0.82	7.9922	0.82	8.9130	1.09	7.9516	25.50	3.9651	57.00	2.2800	15.00	1.0563	57.00	1.5798	57.00	2.0357
0.0860	25.1976	5.20	5.2000	0.83	8.0897	0.83	9.0217	1.11	8.0975	25.75	4.0040	58.00	2.3200	15.25	1.0739	58.00	1.6076	58.00	2.0714
0.0880	25.7836	5.30	5.3000	0.84	8.1871	0.84	9.1304	1.13	8.2434	26.00	4.0429	59.00	2.3600	15.50	1.0915	59.00	1.6353	59.00	2.1071
0.0900	26.3696	5.40	5.4000	0.85	8.2846	0.85	9.2391	1.15	8.3893	26.25	4.0818	60.00	2.4000	15.75	1.1092	60.00	1.6630	60.00	2.1429
0.0920	26.9556	5.50	5.5000	0.86	8.3821	0.86	9.3478	1.17	8.5352	26.50	4.1206	61.00	2.4400	16.00	1.1268	61.00	1.6907	61.00	2.1786
0.0940	27.5416	5.60	5.6000	0.87	8.4795	0.87	9.4565	1.19	8.6811	26.75	4.1595	62.00	2.4800	16.25	1.1444	62.00	1.7184	62.00	2.2143
0.0960	28.1277	5.70	5.7000	0.88	8.5770	0.88	9.5652	1.21	8.8270	27.00	4.1984	63.00	2.5200	16.50	1.1620	63.00	1.7461	63.00	2.2500
0.0980	28.7137	5.80	5.8000	0.89	8.6745	0.89	9.6739	1.23	8.9729	27.25	4.2373	64.00	2.5600	16.75	1.1796	64.00	1.7739	64.00	2.2857
0.1000	29.2997	5.90	5.9000	0.90	8.7719	0.90	9.7826	1.25	9.1188	27.50	4.2761	65.00	2.6000	17.00	1.1972	65.00	1.8016	65.00	2.3214
0.1020	29.8857	6.00	6.0000	0.91	8.8694	0.91	9.8913	1.27	9.2647	27.75	4.3150	66.00	2.6400	17.25	1.2148	66.00	1.8293	66.00	2.3571
0.1040	30.4717	6.10	6.1000	0.92	8.9669	0.92	10.0000	1.29	9.4106	28.00	4.3539	67.00	2.6800	17.50	1.2324	67.00	1.8570	67.00	2.3929
0.1060	31.0577	6.20	6.2000	0.93	9.0643	0.93	10.1087	1.31	9.5565	28.25	4.3928	68.00	2.7200	17.75	1.2500	68.00	1.8847	68.00	2.4286
0.1080	31.6437	6.30	6.3000	0.94	9.1618	0.94	10.2174	1.33	9.7024	28.50	4.4316	69.00	2.7600	18.00	1.2676	69.00	1.9124	69.00	2.4643
0.1100	32.2297	6.40	6.4000	0.95	9.2593	0.95	10.3261	1.35	9.8483	28.75	4.4705	70.00	2.8000	18.25	1.2852	70.00	1.9402	70.00	2.5000
0.1120	32.8157	6.50	6.5000	0.96	9.3567	0.96	10.4348	1.37	9.9942	29.00	4.5094	71.00	2.8400	18.50	1.3028	71.00	1.9679	71.00	2.5357
0.1140	33.4017	6.60	6.6000	0.97	9.4542	0.97	10.5435	1.39	10.1401	29.25	4.5483	72.00	2.8800	18.75	1.3204	72.00	1.9956	72.00	2.5714
0.1160	33.9877	6.70	6.7000	0.98	9.5517	0.98	10.6522	1.41	10.2860	29.50	4.5871	73.00	2.9200	19.00	1.3380	73.00	2.0233	73.00	2.6071
0.1180	34.5737	6.80	6.8000	0.99	9.6491	0.99	10.7609	1.43	10.4319	29.75	4.6260	74.00	2.9600	19.25	1.3556	74.00	2.0510	74.00	2.6429
0.1200	35.1597	6.90	6.9000	1.00	9.7466	1.00	10.8696	1.45	10.5778	30.00	4.6649	75.00	3.0000	19.50	1.3732	75.00	2.0787	75.00	2.6786
0.1220	35.7457	7.00	7.0000	1.01	9.8441	1.01	10.9783	1.47	10.7237	30.25	4.7037	76.00	3.0400	19.75	1.3908	76.00	2.1065	76.00	2.7143
0.1240	36.3317	7.10	7.1000	1.02	9.9415	1.02	11.0870	1.49	10.8696	30.50	4.7426	77.00	3.0800	20.00	1.4085	77.00	2.1342	77.00	2.7500

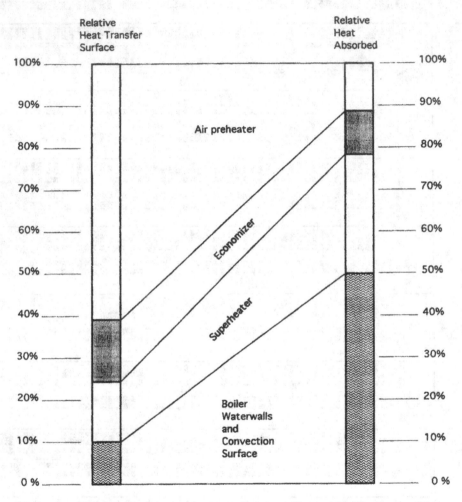

FIGURE 2.2   Heating surface versus heat absorbed. For waterwalls exposed to radiant heat, in the zone of highest temperature, each square foot of waterwalls surface absorbs heat at a rate many times greater than the convection surface.

sumed to remain in the gaseous state, and the water heat of vaporization is not available.

While the high, or gross, heat of combustion can be accurately determined by established (American Society of Testing and Materials; ASTM) procedures, direct determination of the low heat of combustion is difficult. Therefore, it is usually calculated using the following formula:

$$LHV = HHV - 10.30 \, (H_2 \times 8.94)$$

where
LHV = lower heating or net calorific value (Btu/lb)
HHV = higher heating or gross calorific value (Btu/lb)
H_2 = mass percent hydrogen in the fuel (%)

This calculation contains a correction for the difference between a constant volume and a constant pressure process and a deduction for the water vaporization in the combustion products. At 68°F, the total deduction is 1030 Btu/lb of water, including 1055 Btu/lb for the enthalpy of water vaporization.

Note: In the United States fuel is usually purchased on a higher-heating value (HHV) basis. A lower-heating value (LHV) is often used in Europe.

The heating values of gaseous fuels are best calculated from the values listed for the individual constituents. For example, Anadarko Basin natural gas has 8.4% N_2, 84.1% CH_4, and 6.7% C_2H_6. The corresponding HHVs are 0, 1012, and 1773 Btu/ft^3.

Therefore; $(0.084 \times 0) + (0.841 \times 1012) + (0.067 \times 1773) = 969.88$ Btu/ft^3 (HHV)

For more information see Table 2.3 and Figure 2.2.

X. BOILER TUBING [42]

A. Causes and Treatments for Tube Failures

1. Hard Scale

Treatment. Lime and sodium carbonate (soda ash) or zeolite treatment of feed water; phosphate to boiler water.

2. Sludge

Treatment. Concentrates of phosphate, silica, and alkali. Wash out and blowdown. Polymeric conditioners. Chelating agents. Some sludge may be tolerated in low-heat release boilers if dispersive materials are used.

3. Corrosion

Attack by	Treatment
Steam	Increase circulation; baffle changes
Concentrated boiler water	Polymer treatment to reduce deposit of corrosion products, keep boiler clean
Caustic	Sodium nitrate
Acid	Frequent alkalinity checks
Stress corrosion and corrosion fatigue	Mechanical and operational changes
Oxygen	Deaerators—sodium and sulfate
Damage in standby	Proper layup procedures
Acidic sulfur compounds	Fuel conditioners

B. Causes and Prevention of Tube Failures

Boiler generating tubes fail for one of three reasons: they have been overheated; they have been attacked chemically on either the water or fireside; or they have been thinned down from the fireside. Overheating may be caused by deposits on the waterside that prevent adequate cooling of the tube metal, or by unusually hot zones in the furnace. The problem of chemical attack is in some cases related to overheating, for high temperatures greatly speed up certain chemical reactions involving iron. Most chemical attack can be prevented, however, by proper boiler water treatment.

External thinning may be caused by the improper placement of soot blowers and by abrasive particles in the fuel.

C. Scale and Sludge

What is the difference between scale and sludge? Simply stated, *scale* is a coating that forms from a solution when heated; *sludge* is a suspended sediment found in the boiler water that sticks to boiler surfaces if certain precautions are not taken.

When heat transfer rates are high, it is essential that scale formation be prevented. Not only does the insulating effect decrease fuel efficiency, but it also prevents the boiler water from cooling in the tubes. Eventually, this overheating leads to rupture.

Remember, if a large boiler does not have the benefit of a high percentage of condensate return, without proper treatment it will receive many pounds per day of calcium hardness, causing rapid tube failure. The best way to treat calcium scale is pretreatment with lime and sodium carbonate or zeolite for hardness removal followed by phosphate, or phosphate with a sludge conditioner, or treatment with a chelating agent.

Not all scales, however, contain calcium. An equally harmful problem occurs when soluble compounds are deposited in heavily worked areas of the boiler. For instance, steam is generated right at the tube surface, where the temperature is greatest. If this area is not sufficiently scrubbed, boiler water solids will be deposited. Under extreme conditions, highly soluble materials, such as common salt, may precipitate. Under less severe conditions, scrubbing may keep deposits from forming; however, the transfer of heat from metal to water sometimes produces a concentrated film of boiler water that exceeds the solubility of certain compounds of iron, aluminum, sodium, and silicon.

Minimize scale deposits caused by insufficient scrubbing by following these steps: increase liquid velocity to obtain better scrubbing and dissipation of the concentrating film; decrease or redistribute heat input so that lower concentrations will result in the water film and less steam will be generated in the problem areas; and modify boiler water composition to reduce the concentration of scale-forming elements.

Sticky sludge may be avoided by maintaining the proper phosphate, silica, and alkali concentrations in the boiler water. Minimize circulating sludge by pre-treating the feedwater to remove hardness.

When raw water is used or hardness removal is poor, use high blowdown rates to limit sludge concentrations in the boiler. Frequently, high sludge concentrations are tolerated when certain sludge-conditioning materials are fed to the boiler. Lignin derivatives and polymeric sludge conditioners, for example, have effectively reduced boiler sludge.

Chelating agents have also become an important tool in internal treatment. Agents, such as the sodium salts of EDTA or NTA, tie up hardness in a soluble form and avoid the formation of suspended solids. The cost of these chelating agents makes them practical only when feedwater hardness is low. Control and methods of feeding are critical. If misused, serious corrosion results. Especially, with a feedwater pH of less than 8.0 or feedwater dissolved oxygen concentrations greater than 0.01 ppm.

D. Waterside Corrosion

Under normal circumstances, water is the principal reactant for the corrosion of boiler steel. However, oxygen, caustic, and other boiler water solids may chemically affect the rate of reaction or cause it to occur in specific areas.

Caustic damages the boiler tube in two ways. First, it may strip off the protective iron oxide coating and permit the exposed iron to react with the water. Not only is the metal removed by this process, but the remaining metal may be embrittled by the hydrogen given off during the reaction. To compensate, increase the water circulation or modify the furnace baffling to decrease the heat input.

Similar attacks occur beneath deposits in the condensate and feedwater system. Often, the deposits are an accumulation of iron and copper oxides that usually stick on heat-transfer surfaces—especially at the burner level in waterwall tubes and just beyond sharp bends. Because the deposits are porous, they provide a trap where boiler water solids can concentrate.

Damage to the tubes beneath these oxide deposits may be reduced by keeping the amounts of iron and copper returned to the boiler with the condensate at very low levels. Neutralizing amines raise the pH of the condensate of filming amines to provide a nonwettable barrier between the condensate and the metal.

Caustic also damages and embrittles the metal in zones adjacent to minute leaks. Boiler water passes through the opening, flashes to atmosphere, and leaves a concentrated brine residue in the opening. Caustic embrittlement has been successfully prevented, even when leaks have occurred, by the maintenance of a suitable ratio of caustic to other substances, such as sodium nitrate, in the boiler water.

Operation under acidic conditions rapidly removes the oxide coating from the boiler metal and allows the boiler water to react with the iron. Ensure a standard level of alkalinity by frequently checking the boiler water samples.

Corrosion fatigue occurs when water reacts with tube metal during stress fluctuations. Stresses in the tubes may be triggered by nonuniform firing or by a loose fireside refractory that sometimes prevents free movement of the boiler metal as it expands from heating. Repair faulty baffling to prevent local overheating and keep a uniform flow of hot gases through the furnace.

If oxygen in the feedwater causes a scattered pitting in the boiler, reduce the oxygen to a very low level by passing the feedwater through a deaerating heater. If the boiler is susceptible to corrosion even at extremely low levels of oxygen, add oxygen-scavenging chemicals, such as sodium sulfite, to the boiler water.

E.   Corrosion in Standby Boilers

The damage done by oxygen to standby nonoperating boilers and auxiliary equipment cannot be overemphasized. Boiler tubes may develop large pits during either wet or dry standby. Check boilers in wet standby for proper levels of alkali and sodium sulfite in the boiler water. Place trays of quicklime in a boiler during dry standby to prevent damage from oxygen-containing condensate. Enough heat should be applied to keep the metal temperature above the dew point.

F.   Fireside Failure

Some coals and fuel oils often contain significant amounts of sulfur, which decompose to acidic gases while burning. If temperatures become low enough to allow these gases to condense, sulfuric acid results and aggressively destroys the tube metal.

Also, some fuel oils contain vanadium, which corrodes boiler tubes in the high-temperature area of the boiler. Vanadium corrosion can be minimized with fuel additives.

Tubes may also fail on their fireside by the abrasive action of fly ash and dirt particles in the fuel. In some steel mills, where blast furnace gas is used as a fuel, this is a frequent cause of failure. Washing the gas before firing often remedies the problem.

A frequent cause of tube failure is the cutting action of steam from improperly placed soot blowers.

G. Superheater Tube Failure

Superheater tubes rarely fail because of corrosion, unless it has occurred during periods of standby. Failures of these tubes are almost always attributed to overheating, which may be caused by the insulating effect of deposits carried over by the boiler water, or by insufficient steam flow.

Superheater deposits are typically caused by boiler water carryover. Carryover results from excessive concentrations of dissolved solids or alkali or the presence of oil or other organic substances. Foaming caused by high solids and alkalinity may be minimized by increasing the boiler blowdown pressure or by the addition of an antifoam material to the boiler water. The elimination of organic contaminants may require special pretreating equipment or the insertion of an oil separator in the feedwater line. It should be emphasized here that pure hydrocarbon oils do not cause carryover, but that additives in them frequently do.

Carryover in the form of slugs of water is another cause of superheater deposition, and this may be remedied by lowering the water level or by obtaining better water level control. However, large load swings may also contribute to this condition.

Starvation of superheater tubes may result from poor start-up practice or by operating the boiler at undesirably low ratings. If, during start-up, the furnace temperature is brought up too rapidly before full boiler pressure is attained, there will be insufficient steam to cool the tubes. In boilers with a radiant type superheater, this may occur at low loads.

H. Regular Inspection Prevents Tube Failure

With regular, detailed inspection of boiler equipment, many tube failures can be prevented. Written records should note changes in corrosion or deposits. Pay attention to unexpected layers of deposit that flake off tube and drum internal surfaces and accumulate in tube bends or headers. Usually, this indicates reduced boiler water circulation and potential overheating.

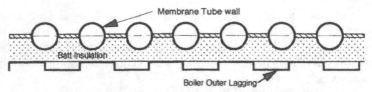

Cross Section of an Outer Membrane Wall

(These walls are water-cooled and constructed of bare tubes
joined by thin membrane bars forming a gas tight wall. This
wall is backed by batt insulation and an outer casing.

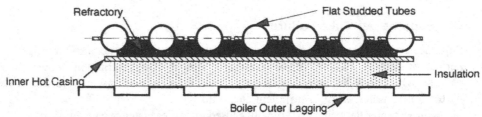

Cross Section of a Flat Stud Tube Wall w/ Inner Casing

(These walls consist of tubes with small, flat bar studs welded at the sides. In current method
of construction, the flat studded tubes are backed with refractory covered with a welded inner
hot casing that is insulated and covered with metal lagging for protection.)

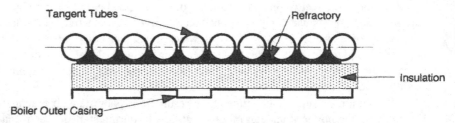

Cross Section of a Tangent Tube Wall w/ Outer Casing

(These walls are constructed of bare tubes placed next to each other with a typical gap
of 0.03125 inch. Backed up with refractory, insulation and casing.)

FIGURE 2.3   Three types of boiler outer wall.

TABLE 2.4

							Boiler House Notes			Page:
						High Heat Input	Other	SH		
					Recommended		Furnace	RH		
	Nominal		Minimum	Minimum	Maximum	Furnace	Walls &	Econ	Drums	
Specification	Composition	Product	Tensile, ksi	Yield, ksi	Use Temp.,F.	Walls	Enclosures			Notes:
ERW (Electric Resistance Welded) Tubing										
SA-178A	Carbon Steel	Tube	.(47.0)	.(26.0)	950	x	x	x		1,2
SA-178C	Carbon Steel	Tube	60.00	37.00	950		x	x		2
SA-178D	Carbon Steel	Tube	70.00	40.00	950	x	x	x		2
SA-250T1A	Carbon-Mo	Tube	60.00	32.00	975		x	x		4,5
SA-250T2	.5Cr-.5Mo	Tube	60.00	30.00	1025	x		x		6,7
SA-250T12	1.0Cr-.5Mo	Tube	60.00	32.00	1050			x		5,7
SA-250T11	1.25Cr-.5Mo-Si	Tube	60.00	30.00	1050			x		5
SA-250T22	2.25Cr-1.0Mo	Tube	60.00	30.00	1115			x		5,7
Seamless Tubing										
SA-192	Carbon Steel	Tube	.(47.0)	.(26.0)	950	x	x	x		1
SA-210-A1	Carbon Steel	Tube	60.00	37.00	950	x	x	x		
SA-210-C	Carbon Steel	Tube	70.00	40.00	950		x	x		
SA-209T1A	Carbon-Mo	Tube	60.00	32.00	975		x	x		4
SA-213T2	.5Cr-.5Mo	Tube	60.00	30.00	1025	x		x		6
SA-213T12	1.0Cr-.5Mo	Tube	60.00	32.00	1050			x		8
SA-213T11	1.25Cr-.5Mo-Si	Tube	60.00	30.00	1050			x		
SA-213T22	2.25Cr-1.0Mo	Tube	60.00	30.00	1115			x		
SA-213T91	9Cr-1Mo-V	Tube	85.00	60.00	1200			x		
SA-213TP304H	18Cr-8Ni	Tube	75.00	30.00	1400			x		
SA-213TP347H	18Cr-10Ni-Cb	Tube	75.00	30.00	1400			x		
SA-213TP310H	25Cr-20Ni	Tube	75.00	30.00	1500			x		
SB-407-800H	Ni-Cr-Fe	Tube	65.00	25.00	1500			x		
SB-423-825	Ni-Fe-Cr-Mo-Cu	Tube	85.00	35.00	1000			x		
Steel Plate										
SA-516-70	Carbon Steel	Plate	70.00	38.00	800				x	
SA-299	Carbon Steel	Plate	75.00	40.00	800				x	

Notes:
1. Values in parentheses are not required minimums, but are expected minimums.
2. Requires special inspection if used at 100% efficiency above e 850 F.
4. Limited to 875 F. maximum for applications outside the boiler setting.
5. Requires special inspection if used at 100% efficiency.
6. Maximum OD temperature is 1025 F. Maximum mean metal temperature for Code calculations is 1000 F.
7. Requires use of a Code Case now. Will not later.
8. 32 ksi minimum yield requires use of Code Case 2070, which is being incorporated into the Code.

re: #1

Ralph L. Vandagriff

But be advised, water treatment is a highly technical science requiring careful water analysis and consideration of boiler design and operating conditions. "Cure all" chemicals can do more harm than good to vital boiler equipment. Many reputable firms specialize in industrial water consulting. So, do not take chances. Consult an expert.

See Figure 2.3 for types of boiler outer walls. Table 2.4 summarizes specifications for boiler tubing and drum materials

XI. REFRACTORY

Nonmetallic refractory materials are widely used in high-temperature applications in which the service permits the appropriate type of construction. The more important classes are described in the following paragraphs.

Fireclays can be divided into plastic clays and hard flint clays; they may also be classified by alumina content.

A. Brick Materials

1. Firebricks

Firebricks are usually made of a blended mixture of flint clays and plastic clays that is formed, after mixing with water, to the required shape. Some or all of the flint clay may be replaced by highly burned or calcined clay, called grog. A large proportion of modern brick production is molded by the dry-press or power-press process, in which the forming is carried out under high pressure and with a low water content. Extruded and hand-molded bricks are still made in large quantities.

The dried bricks are burned in either periodic or tunnel kilns at temperature ranging between 2200°F (1200°C) and 2700°F (1500°C). Tunnel kilns give continuous production and a uniform burning temperature. Fireclay bricks are used in kilns, malleable-iron furnaces, incinerators, and many portions of metallurgical furnaces. They are resistant to spalling and stand up well under many slag conditions, but are not generally suitable for use with high-lime slags or fluid–coal–ash slags, or under severe load conditions.

2. High-Alumina Bricks

These bricks are manufactured from raw materials rich in alumina, such as diaspore. They are graded into groups with 50, 60, 70, 80, and 90% alumina content. When well fired, these bricks contain a large amount of mullite and less of the glassy phase that is present in firebricks. Corundum is also present in many of these bricks. High-alumina bricks are generally used for unusually severe temperature or load conditions. They are employed extensively in lime kilns and rotary cement kilns, in the ports and regenerators of glass tanks, and for slag resistance in some metallurgical furnaces; their price is higher than that of firebrick.

3. Silica Bricks

These bricks are manufactured from crushed ganister rock containing about 97–98% silica. A bond consisting of 2% lime is used, and the bricks are fired in periodic kilns at temperatures of 2700°F (1500°C) to 2800°F (1540°C) for several days until a stable volume is obtained. They are especially valuable when good strength is required at high temperatures. Superduty silica bricks are finding some use in the steel industry. They have a lowered alumina content and often a lowered porosity.

Silica bricks are used extensively in coke ovens, the roofs and walls of open-hearth furnaces, and the roofs and sidewalls of glass tanks, and as linings of acid electric steel furnaces. Although silica brick is readily spalled (cracked by a temperature change) below red heat, it is very stable if the temperature is kept above this range, and consequently, it stands up well in regenerative furnaces. Any structure of silica brick should be heated slowly to the working temperature; a large structure often requires 2 weeks or more.

4. Magnesite Bricks

These bricks are made from crushed magnesium oxide, which is produced by calcining raw magnesite rock to high temperatures. A rock containing several percent of iron oxide is preferable, as this permits the rock to be fired at a lower temperature than if pure materials were used. Magnesite bricks are generally fired at a comparatively high temperature in periodic or tunnel kilns. A large proportion of magnesite brick made in the United States uses raw material extracted from seawater. Magnesite bricks are basic and are used whenever it is necessary to resist high-lime slags, as in the basic open-hearth steel furnace. They also find use in furnaces for the lead-refining and copper-refining industries. The highly pressed unburned bricks find extensive use in linings for cement kilns. Magnesite bricks are not so resistant to spalling as fireclay bricks.

5. Chrome Bricks

Although manufactured in much the same way as magnesite bricks chrome bricks are made from natural chromite ore. Commercial ores always contain magnesia and alumina. Unburned hydraulically pressed chrome bricks are also available.

Chrome bricks are very resistant to all types of slag. They are used as separators between acid and basic refractories, also in soaking pits and floors of forging furnaces. The unburned hydraulically pressed bricks now find extensive use in the walls of the open-hearth furnace. Chrome bricks are used in sulfite-recovery furnaces and, to some extent, in the refining of nonferrous metals. Basic bricks combining various properties of magnesite and chromite are now made in large quantities and, for some purposes, have advantages over either material alone.

6. Insulating Firebrick

This is a class of brick that consists of a highly porous fireclay or kaolin. Such bricks are light in weight (about one-half to one-sixth of the weight of fireclay), low in thermal conductivity, and yet sufficiently resistant to temperature to be used successfully on the hot side of the furnace wall, thus permitting thin walls of low thermal conductivity and low heat content. The low heat content is particularly valuable in saving fuel and time on heating up, allows rapid changes in temperature to be made, and permits rapid cooling. These bricks are made in a variety of ways, such as mixing organic matter with the clay and later burning it out to form pores; or a bubble structure can be incorporated in the clay–water mixture that is later preserved in the fired brick. The insulating firebricks are classified into several groups according to the maximum use limit; the ranges are up to 1600, 2000, 2300, 2600, and above 2800°F.

Insulating refractories are used mainly in the heat-treating industry for furnaces of the periodic type. They are also used extensively in stress-relieving furnaces, chemical-process furnaces, oil stills or heaters, and the combustion chambers of domestic oil-burner furnaces. They usually have a life equal to that of the heavy brick that they replace. They are particularly suitable for construction of experimental or laboratory furnaces because they can be cut or machined readily to any shape. They are not resistant to fluid slag.

7. Miscellaneous Brick

There are several types of special brick obtainable from individual producers. High-burned kaolin refractories are particularly valuable under conditions of severe temperature and heavy load or severe spalling conditions, as in of high-temperature oil-fired boiler settings or piers under enameling furnaces. Another brick for the same uses is a high-fired brick of Missouri aluminous clay.

There are a number of bricks on the market that are made from electrically fused materials, such as fused mullite, fused alumina, and fused magnesite. These bricks, although expensive, are particularly suitable for certain severe conditions.

Bricks of silicon carbide, either recrystallized or clay-bonded, have a high thermal conductivity and find use in muffle walls and as a slag-resisting material.

Other types of refractory that find use are forsterite, zirconia, and zircon. Acid-resisting bricks consisting of a dense–body-like stoneware are used for lining tanks and conduits in the chemical industry. Carbon blocks are used very extensively in linings for the crucibles of blast furnaces in several countries and, to a limited extent, in the United States. Fusion-cast bricks of mullite or alumina are largely used to line glass tanks.

8. Ceramic Fiber-Insulating Linings

Ceramic fibers are produced by melting the same alumina–silica china (kaolin) clay used in conventional insulating firebrick and blowing air to form glass fibers.

The fibers, 2–4 in. long by 3 μm in diameter, are interlaced into a mat blanket with no binders, or chopped into shorter fibers and vacuum-formed into blocks, boards, and other shapes. Ceramic fiber linings, available for the temperature range of 1200–2600°F are more economical than brick in the 1200–2250°F range. Savings come from reduced first costs, lower installation labor, 90–95% less weight, and a 25% reduction in fuel consumption.

Because of the larger surface area (compared with solid-ceramic refractories) the chemical resistance of fibers is relatively poor. Their acid resistance is good, but they have less alkali resistance than solid materials because of the absence of resistant aggregates. Also, because they have less bulk, fibers have lower gas velocity resistance. Besides the advantage of lower weight, because they will not hold heat, fibers are more quickly cooled and present no thermal shock structural problem.

B. Castable Monolithic Refractories

Standard portland cement is made of calcium hydroxide. In exposures above 800°F the hydroxyl ion is removed from portland (water removed); below 800°F water is added. This cyclic exposure results in spalling. Castables are made of calcium aluminate (rather than portland); without the hydroxide, they are not subject to that cyclic spalling failure.

Castable refractories are of three types:

1. *Standard*: 40% alumina for most applications at moderate temperatures.
2. *Intermediate purity*: 50–55% alumina. The anorthite (needle-structure) form is more resistant to the action of steam exposure.
3. *Very pure*: 70–80% alumina for high temperatures. Under reducing conditions the iron in the ceramic is controlling, as it acts as a catalyst and converts the CO to CO_2 plus carbon, which results in spalling. The choice among the three types of castables is generally made for economic considerations and the temperature of the application.

Compared with brick, castables are less dense, but this does not really mean that they are less serviceable, as their cements can hydrate and form gels that can fill the voids in castables. Extralarge voids do indicate less strength regardless of filled voids and dictate a lower allowable gas velocity. If of the same density as a given brick, a castable will result in less permeation.

Normally, castables are 25% cements and 75% aggregates. The aggregate is the more chemically resistant of the two components. The highest-strength materials have 30% cement, but too much cement results in too much shrinkage. The standard insulating refractory, 1:2:4 LHV castable, consists of 1 volume of cement, 2 volumes of expanded clay (Haydite), and 4 volumes of vermiculite.

Castables can be modified by a clay addition to keep the mass intact, thereby allowing application by air-pressure gunning (gunite). Depending on the size and geometry of the equipment, many castable linings must be reinforced; wire and expanded metal are commonly used.

Andalusite	Al_2OSiO_4 A natural silicate of aluminum
Corundum	Al_2O_3 (emery) Natural aluminum oxide sometimes with small amounts of iron, magnesium, silica, or other
Mullite	$3Al_2O_3 \cdot 2SiO_2$ A stable form of aluminum silicate formed by heating other aluminum silicates (such as cyanite, sillimanite, and andalusite) to high temperatures; also found in nature.

C. Refractory Materials in Boilers [1,2,7,8,9]

The furnace and other wall areas of modern fossil fuel boilers are made almost entirely of water-cooled tubes. The increased use of membrane walls has reduced the use of refractory in these areas. However, castable and plastic refractories may still be used to seal flat studded areas, wall penetrations, and door and wall box seals.

Other than membrane wall construction, when tubes are tangent or flat-studded, several types of plastic refractory materials are applied to the outside of tubes for insulation or sealing purposes. When an inner casing is to be applied directly against the tubes, the refractory serves principally as an inert filler material for the lanes between tubes. It has a binder that cements it to the tubes and it is troweled flush with the surface of the wall.

On cyclone-fired and process-recovery boilers, a gunnite grade of plastic refractory is applied to areas having studs welded to the tubes to control burner performance and slagging characteristics. Studded tubes with refractory coating are used in some instances if severe corrosion attack occurs when burning fuels with sulfur, chlorine, and such.

Smaller boilers have furnaces constructed of tubes on wide-spaced centers backed with a layer of brick or tile. Brickwork of this type is supported by the pressure parts and held in place by studs. The brick is insulated and made air tight by various combinations of plastic refractory, plastic or block insulation, and casing. High quality workmanship is mandatory in the application of refractory material. Construction details are clearly outlined on drawings and instructions, which also designate the materials to be used. These materials must be applied to correct contour and thickness without voids or excessive cracking. Skilled mechanics and close supervision are essential.

D. Furnace Refractory: Chemical Considerations [40]

Refractories are affected by the action of the atmosphere in the heating equipment, and by chemical attack on *permeable* materials. In heating equipment, non-oxidizing atmospheres create some problems for insulating refractories. Low oxygen pressure (i.e., concentration) reduces Fe_2O_3 to FeO. Because FeO is less refractory than Fe_2O_3, this condition promotes hot-load deformation of the refractory at temperatures higher than 1800°F. This effect will also reduce SiO_2 to SiO (a gas), which also causes refractory failure. On cooling, the SiO tends to form deposits on surfaces in heat-exchangers, boilers, and reformers.

Refractory fibers should not be used above 900°F, or at a dewpoint lower than −20°F. As the dewpoint is lowered, the service temperature must also be lowered to prevent a particular oxide from being reduced. For example, if the dewpoint for SiO_2 is −60°F, the temperature must be maintained below 2400°F to prevent reduction of SiO_2.

A disintegration triggered by the catalytic decomposition of carbon monoxide or hydrocarbons such as methane, also occurs in prepared atmospheres. Let us review the mechanism for failure because of this reaction. Ferric oxide (FeO_3) is present in the refractory in localized concentrations. This is converted to iron carbide (Fe_3C) that catalyzes the decomposition of carbon monoxide (at 750°– 1,300°F) to carbon dioxide and carbon. The carbon is deposited on the Fe_3C. Carbon builds up on the catalytic surfaces that are under stress, and ultimately caused disintegration of the refractory. Frequently, such stresses are severe enough to burst the steel shell of the furnace.

A similar condition exists in hydrocarbon atmospheres, which persists up to 1700°F. The risk of carbon disintegration can be minimized by using refractories above the carbon-deposition temperature, and by using products having low iron content. This effect is more noticeable in dense-brick or castable refractories.

Furnace atmospheres also have an effect on the thermal conductivity [10] of insulating refractories. Let us review how the thermal conductivity of the furnace gases affects these refractories. The component of thermal conductivity affected by changing the gas constituents in the furnace atmosphere is the gas conduction. It is quite apparent that the gas (normally air) *in the pores* of an insulating refractory can be readily replaced by other gas constituents found in the furnace atmosphere. Replacement or dilution of this air constituent by other gases *will change the insulation's* k *value*. The amount of change is dependent on the k value of the replacement gas, and the *porosity* of the insulation. Most gases involved in a heating atmosphere have essentially the same k value as air. However, hydrogen has a very high k value, causing a significant change in insulation effectiveness. Other compounds can be found in small quantities in many furnace atmospheres. They can originate in the fuel, the refractories, or even the charge in the kiln or furnace. One of these is vanadium pentoxide (V_2O_5) for low-grade fuel oils such as bunker C. Another vanadium compound, sodium vanadate, may be found in oil flames as droplets. It appears to decompose and cause alkali attack at about 1240°F.

Sulfur occurs in fuels and in some clays. Depending on its content and chemical form, it can become part of a furnace atmosphere as sulfur oxides.

Alumina–silica (45–54% Al_2O_3 range) maintains the highest hot strength of the several refractory materials cycled at 1400–1800°F in the presence of an SO_3 atmosphere.

Test data show that disintegration of the refractory may occur if Na_2SO_4, $MgSO_4$, $Al_2(SO_4)_3$, or $CaSO_4$ are formed in a sulfur dioxide atmosphere.

Insulating refractories are seldom used in installations in which chemical attack is expected. Unfortunately, it often arises unexpectedly. Primarily, such attack is due to the *permeability* of insulating refractories. Many times, dense refractories resist chemical attack, not because of chemical resistance but because their very high density and *low permeability prevent damaging materials from entering*.

Slagging is defined by the American Society of Thermal Manufacturers as "the destructive chemical reaction between refractories and external agencies at high temperatures, resulting in the formation of a liquid with the refractory."

Attack from mill scale occasionally causes slagging. As ferrous metal objects are heated, their surfaces oxidize, and the resulting iron oxides flake. These oxides fall onto the hearth, and will readily attack an alumina–silica refractory because the iron oxides act as a fluxing agent. Should the oxides become airborne, they can contaminate refractories on the furnace wall. Other materials heated in a furnace can throw off other oxides, considered to be fluxes for an alumina–silica refractory. Incinerators are the worst of all; they burn or oxidize everything put into them.

Refractory fibers are troublesome in fluxing situations. Experience dictates extreme caution when applying such fibers at temperature, higher than 1800°F; however, these fibers have worked well in incinerator afterburners at higher than 1800°F. Perhaps because the fluxes have by then become so oxidized that they are no longer able to attack the refractory.

XII. CORROSION

(The following material is taken from a book [40] on process plant design written in 1960, please see the latest design information for an update. This is a very good general discussion of corrosion phenomena)

A. Corrosion and Corrosion Phenomena

Corrosion is the destruction or deterioration of a metal or alloy by chemical or electrochemical reaction with its environment. Metals and alloys are affected in many different ways. Corrosion may cause dezincification in brass, pitting in stainless steel, or graphitic attack in cast iron. General corrosion, intergranular attack, and stress-cracking are other forms of this insidious type of metal destruction. It may occur gradually over a period of years or in a matter of hours. The metal always tends to revert to its original or natural state. Metal in the refined form is in an unstable condition, and corrosion is the transition back to the stable state. Thus, in rusting, iron is being converted to an iron oxide, which is one form of iron ore.

The various corrosion phenomena fall into natural divisions: uniform corrosion, galvanic corrosion, concentration-cell corrosion, pitting corrosion, dezincification, graphitic corrosion, intergranular corrosion, stress corrosion, corrosion fatigue, cavitation, and impingement attack.

1. Uniform Corrosion

Uniform attack (general corrosion) is characterized by a chemical or electrochemical reaction that proceeds evenly and at the same rate over the entire exposed area. In liquids, this form of corrosion involves a simple solution of the metal; in gases, it involves scaling or oxidation. Effective measures to overcome this type of corrosion include upgrading the material of construction, using inhibitors, installing cathodic protection, and applying protective coatings and linings.

2. Galvanic Corrosion

Galvanic corrosion occurs when two dissimilar conductors are in contact with each other and exposed to an electrolyte, or when two similar conductors are in contact with each other and exposed to dissimilar electrolytes. An electrical potential exists between the two metals, and a current flows from the less noble (anode) to the more noble (cathode). This current flow produces corrosive attack over and above that which would normally occur if the two metals were not in contact. Corrosion is accelerated on one metal—the anode—and decreased on the other—the cathode.

This form of attack can be recognized by mere visual examination. The accelerated corrosion of the less noble metal is usually localized near the points of contact. It often appears as a groove or deep channel adjacent to the point of coupling. The tendency always exists when two dissimilar metals are in galvanic contact in water or a chemical solution. Galvanic corrosion can be eliminated by breaking or preventing the flow of current.

When dissimilar metal contact is unavoidable, it is possible to minimize the effects of galvanic corrosion by several methods. Materials from adjacent groups should be used. When this is impossible, the part having the smallest area should be fabricated from the most noble metal. Increased thickness of the less noble metal will also increase service life. Threaded fittings should not be used to join dissimilar metals because of the diminished wall thickness in the threaded area.

Design plays a large role in galvanic corrosion. Pockets and crevices should be avoided. Proper drainage and ventilation will reduce corrosion. Free airflow will hasten drying of wetted parts. Magnesium or zinc anodes can be used to protect the dissimilar joint.

3. Concentration–Cell Corrosion

Concentration–cell corrosion occurs when one metal is in contact at two or more points on its surface with different concentrations of the same environment. In other words, current flow and potential differences exist on different areas of the same metal if this metal is in contact with different concentrations of the same solution or liquid. The corrosion that occurs in the areas of weaker concentration may be many times greater than those that would be prevalent in the same liquid of uniform concentration. The corrosion is further accelerated by the large ratio of cathodic areas to anodic areas. This form of attack is associated with crevices, scale, surface deposits, and stagnant solutions. It is sometimes called crevice corrosion.

Concentration–cell corrosion can be minimized and eliminated. The use of butt-welded joints, with complete weld penetration, is recommended; lap joints should be avoided. If lap joints are necessary, they should be sealed by welding or caulking. In design, sharp corners and crevices should be avoided. Cleanliness is the major factor in preventing concentration–cell corrosion. The flow of liquids should be uniform, with minimum turbulence, air entrainment, silt, and other solids. The use of strainers on all pipelines will payoff in decreased downtime and maintenance costs. Wet absorbent packing and gasketing are ideal locations for concentration–cell corrosion; they should be avoided and removed.

4. Pitting Corrosion

Pitting is probably the most destructive and unpredictable form of corrosion. Although pitting is readily recognized, the reasons why it occurs at a certain spot and leaves adjacent areas relatively unattacked are obscure. Apparently, localized corrosion occurs because of a lack of complete homogeneity in the metal surface. The presence of impurities, rough spots, scratches, and nicks may promote the formation of pits; pits also form under deposits and in crevices. If the passive film on corrosion-resistant metals and alloys is broken, the weak spots may become potential pits. Unfortunately, after pitting is started, it tends to progress at an accelerated rate.

5. Dezincification

Dezincification was first observed on brasses, which are copper–zinc alloys. The zinc is selectively removed from the brass to leave a weak, porous mass of copper. Dezincified areas show the red color of copper instead of the distinctive yellow of brass. Usually,dezincification is not accompanied by any significant dimensional changes. The attacked area has a spongy appearance and no physical strength. Normally dezincification occurs in brasses containing more than 15% zinc and no dezincification inhibitor. Admiralty metal, Muntz metal, and high brass are particularly susceptible.

Two types of dezincification are recognized: plug-type and layer-type. Plug-type occurs in highly localized areas and may be as small as pin heads Removal of the plugs reveals hemispherical pits. They can be located through the white, brown, or tan tubercules of zinc-rich salts that form over the plugs. Layer-type dezincification covers larger areas. Sometimes, it is a merger of a large number of small plugs and can be recognized by the nodular appearance of the copper layer. Generally, it proceeds evenly over the metal surface. Both types occur under prolonged exposure in supposedly mild conditions of corrosion.

A minor addition of arsenic, tin, antimony, or phosphorus to brass is an aid in the resistance to dezincification. Alloys, such as inhibited Admiralty metal, are used in mild cases. In more severe cases, red brass is often used because it is practically immune to this attack. Sometimes, inhibitors in the water or chemical solution will diminish the aggressiveness of the solution and hinder dezincification.

6. Graphitic Corrosion

Graphitic corrosion is a peculiar form of disintegration suffered by cast iron when exposure to certain types of mildly acid environments. It is not always recognized because the original shape of the object is retained. Chemical attack removes much of the ferrite phase from the affected zone of the cast iron and leaves behind a black porous structure that is rich in graphite and carbides. In most instances, the graphitized surfaces have the feel of graphite and leave a black smudgy mark or deposit on the fingers. The graphite layer can be removed with a pen knife or other sharp instrument.

Graphitic corrosion is most likely to occur in weak acid solutions. It is prevalent on cast-iron pipelines laid in acidic soils or in slag backfill. The rate is slow and after 10–15 years, the depth of penetration may be only 1/8–1/4 in. Pumps that are handling acid waste or steam condensate may graphitize to the same depth in several months. Pump casings generally outlast impellers in many instances. Occasionally, the life of the cast iron replacement impeller is much shorter than that of the original impeller. The answer may be graphitic corrosion of the casing that causes accelerated galvanic attack on the replacement impeller. Erosion normally prevents accumulation of graphite on the impeller, but not on the casing.

Graphitic corrosion can be limited and controlled. The cast iron may be upgraded by alloying with nickel or nickel and chromium, or the cast iron quality may be refined. Inhibitors in the chemical solution are also very effective.

7. Intergranular Corrosion

Intergranular corrosion consists of localized attack along the grain boundaries of a metal or alloy. There is no appreciable attack on the metal grains or crystals.

Very little weight loss is evident. Occasionally, whole grains are loosened and fall away from the structure. Complete disintegration of the metal or alloy is possible. Intergranular corrosion leads to loss of strength, ductility, and metallic ring.

The austenitic chromium–nickel stainless steels are particularly susceptible to intergranular attack when they are not properly heat treated or stabilized. When the 18-8 alloys are exposed to temperatures in the range 800–1600°F, they become sensitized and susceptible to attack.

Intergranular corrosion can occur on any part of a surface that is held in the sensitizing range or improperly heat treated. In welding, the sensitized zone is not immediately adjacent to the weld, but is located about 1/8 in. away.

Many attempts have been made to control carbide precipitation in a welded austenitic stainless steel article, but only three methods are dependable. Heat-treatment is the oldest and most positive method. The article is heated to 1950–2100°F, depending on the analyses of the stainless steel, and held at that temperature for 30 min. The holding period is followed by rapid quenching to below 800°F. This solution annealing puts the carbides back into solution and rapid cooling keeps them there. Full corrosion resistance is restored.

Stabilized grades of stainless steel were developed to overcome the effect of carbide precipitation. The addition of titanium to type 321, and columbium and tantalum to types 347 and 348, stabilizes these austenitic alloys. If improper welding techniques are employed, some chromium–carbide precipitation may occur, for the stabilized grades are not completely immune to sensitization. When stabilized electrodes are used to weld unstabilized base metal, the stabilized electrodes will not prevent sensitization in the heat-affected zone of the unstabilized base metal.

The extra–low-carbon austenitic stainless steels provide the newest means of combating carbide precipitation. The carbon is held to 0.03% maximum. This will give complete protection during normal welding operations. However, prolonged exposure at 800–1600°F, or the use of improper procedures will cause carbide precipitation.

Intergranular corrosion is by no means confined to stainless steels. There have been several cases of nickel suffering this type of attack in high-temperature steam turbines. "Hastelloy B," "Hastelloy C," and alloy "20" have suffered this phenomenon in certain specific environments.

8. Stress Corrosion

Stress corrosion is a general term for corrosion accelerated by internal or external stresses. Internal stresses are produced by cold-working, welding, unequal cooling, unequal heating, and internal structural changes. External stresses are normally load or operating stresses. Although stresses are additive and complex, surface tension stresses are always required. Almost always, stress corrosion manifests itself in the form of cracks and, consequently, is often known as stress–corrosion cracking. Cracking may be either intergranular or transgranular. Stress and corrosion, acting together, produce failure much more rapidly than either acting alone. No stress level is safe.

There is no universally accepted theory on the mechanism of failure. However, it is generally agreed that corrosive attack initiates small cracks on the surface of the metal parts stressed in tension. As the stress concentration builds up at the base of these tiny cracks, they widen and deepen. Thus, more metal is exposed to the corrosive media and, as stress and corrosion interact, failure occurs. The principal factors in this type of corrosion are stress, corrosive environment, the internal metallurgical structure of the metal, temperature, and time required for failure. No cracking has been known to occur in a vacuum. Time required for failure may vary from a matter of minutes to years. It may occur in one unit, but be absent in other identical units in the same service.

Caustic embrittlement of steel boilers is a well-known example of stress corrosion. Although originally associated with steam boilers, stress corrosion can occur in any strong caustic or alkaline solution and can affect alloy steels as well as carbon steels. The metal away from the cracks is ductile and not brittle, as the name implies.

The initial concentration of caustic may be very low, but under plant-operating conditions, it builds up in crevices and at leaks. In the presence of high tensile stresses, predominantly intergranular cracking occurs. This often takes place along rows of rivets, in welded seams, and around areas of support. Caustic solutions at temperatures below 140°F, can normally be stored or handled in as-welded steel tanks without concern; at higher temperatures, the tanks should be stress-relieved.

Carbon steel is also subject to cracking by nitrate solutions, sulfuric–nitric acid mixtures and calcium chloride brines. Steel equipment in direct contact-scrubbing systems' washing gases that contain hydrogen sulfide or carbon dioxide is likely to stress crack. Cyanides in the gases or water act as poisons and increase the susceptibility.

Stainless steels are susceptible under special conditions. Transgranular cracking is prominent in hot chloride solutions, whereas intergranular cracking appears more prominent in hot caustic solutions. The incidence of cracking is greatest in the austenitic alloys.

Hot, concentrated solutions containing chlorides of aluminum, ammonia, calcium, lithium, magnesium, mercury, sodium, and zinc cause cracking. The use of brackish or seawater for cooling is potentially dangerous because, in the presence of stress, the chlorides may concentrate in crevices and cause cracking. The presence of chlorides in magnesia and silicate insulations is sufficient to cause failure when the insulation is cyclically wetted and dried. Cracking is attributed to the leaching out of water-soluble chlorides and their concentration on hot surface that are under tension. In both brackish water and in wet insulation, the amount of chlorides is small, but the critical value is reached after concentration. Organic chlorides may also be harmful if they decompose in the presence of moisture to form hydrochloric acid and/or chloride salts. Wetted stainless steel will not stress-crack in the absence of oxygen.

The most common and effective method of combating stress corrosion is heat treatment. Internal or residual stresses are relieved by heating to a moderate temperature and holding at that temperature for a short time. Shot or hammer peening of surfaces has sometimes been used to convert the tensile stresses to compressive stresses. Surface coatings have also been used to affect a barrier between the corrosive medium and the metal. However, coatings are thin and porous and in continuous need of repair and maintenance. Their life is relatively short; they must be used with care. Cathodic protection has been tried in several selected applications, but there is always the danger that the evolved hydrogen may accelerate the cracking, addition of inhibitors to the solution may eliminate stress corrosion. The choice of a more resistant alloy may be beneficial; only a minor change may be required. Proper design is another important solution. The equipment should have the stresses spread over a wide area; stress risers should be avoided. The amount of cold deformation should be minimized, and heat treatment should always be considered when there is a history of stress corrosion. High applied stresses should be avoided.

9. Corrosion Fatigue

Corrosion fatigue is closely allied with stress corrosion. Cyclic stresses increase corrosion and may cause pitting and grooving. The applied stresses are concentrated in these areas and failure by cracking may occur. The cracks are normal to the direction of stress. This type of attack is prevalent in shafting. Corrosion fatigue may be guarded against by following the precautions outlined under stress corrosion.

10. Cavitation

Cavitation is defined as deterioration of a metal caused by the formation and collapse of cavities in a liquid. It can occur in all pure hydraulic systems. It exists on agitators and propellers, but is more prevalent in centrifugal pumps. When a liquid enters the eye of the impeller of a centrifugal pump and does not have sufficient head to make the turn into the impeller vane area, a void is formed. This void, which is a partial vacuum, becomes transient, and moves through the vanes of the impeller. When it arrives at the high-pressure area of the casing, the void collapses, and there is a violent inrush of liquid. The impinging force of the inrushing liquid causes a powerful impact and water hammer. The repeated impact causes deformation and fatigue and removes the surface films. The roughened surface is thus more susceptible to corrosion and affords ideal areas for formation and collapse of more voids. This repeated impact causes destruction and irregular thinning of the metal wall or surface. Where cavitation problems exist, they can be solved in several ways. The first one would be a complete redesign of the pump relative to the proper pressure heads. The admission of small amounts of air to the suction side of a centrifugal pump may reduce or eliminate the trouble. The injected air acts as a cushion and absorbs the impinging shock. Sometimes the substitution of an alternative material of construction is beneficial. Austenitic stainless steels and certain high-nickel alloys have given much longer service life than cast iron.

11. Impingement Attack

Impingement attack on the inlet end of condensers may be caused by cavitation. Turbulence and erratic water flow result in low-pressure areas, and some water is vaporized. As vapor bubbles move from a high-velocity to a low-velocity area, the bubbles collapse. This impact removes surface films and accelerated corrosion takes place. In other cases, impingement attack simply occurs as a result of turbulence, velocity, and erratic flow; the surface films are abraded away. Strategically located guide vanes may so direct the water flow that high velocities cannot exist. Perforated plates in the water box and supplementary tube ends have proved useful. The best solution, however, is proper design and selection of material of construction.

XIII. pH VALUES [23]

The pH value of a liquid is the measure of the corrosive qualities of a liquid, either acidic or alkaline. It does not measure the amount of quantity of the acid or alkali; but, instead, the hydrogen or hydroxide ion concentration in gram equivalents per liter of the liquid. pH value is expressed as the logarithm to the base 10 of the reciprocal of the hydrogen ion concentration in gram equivalents per liter. The scale of pH values range from zero through 14. The neutral point is 7. From 6 decreasing to zero denotes increasing acidity. From 8 through 14 denotes increasing alkalinity. It may also be stated that from 6 to zero hydrogen ions predominate; and, from 8 through 14 hydroxide ions predominate. At 7, the neutral point, the hydrogen and hydroxide ions are equal in quantity. The difference in pH numbers is ten-fold. For example, a solution of 3 pH (0.001 hydrogen ion concentration in gram equivalents per liter) has 10 times the hydrogen ion concentration of a 4 pH solution (0.0001 hydrogen ion concentration in gram equivalents per liter). Likewise, a 10 pH solution has 10 times the hydroxide ion concentration of a 9 pH solution. The pH value of a solution can be obtained by colorimetric methods using "universal indicator" or by electric meters designed especially for the purpose.

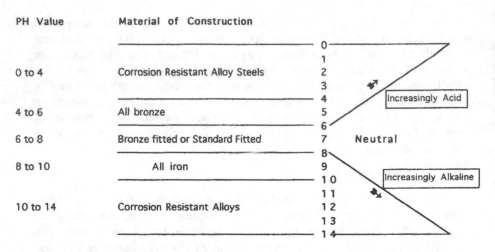

PH Value   Material of Construction

PH Value	Material of Construction
0 to 4	Corrosion Resistant Alloy Steels
4 to 6	All bronze
6 to 8	Bronze fitted or Standard Fitted
8 to 10	All iron
10 to 14	Corrosion Resistant Alloys

FIGURE 2.4 Recommended materials for construction of pumps that handle solutions for which the pH value is known.

Figure 2.4 outlines materials of construction usually recommended for pumps handling solutions for which the pH value is known. Knowing the pH value of a solution does, by no means, answer all questions on the corrosive qualities or characteristics of a solution. Temperatures of the solution affect the pH value. For example, a water solution may have a pH of 7, or neutral, at room temperature but at 212°F, it may have a pH value less than 7 or on the "acid side" of neutral 7. Corrosion effect by dissolved oxygen in a solution and corrosion by electrolysis cannot be predicted by pH values, but knowing the pH value of a liquid to be pumped is an excellent point to start in determining the materials of construction.

XIV. METAL SURFACE PREPARATION [40]

Hot-rolled steel contains a surface layer of mill scale. When intact, this mill scale provides a good base for painting. However, during normal fabrication and erection procedures, the mill scale develops fine cracks and is gradually undercut by rusting. The scale lifts and spalls and corrosion proceeds on the exposed surfaces. As a consequence, mill scale is considered to be a poor base for paints. Cold-rolled steel does not have mill scale, but it does have a tendency to rust. A rusted surface is probably the poorest surface for application of a protective coating. Regardless of advertising claims, there is no known method of painting that has been completely successful when applied over rust. Rust, when painted over, continues to form, and eventually, the paint film is destroyed. Therefore, for adequate protection all mill scale and porous, voluminous rust must be removed to improve paint adhesion. There is no substitute for adequate surface preparation; the most severe services demand the best in surface cleanliness. A paint will not adhere to a metal unless the paint and surface make intimate contact.

There are many methods of surface preparation. All vary in their effectiveness and usefulness. To attain maximum utility, it is often necessary to rely on a combination of procedures. Solvent cleaning is a prerequisite to other methods and is the preferred method for removing oil, grease, dirt, and soluble residues. It is also applicable for removing old paint films. Rust and mill scale are unaffected. Care must be exercised to remove all greasy residues that develop as a result of solvent cleaning. In addition, toxic solvents and those with low flash points should be avoided. Alkaline and steam cleaning are two other acceptable ways to remove oil, grease, and soluble foreign matter. Again, rust and mill scale remain untouched. With alkaline cleaning, the surface must be thoroughly flushed with water to neutralize the cleaners and to remove all residues. The alkaline-cleaned and steam-cleaned surfaces require immediate drying to prevent excessive rusting.

Rust and mill scale may be removed in many ways. The choice of procedure depends on the geometry of the surface, its condition, its accessibility, the degree of preparation required, other specialized considerations, and economics. Weathering has been extensively employed, but it is a controversial method, as it generally results in a surface that contains copious quantities of loose and tight rust, pits of varying sizes, loose mill scale, moisture, and chemical contaminants. By itself it is a poor method and is likely to cause premature failure of the applied paint film. When the surface is further prepared by sandblasting, weathering is deemed beneficial.

For certain items and in some plants, flame cleaning is feasible. The flame dehydrates the surface and the unequal rates of heating set up stresses that cause flaking of the loose mill scale and rust. On new and unpainted steel, one pass of the burner is sufficient; on previously painted or badly rusted steel, multiple slow passes are required. The warm surface should be immediately wire-brushed to remove all loose material, dusted, and primed before moisture condensation occurs. Flame cleaning produces a surface that is better suited for painting than either hand or power cleaning. Naturally, this method must be used with caution. It can induce warpage of the metal and, in certain plant areas, it is a definite fire hazard. Flame cleaning and acid pickling are normally shop techniques, rather than field procedures.

Acid pickling effectively removes all rust and mill scale and, after thorough rinsing and neutralization, leaves a very desirable surface for painting. The procedure involves handling dangerous and very corrosive chemicals and fumes. It is extensively used by steel mills and fabricators of sheets, plates, and shapes. In the small shop, it is useful for parts that can be removed and totally immersed. It is not suitable for cleaning erected structures and operating process equipment.

Hand cleaning is the least effective way of preparing a steel surface for paint. It has its place and is ideal for cleaning small areas or those of complex design. It is not applicable to large flat areas, because only the high spots are touched and only the loosest rust and scale are removed. Hand cleaning with wire brushes, chippers, scrapers, and sanders will not remove intact or firmly adherent scale and rust. At best, only 2 ft²/min can be adequately cleaned by vigorous hand methods. This figure is not a production rate, but rather a work standard that depends on the surface contour and the amount of scale and rust. The method is expensive and is applicable for normal atmospheric exposures. Power cleaning augments hand cleaning. It encompasses the use of electric or pneumatic chippers, descalers, sanders, grinders, and wire brushes. Generally, this method is more economical, effective, and reliable than the hand procedures. The cleaning rate is about the same, but the surface is much better prepared. With power tools, care must be exercised to prevent roughening or burnishing of the steel surface. Paint will not adhere to a slick, smooth surface.

For shop or field preparation of equipment and structures, the most effective method of removing rust and mill scale is blast cleaning. Screened sand, grit, or steel shot may be used. Shot is expensive and is employed when its recovery is feasible. Sand is considered expendable and is often not worthy of reclamation. The depth of the surface profile from peak to valley should be one-third the thickness of a three-coat paint system. For most paints, this averages out to 1.5 mil, and should not be over 2.0 mil. If the surface is too rough, it is difficult to paint and premature failure generally follows. Because sand and shot blasting produce high nozzle forces, they cannot be used on the thin metal without risk of warpage and damage.

The term "blast cleaning" is not too definite, as it may be used to various degrees. The best method is blasting to white metal. This produces a bright clean surface that is free of all rust, scale, old paint, weld flux, and foreign matter. Weld spatter is not removed by blasting. Bright cleaning is most expensive and its cost is warranted only for those exposures involving very corrosive atmospheres or immersion in fresh- or seawater. Maximum paint performance is achieved over a white metal surface. A commonly accepted work standard is 100 ft² of surface area per nozzle per hour. The majority of blast cleaning applications utilize the commercial or gray procedure. Here, the rate of cleaning is roughly 2.5 times that of white metal blasting. This method removes most of the rust and mill scale; some minor residues are tolerable. It does not include the removal of the gray oxide layer between the mill scale and the white steel surface. The surface thus produced is a streaky gray color. Commercial blast-cleaned surfaces are extensively employed for those exposures involving spills, splashes, and severe fumes, but not immersion. The cheapest blast-cleaning method is the blast brush-off. The rate of travel is high—roughly six times that of white blasting or about 600 ft ^{2} of surface per nozzle per hour. All loose mill scale and rust is removed; the tightly adherent material remains on the surface. Metal prepared in this manner is not suitable for severe exposure, but is excellent for the less severe splashes, fumes, and atmospheres. Sand brush-off coupled with suitable primers is preferred to hand- and power-cleaning procedures. With all blast cleaning, the adjacent machinery and equipment must be protected from sand, debris, and dust.

Surface Coverage[a] with Paint or Coating

1 gal = 231 in.^{3} of liquid
1 ft ^{2} = 144 in.^{2}
1 gal will cover 1.6042 ft^2 of surface, 1 in. in depth.
1 mil (coating) = 0.001 in. of coating thickness
1 gal will cover 1604.2 ft^2 of surface, 1 mil thick.
1 gal will cover 802.1 ft^2 of surface, 2 mil thick.
1 gal will cover 530 ft^2 of surface, 3 mil thick.
1 gal will cover 401 ft^2 of surface, 4 mil thick.
1 gal will cover 320 ft^2 of surface, 5 mil thick.

---

[a]All coverage is theoretical. Reality is not the same.

XV. SCREENING [8]

Screening machines may be divided into five main classes: grizzlies, revolving screens, shaking screens, vibrating screens, and oscillating screens. Grizzlies are used primarily for scalping at 2 in. and coarser, whereas revolving screens and shaking screens are generally used for separations above 0.5 in. Vibrating screens cover this coarse range and also down into the fine meshes. Oscillating screens are confined in general to the finer meshes below 4 mesh.

A. Grizzly Screens

Grizzly screens consist of a set of parallel bars held apart by spacers at some predetermined opening. Bars are frequently made of manganese steel to reduce wear. A grizzly is widely used before a primary crusher in rock or ore-crushing plants to remove the fines before the ore or rock enters the crusher. They can be a stationary set of bars or a vibrating screen.

The stationary grizzly is the simplest of all separating devices and the least expensive to install and maintain. It is normally limited to the scalping or rough screening of dry material at 2.0 in. and coarser and is not satisfactory for moist and sticky material. The slope, or angle with the horizontal, will vary between 20 and 50 degrees.

Flat grizzlies, in which the parallel bars are in a horizontal plane, are used on tops of ore and coal bins and under unloading trestles. This type of grizzly is used to retain occasional pieces too large for the following plant equipment. These lumps must then be broken up or removed manually.

Stationary grizzlies require no power and little maintenance. It is difficult to change the opening between the bars, and the separation may not be too complete.

Vibrating grizzlies are simply bar grizzlies mounted on eccentrics so that the entire assembly is given a back-and-forth movement or a positive circle throw.

B. Revolving Screens

Revolving screens or trommel screens, once widely used, are being largely replaced by vibrating screens. They consist of a cylindrical frame surrounded by wire cloth or perforated plate, open at both ends, and inclined at a slight angle. The material to be screened is delivered at the upper end and the oversize is discharged at the lower end. The desired product falls through the wire cloth openings. They revolve at relatively low speeds of 15–20 rpm. Their capacity is not great, and their efficiency is relatively low.

C. Mechanical Shaking Screens

Mechanical shaking screens consists of a rectangular frame, which holds wire cloth or perforated plate and is slightly inclined and suspended by loose rods or cables, or supported from a base frame by flexible flat springs. The frame is driven with a reciprocating motion. The material to be screened is fed at the upper end and is advanced by the forward stroke of the screen while the finer particles pass through the openings. In many screening operations they have given way to vibrating screens. Shaking screens may be used for both screening and conveying. Advantages of this type are low headroom and low power requirement. The disadvantages are the high cost of maintenance of the screen and what the supporting structure does to the vibration, and its low capacity compared with inclined high-speed vibrating screens.

D. Vibrating Screens

Vibrating screens are used as standard practice where large-capacity and high-efficiency are desired. The capacity, especially in the finer sizes, is so much greater than any of the other screens that they have practically replaced all other types when the efficiency of the screen is an important factor. Advantages include accuracy of sizing, increased capacity per square foot, low maintenance cost per ton of material handled, and a saving in installation space and weight. Vibrating screens basically can be divided into two main classes: (1) mechanically vibrated and (2) electrically vibrated.

1. Mechanically Vibrated Screens

The most versatile vibration for medium to coarse sizing is generally conceded to be the vertical circle produced by an eccentric or unbalanced shaft, but other types of vibration may be more suitable for certain screening operations, particularly in the finer sizes.

2. Electrically Vibrated Screens

Electrically vibrated screens are particularly useful in the chemical industry. They very successfully handle many light, fine, dry materials and metal powders from approximately 4 mesh to as fine as 325 mesh. Most of these screens have an intense, high-speed (1500–7200 vibrations per minute) low-amplitude vibration supplied by means of an electromagnet.

3. Oscillating Screens

Oscillating screens are characterized by low-speed (300–400 rpm) oscillations in a plane essentially parallel to the screen cloth. Screens in this group are usually used from 0.5 in. to 60 mesh. Some light free-flowing materials, however, can be separated at 200–300 mesh. Silk cloths are often used.

4. Reciprocating Screens

Reciprocating screens have many applications in chemical work. An eccentric under the screen supplies oscillation, ranging from gyratory at the feed end to reciprocating motion at the discharge end. Frequency is 500–600 rpm; and because the screen is inclined, about 1/10 in. is also set up. Further vibration is caused by balls bouncing against the lower surface of the screen cloth. These screens are used for handling fine separations down to 300 mesh and are not designed for handling heavy tonnages of materials such as rock or gravel.

5. Gyratory Screens

Gyratory screens are a boxlike machine, either round or square, with a series of screen cloths nested atop one another. Oscillation, supplied by eccentrics, or counterweights, is in a circular or near-circular orbit. In some machines a supplementary whipping action is set up. Most gyratory screens have an auxiliary vibration caused by balls bouncing against the lower surface of the screen cloth.

6. Gyratory Riddles

Gyratory riddles are screens driven in an oscillating path by a motor attached to the support shaft of the screen. The gyratory riddle is the least expensive screen on the market and intended normally for batch screening.

More information can be found in Table 2.5 and Figure 2.5.

TABLE 2.5

U. S. Sieve Series & Tyler equivalents						
A.S.T.M. E-11-61						
Ralph L. Vandagriff				Boiler House Notes:		Page No:
File: Sieve						
Sieve Designation		Sieve Opening		Nominal Wire Diameter		Tyler
Standard	Alternate	mm	Inches approx. equiv.	mm	Inches approx. equiv.	Equivalent Designation
107.06 mm	4.24 "	107.06 mm	4.24 "	6.4000	0.2520	
101.06 mm	4.00"	101.06 mm	4.00"	6.3000	0.2480	
90.50 mm	3.50"	90.50 mm	3.50"	6.0800	0.2394	
76.10 mm	3.00"	76.10 mm	3.00"	5.8000	0.2283	
64.00 mm	2.50"	64.00 mm	2.50"	5.5000	0.2165	
53.80 mm	2.12"	53.80 mm	2.12"	5.1500	0.2028	
50.80 mm	2.00"	50.80 mm	2.00"	5.0500	0.1988	
45.30 mm	1.75"	45.30 mm	1.75"	4.8500	0.1909	
38.10 mm	1.50"	38.10 mm	1.50"	4.5900	0.1807	
32.00 mm	1.25"	32.00 mm	1.25"	4.2300	0.1665	
26.90 mm	1.06"	26.90 mm	1.06"	3.9000	0.1535	1.05 inch
24.50 mm	1.00"	24.50 mm	1.00"	3.8000	0.1496	
22.60 mm	.875"	22.60 mm	.875"	3.5000	0.1378	.883 inch
19.00 mm	.75"	19.00 mm	.75"	3.3000	0.1299	.742 inch
16.00 mm	.625"	16.00 mm	.625"	3.0000	0.1181	.624 inch
13.50 mm	.53"	13.50 mm	.53"	2.7500	0.1083	.525 inch
12.70 mm	.50"	12.70 mm	.50"	2.6700	0.1051	
11.20 mm	.4375"	11.20 mm	.4375"	2.4500	0.0965	.441 inch
9.51 mm	.375"	9.51 mm	.375"	2.2700	0.0894	.371 inch
8.00 mm	.3125"	8.00 mm	.3125"	2.0700	0.0815	2.5 mesh
6.73 mm	.265"	6.73 mm	.265"	1.8700	0.0736	3 mesh
6.35 mm	.25"	6.35 mm	.25"	1.8200	0.0717	
5.66 mm	No. 3 1/2	5.66 mm	.223"	1.6800	0.0661	3.5 mesh
4.76 mm	No. 4	4.76 mm	.187"	1.5400	0.0606	4 mesh
4.00 mm	No. 5	4.00 mm	.157"	11.3700	0.0539	5 mesh
3.36 mm	No. 6	3.36 mm	.132"	1.2300	0.0484	6 mesh
2.83 mm	No. 7	2.83 mm	.111"	1.1000	0.0430	7 mesh
2.38 mm	No. 8	2.38 mm	.0937"	1.0000	0.0394	8 mesh
2.00 mm	No. 10	2.00 mm	.0787"	0.9000	0.0354	9 mesh
1.68 mm	No. 12	1.68 mm	.0661"	0.8100	0.0319	10 mesh
1.41 mm	No. 14	1.41 mm	.0555"	0.7250	0.0285	12 mesh
1.19 mm	No. 16	1.19 mm	.0469"	0.6500	0.0256	14 mesh
1.00 mm	No. 18	1.00 mm	.0394"	0.5800	0.0228	16 mesh
841 micron	No. 20	0.841 mm	.0331"	0.5100	0.0201	20 mesh
707 micron	No. 25	0.707 mm	.0278"	0.4500	0.0177	24 mesh
595 micron	No. 30	0.595 mm	.0234"	0.3900	0.0154	28 mesh
500 micron	No. 35	0.50 mm	.0197"	0.3400	0.0134	32 mesh
420 micron	No. 40	0.42 mm	.0165"	0.2900	0.0114	35 mesh
354 micron	No. 45	0.354 mm	.0139"	0.2470	0.0097	42 mesh
297 micron	No. 50	0.297 mm	.0117"	0.2150	0.0085	48 mesh
250 micron	No. 60	0.25 mm	.0098"	0.1800	0.0071	60 mesh
210 micron	No. 70	0.21 mm	.0083"	0.1520	0.0060	65 mesh
177 micron	No. 80	0.177 mm	.0070"	0.1310	0.0052	80 mesh
149 micron	NO. 100	0.149 mm	.0059"	0.1100	0.0043	100 mesh
125 micron	No. 120	0.125 mm	.0049"	0.0910	0.0036	115 mesh
105 micron	No. 140	0.105 mm	.0041"	0.0760	0.0030	150 mesh
88 micron	No. 170	0.088 mm	.0035"	0.0640	0.0025	170 mesh
74 micron	No. 200	0.074 mm	.0029"	0.0530	0.0021	200 mesh
63 micron	No. 230	0.063 mm	.0025"	0.0440	0.0017	250 mesh
53 micron	No. 270	0.053 mm	.0021"	0.0370	0.0015	270 mesh
44 micron	No. 325	0.044 mm	.0017"	0.0300	0.0012	325 mesh
37 micron	No. 400	0.037 mm	.0015"	0.0250	0.0010	400 mesh

MELTING POINTS

TEMPERATURE COLORS

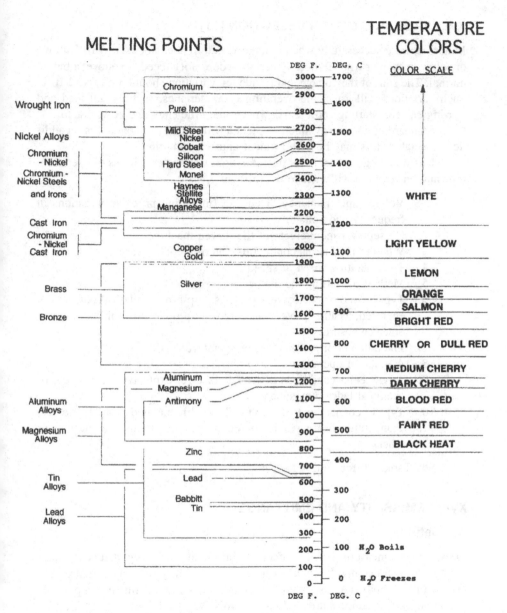

FIGURE 2.5   Melting points and color-temperature relations of metals.

XVI. ELECTRIC MOTOR SELECTION [11]

Motors operate successfully when voltage variation does not exceed 10% above or below normal or when frequency variation does not exceed 5% above or below normal. The sum of the voltage and frequency variation should not exceed 10%. Such variations will affect the operating characteristics, such as full load and starting current, starting and breakdown torque, efficiency and power factor.

Standard motors are available to meet a wide variety of conditions. In addition, special motors may be built to meet unusual conditions.

It is wise to go to the motor manufacturer with the conditions of operation. Information required will include the following:

1. Voltage and frequency of current (including probable variations in frequency and voltage)
2. Horsepower requirement of the driven machine
3. Whether the load is continuous, intermittent, or varying
4. The operating speed or speeds
5. Method of starting the motor
6. Type of motor enclosure—such as drip-proof, splash-proof, totally enclosed, weather protection, explosion proof, dust-ignition proof, or other enclosure
7. The ambient or surrounding temperature
8. Altitude of operation
9. Any special conditions of heat, moisture, explosive, dust laden, or chemical laden atmosphere
10. Type of connection to driven machine (direct,belted, geared, or other)
11. Transmitted bearing load to the motor (overhung load, thrust, or other)

See Table 2.6 for more information.

XVII. EMISSIVITY AND EMITTANCE

A. Definition

Emissivity is a measure of ability of a material to radiate energy; that is, the ratio (expressed as a decimal fraction) of the radiating ability of a given material to that of a black body. (A "black body" emits radiation at the maximum possible rate at any given temperature, and has an emissivity of 1.0.)

Emittance is the ability of a surface to emit or radiate energy (Table 2.7), as compared with that of a black body, which emits radiation at the maximum possible rate at any given temperature, and which has an emittance of 1.0. (Emissivity denotes a property of the bulk material independent of geometry or surface condition, whereas emittance refers to an actual piece of material.)

BOILER HOUSE NOTES:			Page No:	
Subject: Horsepower Worth: Present Worth Analysis				
Ralph L. Vandagriff				
Date: 7/9/96	**Data Input:**			
One (1) Horsepower Installed Cost:	$215.00			
Selling Price: Kw/Hour	$0.055			
Hours / Year Operation:	8,600			
Boiler Life: Years	25			
Cost of Money:	10.00%			
Inflation Rate:	3.00%			
	Conclusions:			
Discount Rate:	13.00%	Discount Rate = Cost of Money + Inflation		
Operating Cost: 1HP/Yr.	$387.60	Operating Cost per Horsepower = ((kW/HP x dollars per Kw per Hour) / by motor efficiency) x hours/year		
Amortization Factor:	0.136425928	Amortization factor = DR/((1+DR)^PL - 1) + DR		
Present Worth Factor:	7.329984978	Present Worth Factor = 1/Amortization Factor		
Present Worth: One (1) Horsepower:	$3,056.10	Present Worth of 1 HP = HP Installed cost + (PWF x Op.Cost)		
Example:	One boiler costs	$100,000.00	more than another boiler	
	But uses	50.0	less fan horsepower	
$3,056.10	times	50.0	equals savings of	$152,805.15
Net Savings over project life = savings minus extra cost =	$52,805.15			
Example:				
One boiler feedwater pump w/ Steam Turbine Costs:	$15,000.00	more than a boiler feedwater pump w/ electric motor drive		
	But uses	200.0	less electric horsepower	
$3,056.10	times	200.0	equals savings of	$611,220.60
Net savings over life of project =	$225,720.60			
savings minus extra cost minus steam cost =				
One HP = 0.7457 kW	Std. Motor Efficiency = 91% @ full load			

TABLE 2.7 Surface Emittances of Metals and Their Oxides

Metal	Condition of surface	Metal temperature (°F)	Emittance
Carbon steel	Oxidized	77	0.80
304A stainless	Black oxide	80	0.30
	Machined	1000	0.15
	Machined	2140	0.73
310 stainless	Oxidized	980	0.97
316 stainless	Polished	450	0.26
	Oxidized	1600	0.66
321 stainless	Polished	1500	0.49
347 stainless	Grit blasted	140	0.47
	Oxidized	600	0.88
	Oxidized	2000	0.92
Nickel	Oxidized	392	0.37
	Oxidized	1112	0.48
	Oxidized	2000	0.86
Inconel X-750	Buffed	140	0.16
	Oxidized	600	0.69
	Oxidized	1800	0.82
Inconel sheet		1400	0.58

Source: Ref. 4.

TABLE 2.8 Normal Emissivities (ϵ) for Various Surfaces

Material	Emissivity (ϵ)	Temp (°F)	Description
Iron	0.21	392	Polished, cast
Iron	0.55–0.60	1650–1900	Smooth sheet
Iron	0.24	68	Fresh emeried
Steel	0.79	390–1110	Oxidized at 1100°F
Steel	0.66	70	Rolled sheet
Steel	0.28	2910–3270	Molten
Steel (Cr-Ni)	0.44–0.36	420–914	18-8 rough, after heating
Steel (Cr-Ni)	0.90–0.97	420–980	25–20 oxidized in service
Brick, fireclay	0.75	1832	

Source: Ref. 1; pp 4–9, 10.

TABLE 2.9

Estimated Relative Properties of Natural and Synthetic Rubber							
Ralph L. Vandagriff						Boiler House Notes Page No:	
Date: 8/7/96							
PROPERTIES	NATURAL RUBBER	BUNA S or GR-S	BUTYL	THIOKOL	NITRILE (BUNA N)	NEOPRENE	SILICONE
		Butadiene styrene	isobutylene isoprene	Organic polysulfide	Butadiene acrylonitrile	Chloroprene	Polysiloxane polymer
Tensile strength (psi) - Pure gum	2500 - 3500	200 - 300	2500 - 3000	250 - 400	500 - 900	3000 - 4000	600 - 1300
Tensile strength (psi) - Black loaded stocks	3500 - 4500	2500 - 3500	2500 - 3000	< 1000	3000 - 4500	3000 - 4000	
Hardness Range (Shore Durometer A)	30 - 90	40 - 90	40 - 75	35 - 80	40 - 95	40 - 95	40 - 85
Specific Gravity (Base Materials)	0.93	0.94	0.92	1.34	1.00	1.23	....
Vulcanizing Properties	Excellent	Excellent	Good	Fair	Excellent	Excellent	....
Adhesion to Metals	Excellent	Excellent	Good	Poor	Excellent	Excellent	....
Adhesion to Fabrics	Excellent	Good	Good	Fair	Good	Excellent	....
Tear Resistance	Good	Fair	Good	Poor	Fair	Good	Poor
Abrasion Resistance	Excellent	Good to excellent	Good	Poor	Good	Excellent	Poor
Compression Set	Good	Good	Fair	Poor	Good	Fair to good	Fair
Rebound - Cold	Excellent	Good	Bad	Fair	Good	Very good	Excellent
Rebound - Hot	Excellent	Good	Very good	Fair	Good	Very good	Excellent
Dielectric Strength	Excellent	Excellent	Excellent	Fair	Poor	Good	Good
Electrical Insulation	Good to excellent	Good to excellent	Good to excellent	Fair to good	Poor	Fair to good	Excellent
Permeability to Gases	Fair	Fair	Very low	Low	Fair	Low	Fair
Acid Resistance - Dilute	Fair to good	Fair to good	Excellent	Fair	Good	Excellent	Excellent
Acid Resistance - Concentrated	Fair to good	Fair to good	Excellent	Fair	Good	Good	Fair
Solvent Resistance - Aliphatic hydrocarbons	Poor	Poor	Poor	Excellent	Excellent	Good	Poor
Solvent Resistance - Aromatic hydrocarbons	Poor	Poor	Poor	Good	Good	Fair	Poor
Solvent Resistance - Oxygenated (ketones, etc.)	Good	Good	Good	Good	Poor	Poor	Fair
Solvent Resistance - Lacquer solvents	Poor	Poor	Poor	Good	Fair	Poor	Poor
Resistance to: Swelling in Lubricating oil	Poor	Poor	Poor	Excellent	Very good	Good	Fair
Resistance to: Oil and Gasoline	Poor	Poor	Poor	Excellent	Excellent	Good	Fair
Resistance to: Animal & vegetable oils	Poor to good	Poor to good	Excellent	Excellent	Excellent	Good	Fair
Resistance to: Water absorption	Very good	Good to very good	Very good	Fair	Fair to good	Good	Good
Resistance to: Oxidation	Good	Good	Excellent	Good	Good	Excellent	Excellent
Resistance to: Ozone	Fair	Fair	Excellent	Excellent	Fair	Excellent	Excellent
Resistance to: Sunlight aging	Poor	Poor	Very good	Good	Poor	Very good	Excellent
Resistance to: Heat aging	Good	Very good	Excellent	Fair	Excellent	Excellent	Outstanding
Resistance to: Flame	Poor	Poor	Poor	Poor	Poor	Good	Fair
Resistance to: Heat	Good	Excellent	Excellent	Poor	Excellent	Excellent	Excellent
Resistance to: Cold	Excellent	Excellent	Good	Fair	Good	Good	Excellent

B. Radiation Properties

Bodies that are good radiation absorbers are equally good emitters (Table 2.8), and Kirchhoff's law states that, at thermal equilibrium, their emissivities are equal to their absorptivities. A *blackbody* is one that absorbs all incident radiant energy while reflecting or transmitting none of it. The absorptivity and emissivity of a blackbody are, by definition, each equal to 1.

C. Emissivity of the Combustion Products

Furnace heat-transfer calculations, must deal on a quantitative basis with the emissivity of the gaseous products of combustion, gas-to-gas absorptivity, and metallic-surface-to-gas absorptivity.

Because about 95% of the heat transfer in large combustion chambers is by radiation, it is important to evaluate the radiative power of the gas/fuel/flame media [13; pp 6–16].

1. Definitions

Absorptivity is the ability of a surface to absorb radiant energy, expressed as a decimal, compared with the ability of a blackbody, absorptivity of which is 1.0.

Emissivity is a measure of ability of a material to radiate energy—the ratio (expressed as a decimal fraction) of the radiating ability of a given material to that of a blackbody. (A blackbody emits radiation at the maximum possible rate at any given temperature, and has an emissivity of 1.0.)

D. Factors Affecting Heat Transfer Rates [4]

Radiation heat flux, qr, Btu/ft^2 hr^{-1} = $0.1713 \times 10^{-8} \times (T^4s - T^4r) \times F_e \times F_a$

(T in degrees Rankine)

Radiation heat flux, qr, kW/m^2 = $0.00567 \times 10^{-8} \times (T^4s - T^4r) \times F_e \times F_a$

(T in degrees Rankine)

where

F_e = emissivity factor
F_a = arrangement factor

See Table 2.9 for more information.

3

Gas and Oil Fuels

Gas and Oil Burners for Boilers: General Data; Gas or Oil Burner Check List: Industrial Quality; Firetube Boilers: Gas and Oil Fired; Waterwall Boilers: Gas and Oil Fired.

I. GAS AND OIL BURNERS FOR BOILERS: GENERAL DATA [53]

The burner is the principal equipment component for the firing of the fuel into the boiler. Its functions include mixing the fuel and air, atomizing and vaporizing the fuel, and providing for continuous ignition of the mixture. Significant burner design characteristics include turndown ratio, stability, and flame shape.

A. Turndown Ratio

The burner turndown ratio is the ratio of the maximum to the minimum fuel-and-air mixture input rates at which the burner will operate satisfactorily. It specifies the range of fuel mixture input rates within which the burner will operate. The maximum input rate is limited by flame blowoff and physical size of the equipment. Flame blowoff is the phenomenon that results when the mixture velocity exceeds the flame velocity. The minimum input rate is limited by flashback and by the minimum flow rate at which the equipment controlling the mixture ratio will function. Flashback occurs when the flame velocity exceeds the mixture velocity.

A high turndown ratio is desirable when a high input is needed during initial heat-up, but cannot be used during the entire heating cycle. It is unnecessary in continuously fired boilers, which seldom have to be started cold.

B. Stability

A burner is considered stable if it will maintain ignition when the unit is cold at normal operating fuel/air ratios and mixture pressures. A burner is not considered

81

stable merely because it is equipped with a pilot. Many burners will not function satisfactorily under adverse conditions, such as cold surroundings, unless the mixture is rich and the flame is burning in free air. With burners of this type, it is necessary to leave the furnace door open during the start-up period. Without an open door, the free air inside the furnace would be quickly used up and the flame extinguished.

C. Flame Shape

For a given burner, changes in mixture pressure or amount of primary air will affect flame shape. In most burners, increasing the mixture pressure will broaden the flame, whereas increasing the amount of primary air will shorten the flame (input rate remaining the same).

Burner design affects flame shape to an even greater degree. Good mixing, resulting from high gas turbulence and velocities, produces a short bushy flame. Poor mixing and low velocities produce long slender flames. High atomizing air pressure tends to throw the fuel farther away from the burner nozzle before it can be heated to its ignition temperature, thereby lengthening the flame.

D. Atomization

Atomization is necessary to burn fuel oil at the high rates demanded of modern boiler units. Atomization exposes a large amount of oil particle surface for contact with the combustion air. This helps assure prompt ignition and rapid combustion. The two most popular types of atomizers are steam (or air) and mechanical.

The steam (or air) atomizer is the most efficient and most commonly used. It produces a steam-and-fuel emulsion (or air-and-fuel mixture) which, when released into the furnace, atomizes the oil through the rapid expansion of the steam (or air). Steam used for atomization must be dry because moisture causes pulsations, which can lead to loss of ignition.

Mechanical atomizers use the pressure of the fuel itself for atomization. One type of mechanical atomizer is the return-flow atomizer. Fuel flows under pressure into a chamber from which it issues through an opening in the sprayer plate as a fine conical mist or spray. Any oil that exceeds the boiler input requirements is returned to the fuel oil system.

Fuel oil must be atomized and vaporized to burn. Fuel oil will not burn as a liquid. It must first be converted into a gas. Good atomization requires that a large amount of air be initially mixed with the oil particles. The air must be turbulent to assure proper mixing. Then heat must be radiated into the spray to vaporize it.

Burner nozzles are designed for oil of a specific viscosity range, and variation from those viscosities will result in poor atomization. With heavy fuel oils, correct oil viscosity may be obtained by preheating the oil. Therefore, it is essential to know the viscosity range for which the burner is designed and determine the correct level of preheat needed to maintain the viscosity within this range. The design atomization viscosity of a burner should be obtained from the manufacturer.

E. Preheat

If the preheat is too high, the viscosity is lower than recommended and poor, fluctuating atomization occurs. This causes the flame to be noisy and unstable. If the preheat is too low, the viscosity is too high. This causes improper atomization, resulting in droplet size being overly large. These larger droplets are more difficult to vaporize and incomplete combustion and soot formation are more likely to occur.

II. GAS OR OIL BURNER CHECK LIST: INDUSTRIAL QUALITY

An industrial quality gas or oil burner should meet the following specifications:

1. Forced draft (FD) fan of industrial quality, mounted on concrete base or on windbox. If FD fan is mounted on windbox, mounting plate is to be of 3/8-in. thick (minimum) with stiffeners. It should run at 1800 rpm for less noise and longer life. Direct drive.
2. FD fan airflow controlled by heavy-duty inlet vortex.
3. Windbox depth: sufficient to ensure even air distribution
4. Windbox plate of 1/4-in. thickness, with angle stiffeners.
5. Access door in windbox.
6. Burner turndown: gas 10:1; oil 8:1.
7. Industrial quality Fireye Flame scanner (or equal): heavy-duty wiring.
8. Flame safeguard system programmed in solid-state programmable controller, of a standard heavy industrial type.
9. Indicating lights should be of heavy-duty industrial quality: low-voltage transformer type.
10. Circuit breakers used for electrical circuit protection.
11. Heavy-duty industrial quality relays and timers used where required.
12. Adjustable register for shaping flame to fit furnace configurations.
13. Hinged burner door allowing easy access to examine register parts.
14. Two (2) flame observation ports per register.
15. Cast alloy diffuser.
16. Stainless steel gas jets, preferably removable.
17. Steam-atomizing oil burner. Option to start with plant air.
18. Easily removable oil gun, preferably screw and yoke connector. Tips to be alloy steel, easy to change.
19. Scavenger pump system to evacuate oil from gun and hose on every oil shutdown. Pump to be positive-displacement gear pump.

20. NEMA 12-flame safeguard panel.
21. Maxon gas valves: standard.
22. 1-in.–diameter (minimum) jackshaft for heavy duty industrial linkage on single point positioning control systems.

III. FIRETUBE BOILERS: GAS AND OIL FIRED [6,7]

A. Scotch Marine

1. Packaged Scotch Marine Boilers

This name applies to horizontal boilers designed around a burner firing either natural gas and/or fuel oil. The flame, in the shape of a candle flame lying on its side, is introduced into a corrugated round metal tube (that is surrounded by water), passes through this tube, hits a rear wall, reverses direction and passes through boiler tubes (fire tubes) surrounded by water, to a front wall, changes direction to a vertical flow and passes out the stack. In this arrangement, the stack is at the front of the boiler. This is a two-pass boiler (Fig. 3.1).

If the hot gases reverse direction at the front wall and pass through firetubes to the back wall, then up and out the stack, you have a three-pass scotch marine packaged boiler. In this arrangement, the stack is at the rear of the boiler.

Four-pass scotch marine packaged boilers have been built, but they are rare. The cost of the fourth pass is just barely justified in the increased heat transfer surface, and this design is seldom seen. (The main reason seems to be that this section of the boiler market is very price competitive, and every effort is made to keep the price of the finished product as low as possible.)

Steaming capacity range	250–40,000 lb/hr
Working pressure limit	300–325 psig (limited by flat heads)
Superheat availability	None
Fuels	Natural gas, no. 2 fuel oil, and rarely no. 6 fuel oil
Water treatment required	Basic
Design types	Dry-back, two-pass (stack on front)
	Dry-back, three-pass (stack on back)
	Dry-back, four-pass (rarely seen—too expensive)
	Wet-back, two-pass (stack on front)
	Wet-back, three-pass (stack on back)
	Wet-back, four-pass (very rare—too expensive)

2. Design Criteria

The scotch marine boiler normally consists of a cylindrical shell, flat heads on each end, a cylindrical corrugated furnace in the lower section running from the

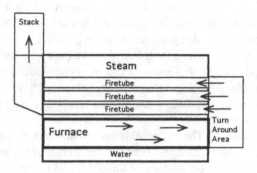

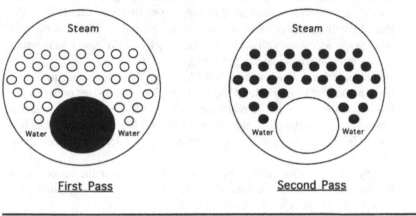

re: (6) & (7)

FIGURE 3.1 Combustion gas passage through a scotch marine packaged boiler.

front head to the back head, and one or more tube passes attached to both heads. The use of flat heads that have to be anchored by staybolts limits the maximum working pressure to approximately 325 psig maximum. It was developed during World War II.

> *Dry-back*:   This term is used to describe a scotch marine boiler that has a rear combustion chamber (turn-around section) that is refractory lined. The rear door is also refractory lined. This is a higher maintenance design than the wet-back.

Wet-back: This term is used to describe a scotch marine boiler that has a rear combustion chamber (turn-around section) that is water-jacketed. The rear door is also designed to be water-jacketed.

Corrugated furnace: This tube is called a Morison tube for the inventor. It is a straight cylinder that has been run through a special roller that puts corrugations at regular intervals, completely around the circumference of the cylinder. This increases strength and heat transfer. The furnace is not refractory lined.

Combustion gas flow: The gas or oil is burned in a burner setting, attached to the front of the furnace. Sometimes, the burner refractory throat extends into the very front section of the furnace. On a two-pass unit, the combustion products proceed through the furnace to the turnaround area at the back, reverse direction and flow through the one pass of tubes to the stack breeching on the front of the boiler. On a three-pass unit, the combustion products make another reversal of direction and flow to the stack breeching, the stack is on the back of the boiler. The heat transfer percentage decreases with each pass, that is why a four-pass unit is rarely seen, the economics just are not there.

Water circulation: Feedwater is introduced to the scotch marine boiler near the bottom of one side of the cylinder. The water rises between the tubes, usually faster in the rear than in the front; cooled water courses downward along the shell, then upward around the furnace to complete the cycle. The large water surface permits steam release with minimum foaming.

Water treatment: Precipitates, scale, and silt collect in the space under the furnace. Blowdown is important. The scotch marine design is well suited for operation with minimum quality feedwater.

3.   Advantages of the Firetube Boiler Versus
the Waterwall Boiler

1.   Cheaper in price than waterwall boilers of the same capacity and pressure.
2.   Respond to the *first* load swing faster, because of about 3.5 times more water at or near the saturation temperature.
3.   Easier installation in low headroom situations.
4.   Low susceptibility to cold-end corrosion. Critical parts of firetube boilers are maintained at or above the saturation temperature throughout the load range.
5.   Require less feedwater treatment than a waterwall boiler.

TABLE 3.1

Boiler House Notes — Page No:
Scotch Marine Boiler Tube Data

ABCO, AMC 2 Pass DryBack

Model Horsepower	No. Tubes 2.5"OD	No.Tubes 3.0"OD	Tube Length-In.
10	20	16	38.875
15	20	16	57.875
20	29	26	50.875
25	29	26	63.875
30	48	37	55.875
40	48	37	74.875
50	52	48 (44)	74.875
60	52	48 (44)	89.875
80	88	78	75.875
100	88	78	94.875
125	123	111 (103)	84.875
150	123	111 (103)	101.875
200	157	141	107.875
250	157	141	135.875
300	173	160	143.875
350	173	160	169.875
400	250	208	142.875
500	250	208	185.875
600	272	226	191.875
700	272	226	239.875

BESCOTCH, BE-H Series, 2PassDryBack

Model Horsepower	No. of Tubes	Tube OD,Inches	Tube Length-In.
20	38	2.0	51
30	38	2.0	78
40	58	2.0	78
50	82	2.0	66
60	82	2.0	80
70	82	2.0	96
80	120	2.0	76
100	120	2.50	95
125	104	2.50	108
150	104	2.50	129
200	106	2.50	165
250	160	2.50	140
300	160	2.50	169
350	160	2.50	197
400	250	2.50	149
500	270	2.50	172
600	270	2.50	206
650	270	2.50	224
700	270	2.50	224
750	270	2.50	224
800	270	2.50	224

Cleaver-Brooks, 4 Pass-Converted to 2 Pass

Model No.	No. of Tubes	Tube OD,Inches	Tube Length Feet-Inches
CB-15	27	2.00	4'-5"
CB-20	38	2.00	4'-5"
CB-30	44	2.00	5'-1"
CB-40	44	2.00	8'-1"
CB-50	49	2.50	6'-10"
CB-60	62	2.50	6'-10"
CB-70	49	2.50	10'-1"
CB-80	57	2.50	10'-1"
CB-100	62	2.50	11'-8"
CB-125	98	2.50	9'-5"
CB-150	98	2.50	11'-5"
CB-200	108	2.50	14'-0"
CB-250	175	2.50	10'-9"
CB-300	175	2.50	13'-0"
CB-350	175	2.50	15'-6"
CB-400	273	2.50	11'-2"
CB-500	273	2.50	14'-0"
CB-600	273	2.50	18'-0"
CB-700	273	2.50	19'-9"

Cyclotherm, 2 Pass Scotch DryBack

Model Horsepower	No. of Tubes	Tube OD,Inches	Tube Length-In
60	44	2.0	84.0
70	46	2.0	102.0
80	50	2.0	94.0
100	56	2.0	108.0
125	44	3.0	116.5
150	48	3.0	128.5
200	57	3.0	152.5
250	87	3.0	120.0
300	87	3.0	144.0

Farrar & Treft, 2 Pass Scotch DryBack

Model Number	No. of Tubes	Tube OD,Inches	Tube Length.Ft-In
1060	74	3.0	8'-10"
1075	74	3.0	11'-0"
1100	92	3.0	12'-0"
1125	116	3.0	12'-0"
1150	154	3.0	11'-2"
1200	179	3.0	13'-0"
1250	198	3.0	14'-6"
1300	219	3.0	16'-0"

Farrar & Treft, Bixon, 2 Pass Series 6000, Type C

Model Number	No. of Tubes	Tube OD,Inches	Tube Length:Ft-In
6050	44	3.0	10'-0"
6060	44	3.0	11'-6"
6075	52	3.0	11'-6"
6080	52	3.0	12'-6"
6100	65	3.0	13'-0"
6125	80	3.0	13'-0"
6150	80	3.0	15'-0"
6200	120	3.0	14'-6"
6250	120	3.0	17'-6"
6300	160	3.0	18'-6"

Federal: FMA & FMB Series, 2 Pass

Model Number	No. of Tubes	Tube OD,Inches	Tube Length: Inch
168	27	3.0	33.625
219	27	3.0	46.125
268	27	3.0	58.625
316	27	3.0	71.125
365	43	3.0	54.125
425	43	3.0	64.125
486	43	3.0	74.125
547	43	3.0	84.125
608	43	3.0	94.125
729	71	3.0	71.125
850	71	3.0	84.125
1033	71	3.0	106.125
1215	71	3.0	123.125
1518	128	3.0	88.000
1822	128	3.0	107.000
2125	128	3.0	126.000
2429	128	3.0	145.000
2429 Spec.	204	3.0	90.000
3036	204	3.0	115.000
3643	204	3.0	140.000
4250	204	3.0	164.000
4250 Spec.	266	3.0	126.000
4857	266	3.0	145.000
5464	266	3.0	164.000
6071	266	3.0	183.000

Fitzgibbons: 2 Pass DryBack

Model Number	No. of Tubes	Tube OD,Inches	Tube Length: Inch
SM-54	43	3.0	83.000
SM-61	43	3.0	93.500
SM-72	59	3.0	85.500
SM-85	59	3.0	101.000
SM-103	76	3.0	97.125
SM-121	76	3.0	115.875
SM-151	97	3.0	115.375
SM-182	97	3.0	140.325
DM-212	107	3.0	148.625
DM-242	107	3.0	171.375
DM-303	124	3.0	186.750
DM-364	154	3.0	182.750
SM-425	188	3.0	177.250

Industrial Combustion: 2 Pass Dry Back

Model Number	No. of Tubes	Tube OD,Inches	Tube Length: Inch
30A	42	2.5	56.000
40A	42	2.5	76.000
50A	56	2.5	74.000
60A	56	2.5	91.000
70A	56	2.5	106.000
80A	68	2.5	102.000
100A	68	2.5	127.000
125A	92	2.5	115.375
150A	92	2.5	140.375
200A	138	2.5	126.500
250A	138	2.5	159.000
300A	138	2.5	191.500
350A	196	2.5	157.375
400A	196	2.5	180.875
450A	196	2.5	204.125
500A	255	2.5	184.625

re: Boiler Tube Data Book, American Fuel Economy, Inc.

TABLE 3.1　Continued

Boiler House Notes											Page No:	
Scotch Marine Boiler Tube Data												

Iron Fireman: D-B, 35 Series, 3 Pass WetBack

Model Number	No. Tubes 3rd Pass	Tube OD Inches	Tube Length:Inch
35-60	14	2.0	33.375
35-80	19	2.0	33.375
35-100	21	2.0	46.375
35-120	24	2.0	46.875
35-160	28	2.0	55.875
35-200	28	2.0	65.375
35-240	46	2.0	57.125
35-280	48	2.0	64.875
35-320	50	2.0	71.025
35-358	61	2.0	65.875
35-400	59	2.0	76.625
35-500	59	2.0	92.125
35-600	62	2.5	86.750
35-800	68	2.5	101.000
35-1000	68	2.5	125.500
35-1250	92	2.5	121.000
35-1500	92	2.5	144.000

Iron Fireman: D-B, 300 Series, 3 Pass WetBack

Model Number	No. Tubes 3rd Pass	Tube OD Inches	Tube Length:Ft-In.
20	20	2.0	4'-7.625"
30	20	2.0	6'-10.25"
40	24	2.0	7'-2.375"
50	40	2.0	6'-0.875"
60	40	2.0	7'-2.625"
70	40	2.0	8'-4.875"
70A	42	2.0	7'-9.75"
80	40	2.0	9'-6.25"
80A	42	2.0	8'-8.25"
100	42	2.0	10'-9.0"
125	52	2.0	11'-3.125"
150	68	2.0	10.9.375"
200	68	2.0	14'-1.875"
250	70	2.50	13'-1.125"
300	70	2.50	15'-6.875"
350	98	2.50	12'-11.125"
400	98	2.50	14'-7.375"
500	112	2.50	15'-8.875"
600	112	2.50	16'-7.625"

Johnston: 3 Pass Wet Back Scotch Marine

Model / H.P.	Number Tubes	Tube OD Inches	Tube Length:Ft-In.
299	32	2.00	7' - 10.5"
358	32	2.00	9' - 4"
510-75	32	2.00	9' - 7"
512-100	44	2.00	9' - 7"
514-125	56	2.00	9' - 7"
516-150	60	2.00	10' - 11"
518-200	80	2.00	10' - 11"
522-250	102	2.00	10' - 11"
524-300	66	2.50	14' - 11.5"
526-320	78	2.50	14' - 11.5"
528-400	88	2.50	14' - 11.5"
530-500	90	2.50	18' - 0.5"
534-600	110	2.50	18' - 0.5"
535-750	138	2.50	18' - 0.5"
536-900	148	2.50	20' - 4.5"
538-1000	148	2.50	22' - 4.5"
540-1200	178	2.50	22' - 4.5"
550-1500	218	2.50	22' - 4.5"

Johnston: 3 Pass Dry Back Scotch Marine

Model Horsepower	No. of Tubes	Tube OD, Inches	Tube Length:Ft-In.
209-3	10	2.0	6' - 7"
298-20	10	2.0	6' - 7"
210-3	12	2.0	7' - 0"
298-25	12	2.0	7' - 0"
211-3	16	2.0	7' - 0"
298-30	16	2.0	7' - 0"
212-3	18	2.0	7' - 0"
298-35	18	2.0	7' - 0"
213-3	18	2.0	8' - 1"
298-40	18	2.0	8' - 1"
214-3	24	2.0	8' - 1"
298-50	24	2.0	8' - 1"
215-3	24	2.0	9' - 7"
298-60	24	2.0	9' - 7"
216-3	32	2.0	8' - 6"
298-70	32	2.0	8' - 6"
217-3	32	2.0	9' - 7"
298-80	32	2.0	9' - 7"
218-3	48	2.0	8' - 6"
298-100	48	2.0	8' - 6"
219-3	48	2.0	10' - 9"
298-125	48	2.0	10' - 9"
220-3	60	2.0	10' - 3"
298-150	60	2.0	10' - 3"
221-3	80	2.0	10' - 9"
298-200	80	2.0	10' - 9"
222-3	96	2.0	10' - 3"
298-250	96	2.0	10' - 3"
223-3	66	2.5	14' - 5.5"
298-300	66	2.5	14' - 5.5"
224-3	78	2.5	14' - 5.5"
298-350	78	2.5	14' - 5.5"
225-3	88	2.5	14' - 5.5"
298-400	88	2.5	14' - 5.5"

re: Boiler Tube Data Book, American fuel Economy, Inc.

Johnston: 2 Pass Wet Back Scotch Marine

Model No.	Number Tubes	Tube OD Inches	Tube Length: Inchs
9599	22	2.5	42.00
9500	22	2.5	54.00
9501	36	2.5	42.00
9502	36	2.5	54.00
9503	36	2.5	66.00
9504	36	2.5	78.00
9505	47	2.5	69.00
9506	47	2.5	81.00
9507	47	2.5	93.00
9508	47	2.5	105.00
9509	65	2.5	93.50
9510	65	2.5	111.00
9510 Spec.	84	2.5	84.00
9511	84	2.5	105.00
9512	105	2.5	102.00
9513	123	2.5	111.00
9514	142	2.5	117.00
9515	180	2.5	108.00
9516	215	2.5	105.25
9516 Spec.	215	2.5	108.75
9-1600	215	2.5	120.75
9517	240	2.5	120.75
9518	286	2.5	120.75

Lookout-Eclipse: 2 Pass Dry Back Scotch Marine
Model SGO and SGOH

Model H.P.	Number Tubes	Tube OD Inches	Tube Length: Inchs
12	14	3.0	42.0
20	17	3.0	68.0
25	17	3.0	72.0
30	22	3.0	75.0
40	27	3.0	84.0
50	36	3.0	84.0
60	44	3.0	84.0
80	57	3.0	90.0
100	68	3.0	96.0

Kieco: 2 Pass Scotch Marine

Model No.	Number Tubes	Tube OD Inches	Tube Length: Inchs
10	24	2.0	40.0
15	24	2.0	60.0
20	33	2.0	60.5
25	33	2.0	75.0
30	43	2.0	72.0
40	43	2.0	96.0
50	62	2.0	87.0
60	62	2.0	106.5
75	70	2.5	94.5
100	70	2.5	126.0
125	104	2.5	108.0
150	104	2.5	126.0
175	126	2.5	126.0
200	126	2.5	144.0

Leffel: 2 Pass Dry Back Scotch Marine

Model H.P.	Number Tubes	Tube OD Inches	Tube Length: Inchs
6	14	3.0	42.0
12	16	3.0	73.0
15	18	3.0	73.0
20	25	3.0	78.0
25	25	3.0	96.0
30	31	3.0	96.0
35	31	3.0	114.0
40	42	3.0	96.0
50	45	3.0	114.0
60	66	3.0	114.0
75	66	3.0	144.0
100	80	3.0	160.0
125	102	3.0	160.0
150	114	3.0	174.0
175	136	3.0	174.0
200	158	3.0	174.0
250	192	3.0	178.0

TABLE 3.1 Continued

Boiler House Notes   Page No:
Scotch Marine Boiler Tube Data

Ocean Shore Iron Works — 2 Pass Scotch Marine Dry Back

Model H.P.	No. of Tubes	Tube OD Inches	Tube Length:inch
10	36	1.5	35.5
16	32	2.0	47.5
28	46	2.0	63.5
40	54	2.0	71.5
56	64	2.0	89.5
80	92	2.0	95.5
107	97	2.0	119.5
126	102	2.5	107.5
162	102	2.5	143.5
254	109	2.5	191.5
336	148	2.5	191.5
400	157	3.0	191.5
500	210	3.0	191.5
640	260	3.0	192.0
748	308	3.0	191.5

Superior - Hutchinson, Kansas — Scotch Marine Dry Back — Series: FDH High Pressure & FDL Low Pressure

Model No.	No. Tubes	Tube OD Inches	Tube Length:inch
FDH - 10	22	2.0	48.0
FDH - 15	26	2.0	57.5
FDH - 20	30	2.0	66.5
FDH - 25	34	2.0	74.0
FDH - 30	44	2.0	74.0
FDH - 40	59	2.0	72.5
FDH - 50	62	2.0	86.0
FDH - 60	67	2.0	103.5
FDH - 75	78	2.0	105.5
FDH - 100	92	2.0	120.5
FDH - 125	114	2.0	123.5
FDH - 150	124	2.0	132.0
FDH - 175	150	2.0	132.0
FDH - 200	178	2.0	133.0

re: Boiler Tube Data Book, American Fuel Economy, Inc.

Orr & Sembower — Powermaster: Models 3 - 5 & J

Model H.P.	No. Tubes 3rd Pass	Tube OD Inches	Tube Length:inchs
15	22	1.5	46.4375
20	22	1.5	46.4375
30	28	1.5	58.25
40	24	2.0	68.25
50	24	2.0	85.875
60	34	2.0	71.0
70	34	2.0	83.25
80	34	2.0	95.5625
100	58	2.0	96.125
125	58	2.0	98.5
150	50	2.5	103.125
200	50	2.5	138.125
250	92	2.5	111.25
300	92	2.5	133.25
350	102	2.5	131.25
400	112	2.5	144.25
500	140	2.5	149.25
600	140	2.5	179.25
100 Ohio	40	2	no data
80 Ohio	34	2	no data

Williams and Davis — Two Pass Scotch Marine DryBack

Model No.	No. of Tubes	Tube OD Inches	Tube Length:inch
10	26	2.0	36.0
15	26	2.0	54.0
20	39	2.0	52.0
25	39	2.0	64.0
30	51	2.0	60.0
40	61	2.0	68.0
50	65	2.0	80.0
60	65	2.0	96.0
80	92	2.0	96.0
100	92	2.0	117.0
125	129	2.0	114.0
150	142	2.0	117.0
175	142	2.0	140.0
200	175	2.0	125.0
250	175	2.0	158.0
300	205	2.0	166.0
400	205	2.0	222.0
500	236	2.5	200.0
600	236	2.5	208.0

Spencer AVCO — Type "A": 3 Pass Wet Back Scotch Marine

Model No./H.P.	No. Tubes	Tube OD Inches	Tube Length:inchs
1 - 18	16	3.0	35.875
1 - 22	16	3.0	47.875
2 - 26	20	3.0	44.875
2 - 30	20	3.0	53.875
3 - 35	27	3.0	42.675
3 - 40	27	3.0	51.375
3 - 45	27	3.0	59.375
4 - 50	34	3.0	53.375
4 - 60	34	3.0	67.375
4 - 70	34	3.0	79.875
5 - 70	44	3.0	58.375
5 - 85	44	3.0	74.375
5 - 100	44	3.0	89.625
6 - 100	64	3.0	60.375
6 - 125	64	3.0	79.375
6 - 150	64	3.0	100.375
7 - 150	82	3.0	73.375
7 - 175	82	3.0	88.375
7 - 200	82	3.0	104.875
8 - 200	104	3.0	79.25
8 - 250	104	3.0	103.25
9 - 300	118	3.0	105.5
9 - 350	118	3.0	125.5

York Shipley — Series 500: 3 Pass DryBack Scotch Marine

Model No.	No. of Tubes	Tube OD Inches	Tube Length:inch
SPLV - 20	24	2.0	33.375
SPWV - 20	24	2.0	33.375
SPHV - 20	24	2.0	33.375
SPLV - 25	24	2.0	42.625
SPWV - 25	24	2.0	42.625
SPHV - 25	24	2.0	42.625
SPLV - 30	24	2.0	51.875
SPWV - 30	24	2.0	51.875
SPHV - 30	24	2.0	51.875
SPLC - 60	34	2.0	60.125
SPWC - 60	34	2.0	60.125
SPHC - 60	34	2.0	60.125
SPLC - 125	56	2.0	106.375
SPWC - 125	56	2.0	106.375
SPHC - 125	56	2.0	106.375
SPLC - 175	68	2.0	122.375
SPWC - 175	68	2.0	122.375
SPHC - 175	68	2.0	122.375
SPLC - 250	68	2.0	140.375
SPLC - 250	66	2.5	141.375
SPWC - 250	66	2.5	141.375
SPHC - 250	66	2.5	141.375
SPLC - 400	72	2.5	207.5625
SPWC - 400	72	2.5	207.5625
SPHC - 400	72	2.5	207.5625
SPLC - 500	92	2.5	207.5625
SPWC - 500	92	2.5	207.5625
SPHC - 500	92	2.5	207.5625

4. Disadvantages of the Firetube Boiler Versus
the Waterwall Boiler

1. Less efficient than the waterwall, thus more expensive to operate.
2. Does not respond to load swings as fast as a waterwall boiler.
3. The firetube design is marginal when used to fire solid fuel.
4. Limited to saturated steam temperature, no superheat.
5. Limited to low to medium steam pressure.
6. Firetube design is not practical over 50,000 lb/hr steaming capacity.
7. Lower steam purity than the waterwall.
8. Higher CO and nitrogen oxides (NO_x) emissions than a waterwall.

See Table 3.1 for further details.

IV. WATERWALL BOILERS: GAS AND OIL FIRED

A. Packaged Waterwall Boilers

This boiler design concept is built around the "waterwall." This refers to the tubes connecting the steam drum to the mud drum. The tubes are attached to each other with a continuous strip of metal (called a membrane), usually this strip is 3/16–1/4 in. thick by 1.0 in. wide and whatever length that is required. This forms a continuous wall of tubes filled with water at the bottom and steam at the top, thus *waterwall*. This waterwall design is used in the outerwalls and also in some cases, the dividing wall between the furnace section and the convection section.

This boiler wall design is very efficient in heat transfer, especially in the furnace area. The majority of the radiant heat that strikes the membrane bar is transferred to the tube on either side. This waterwall design also acts as a gastight wall, which also adds to the efficiency of the boiler.

As you know, boiler walls were originally mostly refractory. The problems were many. The normal furnace temperature is 2200–2400°F when firing natural gas. It is higher when firing fuel oil. This can reach a maximum temperature of 3200°F. This high temperature is very detrimental to refractory. The continuous operation of the furnace at these high temperatures will, over time, cause the disintegration of the refractory. Refractory maintenance was a very high dollar item in the power plant budget.

Waterwalls were added to existing watertube boilers before they were incorporated into the newly designed package units. The waterwalls allowed the boilers to be operated continuously at the maximum firing rate and still realize good operating economics. The waterwalls cut down on outages, greatly reduced or eliminated refractory maintenance, and also permitted the efficient firing of lower grade fuels.

A secondary effect of waterwalls is the lowering of the furnace temperature. Most of this is due to the increased absorption of radiant heat by the waterwall. This lowering of furnace temperature directly affects the formation of NO_x. As the furnace temperature decreases, the formation of combustion NO_x decreases.

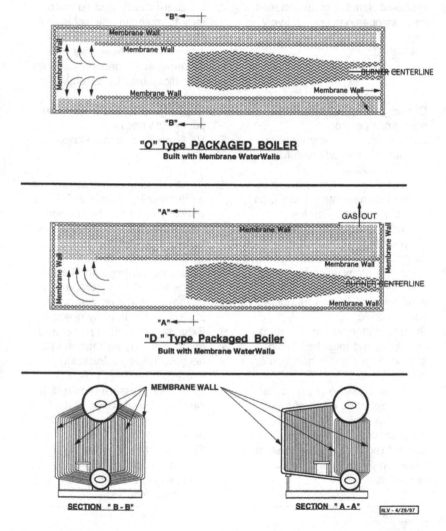

FIGURE 3.2   Two types of packaged boilers built with membrane waterwalls.

TABLE 3.2 Fuel Oil Atomization

Steam atomization	Mechanical atomization
1. Best suited to variable load.	1. Best suited to steady load and high capacity.
2. Wide capacity range without changing tip or gun assembly.	2. Limited capacity range for any given tip size; wide-range system overcomes this somewhat.
3. Frequent cleaning of tip unnecessary, as openings are relatively large and can be quickly blown out with steam.	3. Frequent cleaning of tip necessary to maintain efficient spray. Owing to relatively small opening, entire gun assembly must be removed so sprayer plate and tip orifices may be carefully cleaned.
4. Capacity up to 120 million Btu per nozzle per hour.	4. Capacity up to 100 million Btu per nozzle per hour.
5. Considerable flexibility for shaping flame to conform with furnace conditions.	5. No flexibility in flame shape.
6. Relatively low oil temperature required (approximately 185°F), as viscosity need only be low enough (40 SSU) to readily permit pumping.	6. Relatively high oil temperature required (approximately 220°F), as viscosity must be low enough (180–220 SSU) to produce satisfactory atomization.
7. Oil pressure 2–125 psi.	7. Oil pressure 50–250 psi.
8. Steam for atomization may vary from 0.7 to 5.0%. The approximate average for careful operation is 1.25%.	8. Steam for pumping and heating varies from 0.5 to 1.0%, and is governed by the equipment installed, rather than by operation.
9. Steam for atomization is lost up the stack, and must be considered when there is a question of makeup water.	9. Exhaust steam from pump and heater set may be returned to hotwell, thereby minimizing makeup.
10. Lower air pressure required, because aspirating effect of the steam jets makes up for some of the pressure drop in the register.	10. Higher air pressure required, because of the absence of aspirating effect with the mechanically produced spray.
11. Lower fixed charges, because of lower temperature and furnace requirements.	11. Fixed charges high, owing to cost of equipment to provide for high-pressure and high-temperature requirements.

Source: Ref. 15.

TABLE 3.3

Boiler House Notes					Page No.	
Fuels: Oil & Gas Analysis						
Ralph L. Vandagriff						
NATURAL GAS	Pennsylvania	South Carolina	Ohio	Louisiana	Oklahoma	Kansas
Analysis:						
Constituents, % by Volume						
H2, Hydrogen	0.00	0.00	1.82	0.00	0.00	0.00
CH4, Methane	83.40	84.00	93.33	90.00	84.10	84.10
C2H4, Ethylene	0.00	0.25	0.00	0.00	0.00	0.00
C2H6, Ethane	15.80	14.80	0.00	5.00	6.70	6.70
CO, Carbon Monoxide	0.00	0.00	0.45	0.00	0.00	0.00
CO2, Carbon Dioxide	0.00	0.70	0.22	0.00	0.80	0.80
N2, Nitrogen	0.80	0.50	3.40	5.00	8.40	8.40
O2, Oxygen	0.00	0.00	0.35	0.00	0.00	0.00
H2S, Hydrogen sulfide	0.00	0.00	0.18	0.00	0.00	0.00
Ultimate Analysis, % by Weight						
S, Sulfur	0.00	0.00	0.34	0.00	0.00	
H2, Hydrogen	23.53	23.30	23.20	22.68	20.85	
C, Carbon	75.25	74.72	69.12	69.26	64.84	
N2, Nitrogen	1.22	0.76	5.76	8.06	12.90	
O2, Oxygen	0.00	1.22	1.58	0.00	1.41	
Specific Gravity (relative to air)	0.6360	0.6360	0.5670	0.6000	0.6300	0.6300
Lb.m/Cu.Ft.	0.04873	0.04873	0.04367	0.04683	0.04831	0.04831
HHV, Btu/cu.ft. @ 60 deg. F.	1,129	1,116	964	1,022	974	970
HHV, Btu/lb.	23,170	22,904	22,077	21,824	20,160	20,079

Fuel Oil	#1	#2	#4	#5 (Light)	#5 (Heavy)	#6
Flash Point, Deg. F.	100 or legal	100 or legal	130 or legal	130 or legal	130 or legal	150
Viscosity, SSU @ 100 deg. F.						
Minimum		32.60	45.00	150.00	350.00	900
Maximum		37.93	125.00	300.00	750.00	9000
Viscosity, Centistokes @ 100 deg. F.						
Minimum	1.40	2.00	5.80	32.00	75.00	198
Maximum	2.20	3.60	26.40	65.00	162.00	1980
Specific Gravity	0.825 to 0.806	0.887 to 0.825	0.986 to 0.876	0.972 to 0.922		1.022 to 0.922
Lb. per gallon	6.87 to 6.71	7.39 to 6.87	8.04 to 7.3	8.10 to 7.68		8.51 to 7.68
Heating Value, Btu/lb. (gross)	19,570 to 19,860	19,170 to 19,750	18,280 to 19,400	18,100 to 19,020		17,410 to 18,990
Heating Value, Btu/gal. (gross)min.	131,986	131,698	133,444	139,008		133,709
Heating Value, Btu/gal. (gross)max.	136,438	145,953	155,976	154,062		161,605
Percent by Weight						
Sulfur	0.01 to 0.5	0.05 to 1.0	0.2 to 2.0	0.5 to 3.0		0.7 to 3.5
Hydrogen	13.3 to 14.1	11.8 to 13.9	10.6 to 13.0	10.5 to 12.0		9.5 to 12.0
Carbon	85.9 to 86.7	86.1 to 88.2	86.5 to 89.2	86.5 to 89.2		86.5 to 90.2
Nitrogen	nil to 0.1	nil to 0.1	nil	nil		
Oxygen	nil	nil	nil	nil		
Ash	nil	nil	0.0 to 0.1	0.0 to 0.1		0.01 to 0.5

re: #1, #3, & #4

TABLE 3.4

Boiler House Notes						Page No:

Subject: Combustion constants

Adapted from "Fuel Flue Gases", combustion Flame & Explosions of Gases, American Gas Association, 1951

(1) Gas volumes corrected to 60 deg. F & 30 in.Hg dry.

Compound	Density Lb/cu.ft.	Specific Volume(1) cu.ft./lb.	Heat of Combustion			
			Btu/cu.ft.		Btu/lb.	
			Gross	Net	Gross	Net
Carbon, C					14,093	14,093
Hydrogen, H2	0.0053	187.7230	325	275	61,100	51,623
Oxygen, O2	0.0846	11.8190				
Nitrogen, N2	0.0744	13.4430				
Carbon monoxide, CO	0.0740	13.5060	322	322	4,347	4,347
PARAFFIN SERIES						
Methane, CH4	0.0424	23.5650	1,013	913	23,879	21,520
Ethane, C2H6	0.0803	12.4550	1,792	1,641	22,320	20,432
Propane, C3H8	0.1196	8.3650	2,590	2,385	21,661	19,944
n-Butane, C4H10	0.1582	6.3210	3,370	3,113	21,308	19,680
Isobutane, C4H10	0.1582	6.3210	3,363	3,105	21,257	19,629
n-Pentane, C5H12	0.1904	5.2520	4,016	3,709	21,091	19,517
Isopentane, C5H12	0.1904	5.2520	4,008	3,716	21,052	19,478
Neopentane, C5H12	0.1904	5.2520	3,993	3,693	20,970	19,396
n-Hexane, C6H14	0.2274	4.3980	4,762	4,412	20,940	19,403
n-Heptane, C7H16	0.2641	3.7870	5,502	5,100	20,681	19,314
OLEFIN SERIES						
Ethylene, C2H4	0.0746	13.4120	1,614	1,513	21,644	20,295
Propylene, C3H6	0.1110	9.0070	2,336	2,186	21,041	19,691
I-Butene, C4H8	0.1480	6.7560	3,084	2,885	20,840	19,496
Isobutene, C4H8	0.1480	6.7560	3,068	2,869	20,730	19,382
I-Pentene, C5H10	0.1852	5.4000	3,836	3,586	20,712	19,363
AROMATIC SERIES						
Benzene, C6H6	0.2060	4.8520	3,751	3,601	18,210	17,480
Toluene, C7H8	0.2431	4.1130	4,484	4,284	18,440	17,620
Xylene, C8H10	0.2803	3.5670	5,230	4,980	18,650	17,760
MISCELLANEOUS						
Acetylene, C2H2	0.0697	14.3440	1,499	1,448	21,500	20,776
Napthalene, C10H8	0.3384	2.9550	5,854	5,654	17,298	16,708
Methyl alcohol, CH3OH	0.0846	11.8200	868	768	10,259	9,078
Ethyl alcohol, C2H5OH	0.1216	8.2210	1,600	1,451	13,161	11,929
Ammonia, NH3	0.0456	21.9140	441	365	9,668	8,001
Sulfur, S					3,983	3,983
Hydrogen Sulfide, H2S	0.0911	10.9790	647	596	7,100	6,545

TABLE 3.5

		MINIMUM AUTO-IGNITION TEMPERATURE				
Ralph L. Vandagriff				**Boiler House Notes:**		**Page:**
Date: May 6, 1996		Ignition				
		Temperature	Molecular	HHV	LHV	
	Name	Formula	deg. F.	Weight	Btu/lb.	Btu/lb.
1	Acetone	C3 H6 O	1,042			
2	Acetylene	C2 H2	581	26.04	21,460	20,734
3	Amyl Alcohol	C5 H12 O	801			
4	Benzene	C6 H6	1,078	78.11	17,986	17,259
5	Butadiene	C4 H6	804			
6	Butane	C4 H10	826	58.12	21,293	19,665
7	Isobutane	C4 H10	1,010	58.12	21,242	19,614
8	Butyl Alcohol	C4 H10 O	653	74.10	15,500	14,284
9	Isobutyl Alcohol	C4 H10 O	813			
10	Butylene	C4 H8	829			
11	Carbon,fixed: Bituminous Coal	C	766	12.01	14,087	
12	Carbon,fixed: Semi-bituminous coal	C	870			
13	Carbon,fixed: Anthracite	C	925			
14	Carbon Monoxide	CO	1,128	28.01	4,344	
15	Creosote		637			
16	Cyclohexane	C6 H12	565	84.16	20,026	18,676
17	Cyclopropane	C3 H6	928			
18	Decane	C10 H12	482	142.28	20,483	19,020
19	Dodecane	C12 H26	993			
20	Ethane	C2 H6	882	30.07	22,304	20,416
21	Ethyl Alcohol	C2 H6 O	738	46.00	12,780	11,604
22	Ethyl Bromide	C2 H5 Br	952			
23	Ethylene	C2 H4	914	28.05	21,625	20,276
24	Ethylene Glycol	C2 H6 O2	775			
25	Ethyl Ether	C4 H10 O	379			
26	Gasoline	73 octane	570			
27	Gasoline	92 octane	734			
28	Gasoline	100 octane	804			
29	Glycerine	C3 H8 O	739			
30	Heptane	C7 H16	451	100.20	20,668	19,157
31	Hexane	C6 H14	478	86.17	20,771	19,233
32	Isohexane	C6 H14	543			
33	Hydrogen	H2	1,065	2.02	60,958	51,571
34	Kerosene		491			
35	Methane	C H4	1,170	16.04	23,861	21,502
36	Methyl Alcohol	C H4 O	878	32.00	9,770	8,644
37	Naphtha		450 to 531			
38	Naphthalene	C10 H8	1,038			
39	Nonane	C9 H20	545	128.25	20,531	19,056
40	Octane	C8 H18	446	114.22	20,591	19,100
41	Oil, castor		840			
42	Oil, cottonseed		660			
43	Oil, gas		640			
44	Oil, lard		650			
45	Oil, linseed		650			
46	Oil, lubricating		711			
47	Oil, cylinder		783			
48	Oil, turbine lube		700			
49	Oil, peanut		833			
50	Oil, rosin		648			
51	Oil, sperm		586			
52	Oil, tung		855			
53	Oil, whale		878			
54	Paraffin Wax	C36 H74	473			
55	Pentane	CH3(CH2)3CH3	527	71.15	20,914	19,340
56	Petroleum Ether		624			
57	Propane	C3 H8	898	44.09	21,646	19,929
58	Propyl Alcohol	C3 H8 O	822	60.00	14,500	13,300
59	Isopropyl Alcohol	C3 H8 O	853			
60	Propylene	C3 H6	856	42.08	21,032	19,683
61	Styrene	C8 H8	914			
62	Toluene	C7 H8	1,026	92.13	18,245	17,424
63	Turpentine	C10 H16	464			
64	Sulfur	S	470	32.06	3,980	
65	ref #3					

TABLE 3.6

Spreadsheet Example:	This spreadsheet is set up to calculate one (1) gas fuel.										
	Variable to be entered. Percent by Volumn for each component. (gas origin data is optional)								re: # 49, pg. 3.190		
			Gas Combustion								
	(Note: This method of combustion calculations may be used for any gas.)								**BTU PER CUBIC FOOT OF GAS**		
Gas Origin:	Louisiana		Gas Type:	Natural					(re: #3, pg. 91)		
	Methane	Ethane	Propane	n-Butane						HHV	LHV
Components:	CH4	C2H6	C3H8	C4H10	N2	S	CO2	Total:		Heat of	Heat of
Percent by Volume	90.00%	5.00%	0.00%	0.00%	5.00%	0.00%	0.00%	100.00%		Reaction:	Reaction:
Density:									Component	Btu/Mole	Btu/Mole
Cu.Ft./Lb.	23.6500	12.6200	8.6060	6.5290	13.5318		8.6133		CH4	382,730	344,892
Lb./cu.ft.	0.0423	0.0792	0.1162	0.1532	0.0739		0.1161		C2H6	870,681	813,909
Kg/cu.meter	0.6770	1.2680	1.8613	2.4534	0.0940		1.8590		C3H8	954,372	878,870
									C4H10	1,234,585	1,142,930
										Btu per cubic foot:	
Component Wt.										996.21	898.97
Lb./cu.ft.	0.0381	0.0040	0.0000	0.0000	0.0037	0.0000	0.0000	0.04573			
Kg/cu.meter	0.6093	0.0634	0.0000	0.0000	0.0047	0.0000	0.0000	0.67740		Btu per pound.	
										21,786.91	19,660.44
Percent by Weight:	83.26%	8.66%	0.00%	0.00%	8.08%	0.00%	0.00%	100.00%			
Lb. O2 Required	3.3303	0.3233	0.0000	0.0000	0.0000	0.0000	0.0000	3.6537			
for Combustion											
Lb. of Air Required	14.3549	1.3936	0.0000	0.0000	0.0000	0.0000	0.0000	15.7486			
(N2=76.8%)											
Products of Combustion	3.3684	0.3273	0.0000	0.0000	0.0000	0.0000	0.0000	3.6957			
Lbs:											
Plus Nitrogen:	11.0246	1.0703	0.0000	0.0000	0.0037	0.0000	0.0000	12.0986			
POUNDS OF FLUE GAS PER POUND OF GAS FUEL - ZERO EXCESS AIR							TOTAL	15.7943			

Products of Combustion: by weight									Percent	
CO2	0.1047	0.0116	0.0000	0.0000	0.0000		0.0000	0.1163	15.18%	
H2O	0.0857	0.0071	0.0000	0.0000	0.0000			0.0928	12.11%	
N2							0.0037		0.5569	72.70%
SO2							0.0000		0.0000	0.00%
							Total:	0.7660	100.00%	

PRODUCTS OF COMBUSTION: VOLUMN AT SPECIFIED TEMPERATURES											
DEGREES Fahr.	300	350	600	850	900	950	1000	1050	1250	1500	1800
CO2: Cu.Ft.	1.5476	1.6494	2.1584	2.6675	2.7693	2.8711	2.9729	3.0748	3.4820	3.9911	4.6019
H2O: Cu.Ft.	3.0178	3.2164	4.2091	5.2018	5.4003	5.5989	5.7974	5.9960	6.7901	7.7828	8.9741
N2: Cu.Ft.	11.6442	12.4103	16.2406	20.0709	20.8370	21.6030	22.3691	23.1352	26.1994	30.0298	34.6262
SO2: Cu.Ft.	0.0000	0.0000	0.0000	0.0000	0.0000	0.0000	0.0000	0.0000	0.0000	0.0000	0.0000
Total:	16.2096	17.2760	22.6081	27.9402	29.0066	30.0730	31.1395	32.2059	36.4716	41.8037	48.2022
Cubic Feet of Flue Gas per cubic foot of gas fuel fired - ZERO EXCESS AIR.											
CO2, wet basis	9.55%		CO2, dry basis	11.73%							

EXCESS AIR											
PERCENT:	0.00%	2.50%	5.00%	7.50%	10.00%	12.50%	15.00%	17.50%	20.00%	22.50%	25.00%
Lb. of Combustion Air Required per lb. of fuel gas with Stated Excess Air:											
	15.7486	16.1423	16.5360	16.9297	17.3234	17.7171	18.1109	18.5046	18.8983	19.2920	19.6857
Lb. of Combustion Air Required per cu.foot of fuel gas with Stated Excess Air:											
	0.7201	0.7381	0.7561	0.7741	0.7921	0.8101	0.8281	0.8461	0.8641	0.8821	0.9001
Pounds of Flue Gas	15.7943	16.1880	16.5817	16.9754	17.3692	17.7629	18.1566	18.5503	18.9440	19.3377	19.7314
per pound of fuel gas											
with stated excess air.											
	Cu. Ft of Flue Gas per Cu.Foot of Fuel Gas w/ Stated Excess Air & Stated Temperature.										
300 deg. F.	16.2096	16.5736	16.9377	17.3018	17.6658	18.0299	18.3940	18.7580	19.1221	19.4861	19.8502
600 deg. F.	22.6081	23.1159	23.6236	24.1314	24.6392	25.1470	25.6547	26.1625	26.6703	27.1780	27.6858
900 deg. F.	29.0066	29.6581	30.3096	30.9611	31.6125	32.2640	32.9155	33.5670	34.2184	34.8699	35.5214
1800 deg. F.	48.2022	49.2848	50.3674	51.4500	52.5326	53.6152	54.6978	55.7804	56.8630	57.9456	59.0282
CO2 (wet basis)	9.55%	9.34%	9.14%	8.94%	8.76%	8.58%	8.41%	8.25%	8.09%	7.94%	7.80%
CO2 (dry basis)	11.73%	11.42%	11.12%	10.83%	10.56%	10.31%	10.06%	9.83%	9.61%	9.40%	9.19%
O2 (dry basis)	0.00%	0.57%	1.10%	1.62%	2.11%	2.58%	3.03%	3.46%	3.87%	4.26%	4.64%

TABLE 3.7

	A	B
1	Spreadsheet Example:	Spreadsheet software used: Excel ® by Microsoft®
2	FORMULAS:	
3		**Gas Combustion**
4	Gas Origin:	Louisiana
5		Methane
6	Components:	CH4
7	Percent by Volume	0.9
8	Density:	
9	Cu.Ft./Lb.	23.65
10	Lb/cu.ft.	0.0423
11	Kg/cu.meter	0.677
12		
13		
14		
15	Component Wt.	
16	Lb/cu.ft.	=B7*B10
17	Kg/cu.meter	=B7*B11
18		
19	Percent by Weight:	=B16/$I$16
20		
21	Lb. O2 Required	=B19*(64/16)
22	for Combustion	
23		
24	Lb. of Air Required	=(B21*(0.768/0.232))+B21
25	(N2=76.8%)	
26		
27	Products of Combustion	=B16+B21
28	Lbs:	
29	Plus Nitrogen:	=B21*(0.768/0.232)
30		
31		
32		
33	CO2	=B16*(44/16)
34	H2O	=B16*(36/16)
35	N2	
36	SO2	
37		
38		
39		
40	DEGREES Fahr.	300
41	CO2: Cu.Ft.	=(379/44)*((B40+460)/(32+460))*$I$33
42	H2O: Cu. Ft.	=(379/18)*((B40+460)/(32+460))*$I$34
43	N2: Cu.Ft.	=(379/28)*((B40+460)/(32+460))*$I$35
44	SO2: Cu.Ft.	=(379/32)*((B40+460)/(32+460))*$I$36
45		
46	Total:	=SUM(B41:B44)
47		
48		
49	CO2, wet basis	=B41/B46
50		
51		
52	PERCENT:	0
53	Lb. of Combustion Air Required per lb. of	
54		=$I$24*(1+B52)
55	Lb. of Combustion Air Required per cu.fo	
56		=B54/(1/$I$16)
57		
58	Pounds of Flue Gas	=$I$30+($I$24*B52)
59	per pound of fuel gas	
60	with stated excess air.	
61		
62	300 deg. F.	=((($I$24/(1/$I$16))*B52))*((379/28.95)*((300+460)/(32+460)))+$B$46
63	600 deg. F.	=((($I$24/(1/$I$16))*B52))*((379/28.95)*((600+460)/(32+460)))+$D$46
64	900 deg. F.	=((($I$24/(1/$I$16))*B52))*((379/28.95)*((900+460)/(32+460)))+$F$46
65	1800 deg. F.	=((($I$24/(1/$I$16))*B52))*((379/28.95)*((1800+460)/(32+460)))+$L$46
66		
67	CO2 (wet basis)	=$B$41/B62
68	CO2 (dry basis)	=$B$41/(B62-$B$42)
69	O2 (dry basis)	=((($I$23*(1+B52))-$I$23)/B54)

TABLE 3.8

	A	B
1	Spreadsheet Example	This spreadsheet is set up to calculate one fuel oil.
2	Fuel Oil Combustion	Variables to be entered are:
3		Fuel moisture content, fuel elements and
4		excess air.
5		
6	By: R.L.Vandagriff	Program: OilFormulas
7	Fuel Number:	1
8	Fuel:	#2 Fuel Oil
9	Fuel State	Liquid
10	Fuel Temperature:Deg.F.	80
11	Fuel Moisture Content:%/lb.	0
12	Pounds/Hour Input	1
13		
14	Fuel Elements: % mass	
15	Carbon	85.43
16	Hydrogen	11.31
17	Oxygen	2.70
18	Nitrogen	0.22
19	Sulfur	0.34
20	Chlorine	0.00
21		0.00
22		0.00
23	Ash	0.00
24	Fuel Mol Wt.	12.82103371
25	Gross Heat Value:BTU/lb	19,244.7315
26	Lb. of Water per lb of fuel	0.0000
27	Net Heat Value:BTU/lb	19,244.7315
28	Net BTU/Hr. Input	19,245
29		
30	Excess Air: %	15
31		
32	Combustion:	
33	Lb. O2 required / Lb. Fuel	3.1593
34	Lb. Air required/ Lb. Fuel	13.6178
35	Lb. Excess Air/Lb. Fuel	2.0427
36	Lb. Excess Air per Hour	2.0427
37	Lb. Air Per Hour Req'd.	15.6605
38		
39	Products: Lb. per Lb. Fuel	
40	Carbon Dioxide (CO2)	3.132433
41	Water Vapor (H2O)(Fuel)	1.017900
42	Water Vapor (H2O)(Free)	0.000000
43	Sulfur Dioxide (SO2)	0.006800
44	Hydrogen chloride (HCl)	0.000000
45	Nitrogen (N) (Air)	13.831343
46	Nitrogen (N) (Fuel)	0.002200
47	Excess O2	0.473900
48	Total:Lb.Gas/Lb.Fuel	18.46457678
49		
50	Volume of Flue Gas	
51	Flue Gas Temperature, deg. F.	600
52	CO2, cu.ft. per lb.fuel oil	55.06
53	H2O, cu.ft. per lb. fuel oil	43.74
54	N2, cu.ft. per lb. fuel oil	382.13
55	SO2, cu.ft. per lb.fuel oil	0.08
56	Excess Air, cu.ft. per lb.fuel oil	54.57
57	Cu.Ft. of Flue Gas	535.59
58	per lb. of oil burned	

TABLE 3.9

	A	B
1	Fuel Oil Combustion	Spreadsheet software used: Excel® by Microsoft®.
2	FORMULAS	Formulas shown for Column B are same for other columns. Column designation changes.
3		
4		
5		
6	By: R.L.Vandagriff	Program: OilFormulas
7	Fuel Number:	1
8	Fuel:	#2 Fuel Oil
9	Fuel State	Liquid
10	Fuel Temperature:Deg.F.	80
11	Fuel Moisture Content:%/lb.	0
12	Pounds/Hour Input	1
13		
14	Fuel Elements: % mass	
15	Carbon	85.43
16	Hydrogen	11.31
17	Oxygen	2.7
18	Nitrogen	0.22
19	Sulfur	0.34
20	Chlorine	0
21		0
22		0
23	Ash	0
24	Fuel Mol Wt.	=(B15/12)+(B16/2.02)+(B17/32)+(B18/28)+(B19/32)+(B20/35.5)
25	Gross Heat Value:BTU/lb	=((14544*(B15/100))+(62028*((B16/100)-(B17/100)/8))+(4050*(B19/100)))-(760*(B20/100))
26	Lb. of Water per lb of fuel	=B11/100
27	Net Heat Value:BTU/lb	=(B25*(1-B26))-((1150.4-(B10-32))*(B26))
28	Net BTU/Hr. Input	=B12*B27
29		
30	**Excess Air: %**	15
31		
32	Combustion:	
33	Lb. O2 required / Lb. Fuel	=(((B15/100)*(32/12))+(((B16/100)*(16/2)))+(B19/100))-(B17/100)
34	Lb. Air required/ Lb. Fuel	=(B33*(0.768/0.232))+B33
35	Lb. Excess Air/Lb. Fuel	=(B30/100)*B34
36	Lb. Excess Air per Hour	=B12*B35
37	Lb. Air Per Hour Req'd.	=B12*(B34+B35)
38		
39	Products: Lb. per Lb. Fuel	
40	Carbon Dioxide (CO2)	=(B15/100)+((B15/100)*(32/12))
41	Water Vapor (H2O)(Fuel)	=(B16/100)+((B16/100)*(16/2))
42	Water Vapor (H2O)(Free)	=(B26)-(B44/2)
43	Sulfur Dioxide (SO2)	=(B19/100)*2
44	Hydrogen chloride (HCl)	=(B20/100)*2
45	Nitrogen (N) (Air)	=((B33*(0.768/0.232))*(1+(B30/100)))*(1+(B30/100))
46	Nitrogen (N) (Fuel)	=B18/100
47	Excess O2	=(B30/100)*B33
48	Total:Lb.Gas/Lb.Fuel	=SUM(B40:B47)
49		
50	Volume of Flue Gas	
51	Flue Gas Temperature, deg. F.	600
52	CO2, cu.ft. per lb.fuel oil	=(359/44)*B40*((B51+460)/492)
53	H2O, cu.ft. per lb. fuel oil	=(359/18)*(B41+B42)*((B51+460)/492)
54	N2, cu.ft. per lb. fuel oil	=(359/28)*(B45+B46)*((B51+460)/492)
55	SO2, cu.ft. per lb.fuel oil	=(359/64)*B43*((B51+460)/492)
56	Excess Air, cu.ft. per lb.fuel oil	=B35*(359/28.95)*((B51+460)/492)
57	Cu.Ft. of Flue Gas	=SUM(B52:B56)
58	per lb. of oil burned	

As the prices of natural gas and fuel oil slowly edge upwards, the efficient operation of a boiler becomes more and more important. This leads to the following considerations in packaged boiler design.

1. All boiler outer walls are to be of the membraned waterwall design.
2. All outer tube walls are to be connected by membrane with no corner joints.

3. The complete packaged membrane waterwall boiler to be 100% gas tight.
4. The dividing wall between the furnace and convection sections is to be a gas-tight membrane waterwall.
5. A refractory is to be used only to protect the steam drum and mud drum from flame impingement and thus disruptive heat transfer affecting water circulation.

When a boiler manufacturer follows these design parameters, a highly efficient, compact boiler can be shop-assembled and shipped. Size is limited only by rail or road clearance. If shipment by water is available to the plant, then the size limit to a packaged waterwall boiler seems to be about 350,000 lb of steam per hour. Packaged waterwall boilers of over 3000 psig operating pressure have been built and are in operation. Steam temperature up to 1000°F is standard design.

A large packaged boiler, say 150,000 lb of steam per hour with steam at 650 psig and 710°F, built to the foregoing parameters, operating at capacity around the clock, using the best controls, using the best low NO_x burner, burning natural gas, will operate with an efficiency of 84.0–85.0%.

Figure 3.2 shows for two typical packaged waterwall boiler designs. Tables 3.2–3.9 give fuel oil atomization, oil and gas analysis, combustion constants, minimum autoignition temperatures, and natural gas and fuel oil examples and formulas.

4

Solid Fuels

Combustion in Solid Fuel Beds; Biomass; Water in the Fuel; Solid Fuels;
Miscellaneous Fuel.

I. COMBUSTION IN SOLID FUEL BEDS [15,28]

When fresh fuel is charged into a hot furnace the moisture and volatile matter
are first distilled off. The combustible matter in the volatile matter, along with
the carbon monoxide (CO) and hydrogen formed by reactions of hot carbon with
carbon dioxide (CO_2) and water, burns in the free space above the fuel bed that,
in the final analysis, consists chiefly of carbon. Combustion within fuel beds is
concerned largely with reactions involving oxygen of the air and hot carbon.

Hot carbon is very active in combining with oxygen. It will combine not
only with free oxygen, but also will take oxygen away from water vapor and
from CO_2, reducing them to H_2 and CO. Because fuel beds in furnaces consist
mostly of hot carbon, the gas in the fuel bed and immediately above it contains
a considerable percentage of CO and H_2. This gas, in turn, combines with free
oxygen, or even takes oxygen away from other compounds, such as iron oxides
in ash, and reduces them to lower oxides. In some instances it may even reduce
compounds such as iron pyrites to lower stages of oxidation, such as ferrous
sulfide. This behavior is an important factor in clinker formation.

Experiments indicate that combustion takes place in two zones: (1) an oxidation zone giving carbon dioxide and consuming all but a negligible amount of
oxygen, and (2) a reduction zone in which the carbon dioxide is reduced to carbon
monoxide. Later investigators concluded from experiments on small gas producers and small furnaces at high rates of combustion that carbon monoxide might be
a primary product in the reaction of oxygen with carbon under these conditions.

Other investigators have recently reported studies on the combustion of
coke that demonstrate the validity of the earlier conclusion as far as the overall

effects are concerned. These experiments show that the oxygen of air supplied to the fuel bed is virtually consumed within 0.5–9 in. of the point of entry at rates of 1–65 lb/ft^2hr^{-1}. The height of this oxidation zone is little affected by the rate of air supply. The predominant product in this zone is carbon dioxide, but some carbon monoxide is also formed.

When the oxygen is all consumed, the predominant reaction is the reduction of the carbon dioxide to carbon monoxide by the hot carbon. For coke, this reaction can be encouraged or restricted by controlling the size of the particles, their reactivity, or the depth of the bed.

The maximum temperature in the fuel bed occurs at the top of the oxidation zone and depends on the ratio of CO_2 to CO. High temperatures are favored by high air rates, large sizes of unreactive coke, and low external heat losses (use of large grate areas). A general deduction can be made, that all fuel beds consist of hot carbon and have a reducing zone.

In very simple terms, the burning of carbon occurs in this sequence. First, the carbon and oxygen react to form carbon monoxide and some carbon dioxide. Next, the CO burns with oxygen to form CO_2. In the next stage, the CO_2 is reduced by the fuel bed carbon to form CO again. The primary (undergrate) air should by fairly depleted within the fuel bed, and an excess of CO gas develops. The secondary air is then added above the bed to provide the oxygen needed to burn the CO and produce CO_2 again.

Complete combustion in the fuel bed cannot be obtained by increasing the supply of air through the fuel bed. Increased air supply proportionately increases the rate of combustion of gasification, but the composition of gas rising from the fuel bed remains the same. This is true as long as the fuel bed is free from holes. When holes are formed, large excess of air may be blown through them into the furnace.

For efficient combustion the following conditions are required: a uniform, thin fuel bed, free from blow holes; an adequate supply of secondary air; and a sufficient furnace volume to enable the gas to burn.

II. BIOMASS

A. Wood Fuel Characteristics [15,29,30]

The principal characteristics of wood fuels are high moisture, high volatile matter, and high oxygen content. About four-fifths of the fuel on a dry basis comes off as volatile matter and must be burned in the furnace space above the grates. Only one-fifth is fixed carbon, which must be burned on the grate.

The process of combustion takes place in three consecutive, somewhat overlapping, steps: the evaporation of moisture; the distillation and burning of the volatile matter; and the combustion of the fixed carbon.

The evaporation of the moisture absorbs about 1000 Btus per pound of moisture. At temperatures below 500°F, distillation of the volatile matter also absorbs heat. Beyond 540°F, exothermic reaction takes place and the distillation of the volatile matter continues with the evolution of some heat, even if no additional air is supplied.

In small pieces of wood, such as sawdust, the different phases of burning occur in rapid succession. With large chunks of wood, the processes are overlapping because wood is a very poor conductor of heat. To obtain complete combustion of the volatile gases formed, it is necessary to supply about 80% of the total air required close to the surface of the fuel bed where it can readily mix with the volatilized gases. Because these gases do not ignite below 1100°F, it is best to burn dry wood to obtain a higher maximum temperature.

As long as the fuel contains any appreciable amount of moisture, it cannot be brought to sufficiently high temperature to drive off the volatile matter and ignite it, because any heat imparted is used in evaporating the water. Therefore, hogged fuel can be burned only as fast as moisture can be evaporated from it. To speed up the evaporation, the dried fuel and the distilled gas should be burned in close proximity to the incoming wet wood. This is true of all wet fuels.

Wood is composed of 50–54% cellulose, 15–18% hemicellulose, 26–28% lignin, and minor quantities of other constituents.

The heating value of chemically isolated hemicellulose and cellulose have been measured to be about 8000 Btu/lb. The heating value of various lignins range from 10,000 to 11,000 Btu/lb. The resinous material from softwood species has a heating value of about 16,000 Btu/lb. Charcoal formed after most of the volatiles have been distilled off has a heating value of about 12,000 Btu/lb.

The most common constituents of the ash in wood are calcium, potassium, phosphorus, magnesium, and silica. Ashes recovered from burned wood are about 25% water-soluble and the extract is highly alkaline. The ash fusion temperature is in the range of 1300–1500°C. (2372–2732°F).

B. Furnaces for Wood Refuse up Through 1948 [15]

Woodworking plants, such as furniture factories, box factories, planing mills, and other similar industries are the principal sources of dry wood for steam-generation purposes. Although the refuse from these plants may contain as high as 25% moisture, the average will generally be in the neighborhood of 20%. The wood to be burned consists of large percentages of sawdust and shaving, with considerable lesser amounts of edgings, blocks, slabs, and sticks. Because dry wood burns readily, it is necessary to apply different principles in the design of furnaces for this fuel. Furthermore, the problem of providing suitable furnace cooling is of great importance, as high flame temperatures are developed when burning wood with low excess-air quantities. Under these conditions, the silica and alkaline constituents of the wood ash are combined to form a low-fusion–temperature slag, which fluxes with the silica in the refractories. As a result, there is considerable wall erosion. Even though air-cooled refractory walls have served to reduce somewhat the extent and penetration of this erosion, the use of substantial amounts of watercooled surface over the furnace walls minimizes maintenance and considerably lengthens the time between outages for necessary repair. The location and area of these furnace watercooling surfaces must be carefully studied, so that sufficiently high furnace temperatures, as required for smokeless combustion, may be maintained and, at the same time, fusion of any exposed refractories may be avoided.

Dry wood furnaces may be divided into two general types: one for burning the wood partly in suspension and partly on flat or sloping grates, and the other for burning sawdust, shavings, and other hogged wood in suspension.

1. Flat Grate Furnace

The flat grate-type furnace was formerly commonly used in many furniture factories and planing mills. Cyclone collectors supplied the dry wood, simultaneously with large quantities of excess air, through chutes to the furnace. The wood was burned as produced, and the quantity was therefore irregular. Some of it burned in suspension, while the remainder smoldered in uneven piles that spotted the grate because distribution and supply lacked uniformity. The furnace volume provided was exceedingly small, and this lack of sufficient combustion space resulted in incomplete combustion of the large amount of volatile matter in the wood, notwithstanding the presence of large quantities of excess air, and caused the production of dense smoke at practically all burning rates. Furthermore, flame impingement on the boiler surfaces, because of the short distance between them and the fuel bed, resulted in chilling of the burning gas to produce additional smoke, along with deposits of soot, in large quantities, throughout the boiler passes. Many of these older installations are now being replaced with properly designed furnaces in which modern feeding and burning equipment are used.

2. Fuel Feeding Equipment

An important requirement for these newer installations is the use of suitable equipment to feed properly sized fuel to the furnace. The regulation of dry wood supply is important, because the steam demands for plant operation may bear only a small relation to the simultaneous fuel production cycle. Therefore, it is necessary to control the wood fed to the furnace to avoid flooding with fuel at times of low steam demand, or starving at times of high demand. The use of a fuel storage bin, equipped with some form of feeding device, provides the means for control of fuel flow. These bins are usually of the flat-bottom type, with slightly tapering sides. In the bottom are several helicoid screws used to agitate the fuel, to overcome any tendency to arching, and at the same time slip it forward to a horizontal screw conveyor. Operating in synchronism with this screw conveyor is a star wheel feeder to control fuel supply and provide sealing against any sparks or backfiring into the storage bin, or against needless infiltration of air into the furnace through the feed openings.

3. Inclined Grate Furnace

An inclined grate, similar to that for wet wood, may be used in burning hogged dry wood. The slope of the fuel supporting surface, however, is decreased to approximately 30 degrees. The upper section is composed of stationary elements, with horizontal air spaces formed by a series of ledges, which also act as retarders to fuel slippage. This construction provides a nonsifting feature to prevent wood particles from falling into the windbox. Alternate longitudinal sections of the lower grate are equipped with pushers to move the fuel gradually, as it burns, down the grate. Retarders are located at the end of the grate, so that accumulated refuse can be dumped without danger of the entire fuel bed slipping into the ashpit. These inclined grates are applicable to both large and small boiler units.

4. Furnaces for Suspension Burning

Furnaces for burning sawdust, shavings, and hogged dry wood in suspension find their application primarily in those industries where the steam demands are such that large units are required. In most of these it is also necessary to provide for an auxiliary fuel that is used when the wood supply is low. Pulverized coal, oil, and gas are well adapted to these applications because design requirements and disposition of furnace volume are practically the same as for the wood.

Because of the ease with which dry wood is kindled, temperature in a refractory furnace is sufficiently high to maintain ignition at all loads, and arches are not required. The wood is supplied to the furnace through openings in the upper part of the frontwall. As it falls, the major portion is burned in suspension, while the larger particles drop to the hearth and are burned in the same manner as on a flat grate. Air for combustion is supplied through a series of tuyeres located in the lower portion of the furnace walls. The air streams from these are directed to sweep the pile of accumulated wood, and also to set up a zone of turbulence that breaks up any stratification and produces uniform mixture of the gas leaving the furnace. The earlier design used air-cooled refractory walls, the lanes of which were connected to the wood-burning tuyeres or the auxiliary fuel burners. For reasonable maintenance, the heat liberation rate in these refractory furnaces is limited to approximately 15,000 Btu/ft^3 hr^{-1}.

5. Watercooled Wall Construction

The application of watercooled wall construction resulted from the necessity for overcoming excessive outage, caused by fluxing of refractories by the wood ash, when furnace volume is otherwise insufficient to develop the required rate of steam output. At first there was a feeling that the cooling effect of bare metallic walls, capable of high rates of radiant heat absorption, would chill the furnace to a point where ignition would be impaired. Nevertheless, a number of units were installed in several plants. In these a large portion if the furnace sidewalls were watercooled, whereas the front and rearwalls were of refractory construction. Operation was successful, and availability increased to the extent that outage, owing to furnacewall failure, became practically nonexistent. In addition, it was possible to maintain combustion rates of 20,000 Btu/ft^3 hr^{-1} for long periods.

The final step came with the use of fully watercooled furnaces, in which the liberation rates were 25,000 Btu/ft^3 hr^{-1}, and even higher in some instances.

In some industries refuse wood supply is small and erratic; therefore, it does not warrant the use of special furnace designs. A satisfactory solution for disposing of this refuse is, then, to burn it on stoker fuel beds. When this is done, however, provision must be made for supplying controlled amounts of overfire air to burn the wood quickly and thus prevent it from eventually blanketing the stoker fuel bed.

C. Wood Fuel and Furnace Design in the 1920s [31]

Wood is vegetable tissue that has undergone no geological change. When newly cut, wood contains from 30 to 50% of moisture. When dried in the atmosphere for approximately 1 year, the moisture content is reduced to 18 or 20%.

Wood is ordinarily classified as hardwood, including oak, maple, hickory, birch, walnut and beech, and softwood, including pine, fir, spruce, elm, chestnut, poplar, and willow. While, theoretically, equal weights of wood substance should generate the same amount of heat, regardless of species, practically the varying form of wood tissue and the presence of rosins, gums, tannin, oils, and pigments result in different heating values, and, more particularly, in a difference in the ease with which combustion can be accomplished. Rosin may increase the heating value as much as 12%. Contrary to general opinion, the heat value per pound of softwood is slightly greater than that of hardwood.

The heat values of wood fuels are ordinarily reported on a dry basis. It is to be remembered, however, that because of the high moisture content, the ratio of the amount of heat available for steam generation to that of the dry fuel is much lower than that of practically all other solid fuels. Even woods that are air dried contain approximately 20% moisture, and this moisture must be evaporated and superheated to the temperature of the escaping gases before the heat evolved, for absorption by the boiler, can be determined.

In industrial wood refuse from lumber mills and sawmills, the moisture content may run as high as 60% and the composition of the fuel may vary over a wide range during different periods of mill operation. The fuel consists of sawdust, "hogged" wood, and slabs, and the proportions of these may vary widely. Hogged wood is mill refuse and logs that have been passed through a "hog" machine or macerator that cuts or shreds the wood with rotating knives to a state in which it may be readily handled as fuel.

1. Furnace Design

The principal features of furnace design for the satisfactory combustion of wood fuel are ample furnace volume and the presence of a large area of heated brickwork to radiate heat to the fuel bed. The latter factor is of particular importance in the case of wet wood, and ordinarily necessitates the use of an extension furnace. A furnace of this form not only gives the required amount of heated brickwork for proper combustion, but enables the fuel, in the case of hogged wood and sawdust, to be most readily fed to the furnace. With wet mill refuse, the furnace should be "bottled" at its exit to maintain as high a temperature as possible, the extent to which the bottling effect is carried being primarily dependent on the moisture content of the fuel and being greater as the moisture content is higher. The bottling effect, which is ordinarily secured by a variation in the height of the extension furnace bridge wall, has, in several recent installations, been accomplished by the use of a "drop-nose" arch at the rear of the furnace combustion arch.

Secondary air for combustion is of assistance in securing proper results and may be admitted through the bridge wall to the furnace or, where there is a secondary combustion space behind the bridge wall, into that space.

For hogged wood and sawdust, the fuel is fed through fuel chutes in the roof of the extension furnace, ordinarily being brought from the storage supply to the chutes by some type of conveyor system. With this class of wood fuel, in-swinging fire doors are placed at the furnace front for fire-inspection purposes. Where slabs are burned in addition to hogged wood and sawdust, large side-hinged slab firing doors are usually installed above the in-swinging doors.

Fuel chutes should be circular on the inside and square outside, such design enabling them to be installed most readily in the furnace roof. For ordinary mill refuse, the chute should be 12 in. in diameter, although for shingle mill refuse the size should be 18 in.

Each fuel chute should handle a square unit of grate surface, the dimensions of such units varying from 4 × 4 to 8 × 8 ft, depending on the moisture content and nature of the fuel.

Dry sawdust, chips, blocks, and veneer are frequently burned in plants of the woodworking industry. With such fuel, as with wet wood refuse, an ample furnace volume is essential, although because of the lower moisture content, the presence of heated brickwork is not as necessary as with wet wood fuel.

In a few localities cord wood is burned. With this as with other classes of wood fuel, a large combustion space is an essential feature. The percentage of moisture in cord wood may make it necessary to use an extension furnace, but ordinarily this is not required. Cord wood and slabs form an open fire through which the frictional loss of the air is much less than for sawdust or hogged material. The combustion rate with cord wood is, therefore, higher, and the grate surface may be considerable reduced. Such wood is usually cut in lengths of 4 ft or 4 ft 6 in., and the depth of the grates should be kept approximately 5 ft to obtain the best results.

D. Bagasse [31, 32]

Bagasse is the refuse of sugar cane from which the juice has been extracted, and from the beginning of the sugar industry, it has been the natural fuel for sugar plantation power plants. Physically it consists of matted cellulose fibers and fine particles, the percentage of each varying with the process. Bagasse generally contains about 50% moisture and has a heat content of 3600–4200 Btu/lb as fired. It is used chiefly as a fuel to generate steam and power for the plant. Other by-product uses are for cellulose, for paper and paperboard manufacture, and for furfural production.

In the early days of sugar manufacture, the cane was passed through a single mill and the defecation and concentration of the saccharine juice took place in a series of vessels mounted over a common flue with a fire at one end and a stack at the other. This method required an enormous amount of fuel, and it was frequently necessary to sacrifice the degree of extraction to obtain the necessary amount of bagasse and a bagasse that could be burned. In the primitive furnaces of early practice, it was necessary to dry the bagasse before it could be burned, and the amount of labor involved in spreading and collecting it was great.

With the general abolition of slavery and resulting increased labor cost of production, and with growing competition from European beet sugar, it was necessary to increase the degree of extraction, the single mill being replaced by the double mill, and the open wall or Jamaica train method of extraction as just described was replaced by vacuum-evaporating apparatus and centrifugal machines. Later a third grinding was introduced, and the maceration and dilution of the bagasse were carried to a point where the last trace of sugar in the bagasse was practically eliminated. The amount of juice to be treated was increased by these improved manufacturing methods from 20 to 30%, but the amount of bagasse available for fuel and its calorific value as fuel were decreased to an extent that the combustion capacity of the furnaces available could not meet. In the older plants the raw cane was ground by passing it in series through sets of grooved rolls, each set comprising a mill having finer groves than the preceding one. Modern practice incorporates a shredder that cuts the cane with revolving knives before the tandem milling previously described. The end product has a higher percentage of fines and short fibers. For the steam-generation end of manufacture to keep pace with the process end, it was necessary to develop a more efficient method of burning the bagasse commercially than that employed in the drying of the fuel.

During the transition period of manufacture may furnaces were "invented" for burning green bagasse, the saving in labor by this method over that necessary in spreading, drying, and collecting the fuel obviously being the primary factor in reduction of the cost of steam generation. None of these furnaces, however, gave satisfactory results until the hot-blast bagasse furnace was introduced in 1888. Although furnaces of this design operated satisfactorily, their construction was expensive and, because of the cost to the planters in changing to improved sugar manufacture apparatus, they were difficult to introduce.

1. Composition and Calorific Value of Bagasse

The proportion of fiber contained in the cane and the density of the juice are important factors in the relation the bagasse fuel will have to the total fuel necessary to generate the steam required in a mill's operation. A cane rich in wood fiber produces more bagasse than a poor one, and a thicker juice is subjected to a higher degree of dilution than one not so rich.

Besides the percentage of bagasse in the cane, its physical condition has a bearing on its caloric value. The factors that enter here are the age at which the cane must be cut, the locality in which it is grown, and so on. From the analysis of any sample of bagasse its approximate caloric value may be calculated from the formula

$$\text{Btu/lb bagasse} = \frac{8550F + 7119S + 6750G - 972W}{100}$$

Where F = percentage of fiber in cane, S = percentage of sucrose, G = percentage of glucose, and W = percentage of water.

This formula gives the total available heat per pound of bagasse, that is, the heat generated per pound less the heat required to evaporate its moisture and superheat the steam thus formed to the temperature of the stack gases.

A sample of Java bagasse having $F = 46.5$, $S = 4.5$, $G = 0.5$, $W = 47.5$ gives Btu of 3868. These figures show that the more nearly dry the bagasse is, the higher the caloric value, although this is accompanied by a decrease in sucrose. The explanation is that the presence of sucrose in an analysis is accompanied by a definite amount of water, and that the residual juice contains sufficient organic substance to evaporate the water present when a fuel is burned in a furnace.

A high percentage of silica or salts in bagasse has sometimes been ascribed as the reason for the tendency to smoulder in certain cases of soft fiber bagasse. This, however, is due to the large moisture content of the sample resulting directly from the nature of the cane. Soluble salts in the bagasse have also been given as the explanation of such smoldering action of the fire, but here too, the explanation lies solely in the high moisture content, this resulting in the development of only sufficient heat to evaporate the moisture.

2. Furnace Design and the Combustion of Bagasse

With the advance in sugar manufacture there came, as described, a decrease in the amount of bagasse available for fuel. As the general efficiency of a plant of this description is measured by the amount of auxiliary fuel required per ton of cane, the relative importance of the furnace design for the burning of this fuel is apparent.

In modern practice, under certain conditions of mill operation and with bagasse of certain physical properties, the bagasse available from the cane ground will meet the total steam requirements of the plant as a whole; such conditions prevail, as described, in Java. In the United States, Cuba, Puerto Rico, and like countries, however, auxiliary fuel is almost universally a necessity. The amount will vary, largely depending on the proportion of fiber in the cane, which varies widely with the locality and with the age at which it is cut, and to a lesser extent on the degree of purity of the manufactured sugar, the use of the maceration water, and the efficiency of the mill apparatus as a whole.

In general, it may be stated that this class of fuel may be best burned in large quantities. Because of this fact, and to obtain the efficient combustion resulting from burning a bulk of this fuel, a single large furnace is frequently installed between two boilers, serving both, although there is a limit to the size of boiler units that may be set in this manner. A disadvantage of this type of installation results from the necessity of having two boiler units out of service when it is necessary to take the furnace down for repairs, requiring a greater boiler capacity than if single furnaces are installed to assure continuity of service. On the other hand, the lower cost of one large furnace as against that of two individual smaller furnaces, and the increased efficiency of combustion with the former, may more than offset this disadvantage.

As with wet wood refuse and, as a matter of fact, for all fuels containing an excessive moisture content, the essential features of furnace design for the proper combustion of green bagasse are ample combustion space, a large mass of furnace brickwork for maintaining furnace temperature, and a length of gas travel sufficient to enable combustion to be completed before the boiler-heating surfaces are encountered. The fuel is burned either on a hearth or on grates. The objection to the latter method, particularly where blast is used, is that the air for combustion enters largely around the edges of the fuel pile where the bed is thinnest. Furthermore, when the fuel is burned on grates, the tendency of the ash and refuse to stop the air spaces does not allow a constant combustion rate for a given draft, and because there is a combustion rate that represents the best efficiency with this class of fuel, such efficiency cannot be maintained throughout the entire period between cleaning intervals. If the bagasse is burned on a hearth, the ash and refuse form on the hearth, do not affect the air supply, and allow a constant combustion rate to be maintained. When burned on a hearth, the air for combustion is admitted through a series of tuyeres extending around the furnace and upward from the hearth. In some cases a combination of grates and tuyeres has been used. When air is admitted through tuyeres, it impinges on the fuel pile as a whole and gives a uniform combustion. The tuyeres are connected to an annular space in which, where blast is used, the pressure is controlled by a blower.

As stated, bagasse is best burned in large quantities, with corresponding high combustion rates. When burned on grates with a natural draft of 0.3 in. of water in the furnace, a combustion rate of from 250 to 300 lb/ft^2 of grate surface per hour may be obtained, whereas with a blast of 0.5 in. this rate may be increased to approximately 450 lbs. When burned on a hearth with a blast of 0.75 in. a combustion rate of approximately 450 lb/ft^2 of hearth per hour may be obtained, whereas with the blast increased to 1.6 in., this rate may be increased to approximately 650 lbs. These rates apply to bagasse containing about 50% moisture. It would appear that when burned on grates the most efficient combustion rate is approximately 300 lb/ft^2 of grate per hour, and as stated this rate is obtainable with natural draft. When burned on a hearth, and with blast, the most efficient rate is about 450 lb/ft^2 of hearth per hour, which rate requires a blast of approximately 0.75 in.

The hearth on which the bagasse is burned is ordinarily elliptical. Air for combustion is admitted through a series of tuyeres above the hearth line. The supply of air is controlled by the amount and pressure of the air within the annular space to which the tuyeres are connected. Secondary air for combustion is admitted at the rear of the bridge wall, as indicated. The roof of the furnace is ordinarily spherical, with its top from 11 to 13 ft above the grate or hearth. The products of combustion pass from the primary combustion chamber under an arch to a secondary combustion chamber. A furnace of this design embodies the essential features of ample combustion space, the mass of heated brickwork necessitated by the high moisture content of the fuel, and a long travel of gases before the boiler-heating surfaces are encountered. The fuel is fed through the roof of the furnace, preferably by some mechanical method that will assure a constant fuel supply and at the same time prevent the inrush of cold air into the furnace.

This class of fuel deposits an appreciable quantity of dust and ash which, if not removed promptly, fuses into a hard, glass-like clinker. Ample provision should be made for the removal of this material from the furnace, the gas ducts, and the boiler setting and heating surfaces.

As a fuel for the production of steam, bagasse has been burned in several types of furnaces, the oldest being a Dutch oven with flat grates. Since it was difficult to distribute the bagasse evenly on the grates, the latter were subject to high maintenance costs from burning. Therefore, a new type of furnace was developed to burn the bagasse in a pile on a refractory hearth. Air was admitted to the pile around its circumference through tuyeres. The most popular of these extension furnaces was the Cooke, but it also suffered from high-cost maintenance because of excessive radiation and cleaning difficulties. To overcome these problems the Ward furnace was designed. The Ward furnace has been very successfully used under sugar-mill boilers. It is easy to operate and maintain. Bagasse is gravity fed through chutes to the individual cells, where it burns from the surface of the pile with approximately 85% of the air that is injected into the sides of the pile adjacent to the hearth. This causes local incomplete combustion, but there is sufficient heat released to partially dry the entering raw fuel. Additional drying is accomplished by radiant heat reflected from the hot refractory to the cells. Combustion is completed in the secondary furnace above the arch. Ward furnaces are now equipped with dumping hearths, which permit the ashes to be removed while the unit is in operation.

Mechanical harvesting of sugar cane increases the amount of dirt in the bagasse to as much as 5–10%. To overcome the resultant slagging tendency of the ash, watercooling is incorporated in the furnaces. In the older mills, the drives for the milling equipment were large reciprocating steam engines, which used steam at a maximum of 150 psi and with a few degrees of superheat, exhausting at 15 psi to the boilinghouse steam supply. In more modern mills the drives are either turbines or electric motors with reducing gears. Both the turbines and the turbogenerators use steam at pressures of 400–600 psi and with temperatures up to 750°F.

Raw-sugar mills produce sufficient bagasse to meet all their steam requirements and in some cases an excess. Sugar mills that also refine usually generate from 80–90% of their steam requirements with bagasse, the remainder with supplementary fuel oil. Because of the high moisture content of the gas, the weight of the gaseous combustion products is about twice that from oil and one and one-half times that from coal. This high gas weight causes excessive draft loss and requires either extremely high stacks or fans to obtain the required steam capacity from the boilers. A thermal efficiency of 65% may be obtained by the addition of an air heater and an induced-draft fan.

In recent years bagasse has been burned on stokers of the spreader type. This method of burning, however, requires bagasse with a high percentage of fines, a moisture content not over 50%, and a more experienced operating personnel. Because of such limitations, the Ward furnace is considered the most reliable, flexible, and simple method of burning bagasse.

See Tables 4.1–4.3 for information on biomass fuel combustion.

E. Burning Residential Solid Waste [28]

A paper presented in 1969 at the Breighton Conference (London, England) described in detail the principles of burning residential solid waste [28]. As discussed in that paper, temperatures of approximately 1340°F (1000 K) are required to destroy odors that exist in the garbage. It is, therefore, necessary to bring the waste up to this temperature. During heating, the stages through which the refuse passes are drying, devolatilizing, and igniting.

1. Drying

The drying stage principally involves heat transfer to the refuse to drive out the moisture. The value of moisture typically assumed in solid waste is in the neighborhood of 20%. However, garbage collected after a heavy rain will have a much

TABLE 4.1

Spreadsheet Example:	This spreadsheet is set up to calculate eight(8) different solid fuels.						Page:	
	Variables to be entered are: Fuel moisture content, Fuel Elements, Theoretical Excess Air and Estimated Flue Gas Temperature.							
	Answers calculated are: Fuel Molecular Wt. and following. Each column is a separate calpulation.							
Boiler House Notes:				BIOMASS FUEL COMBUSTION				
re: # 51. pg. 3.195								
Fuel Number:	1	2	3	4	5	6	7	8
Fuel:	Pine Bark	Pine Bark	Pine Bark	Pine Bark	Pine Bark	Pine Bark	Pine Bark	Pine Bark
Fuel State	solid	solid	solid	solid	solid	solid	solid	solid
Fuel Temperature:Deg.F.	80	80	80	80	80	80	80	80
Fuel Moisture Content:	0%	20%	25%	30%	35%	40%	45%	50%
Water: Lb./Hr. Input	0.00	0.20	0.25	0.30	0.35	0.40	0.45	0.50
Bone Dry Fuel,Lb./Hr. Input	1.00	0.80	0.75	0.70	0.65	0.60	0.55	0.50
Fuel Elements: % mass								
Carbon	52.60	52.60	52.60	52.60	52.60	52.60	52.60	52.60
Hydrogen	7.02	7.02	7.02	7.02	7.02	7.02	7.02	7.02
Oxygen	40.07	40.07	40.07	40.07	40.07	40.07	40.07	40.07
Nitrogen	0.17	0.17	0.17	0.17	0.17	0.17	0.17	0.17
Sulfur	0.08	0.08	0.08	0.08	0.08	0.08	0.08	0.08
Chlorine	0.00	0.00	0.00	0.00	0.00	0.00	0.00	0.00
	0.00	0.00	0.00	0.00	0.00	0.00	0.00	0.00
	0.00	0.00	0.00	0.00	0.00	0.00	0.00	0.00
Ash	1.20	1.20	1.20	1.20	1.20	1.20	1.20	1.20
Fuel Mol Wt.	9.1193	9.1193	9.1193	9.1193	9.1193	9.1193	9.1193	9.1193
Gross Heat Value:BTU/lb	8900.9222	8900.9222	8900.9222	8900.9222	8900.9222	8900.9222	8900.9222	8900.9222
Lb. Water per lb of fuel	0.00	0.20	0.25	0.30	0.35	0.40	0.45	0.50
Btu req'd.: water to steam@1750F.	0.00	388.80	486.00	583.20	680.40	777.60	874.80	972.00
Net Heat Value:BTU/lb	8,900.92	8,731.94	6,189.89	5,647.45	5,105.20	4,562.95	4,020.71	3,478.46
Net BTU/Hr. Input	8,901	6,732	6,190	5,647	5,105	4,563	4,021	3,478
Theoretical Excess Air:%	30.0000	30.0000	30.0000	30.0000	30.0000	30.0000	30.0000	30.0000
Combustion:								
Lb. O2 required / Lb. Fuel	1.5644	1.2515	1.1733	1.0951	1.0168	0.9386	0.8604	0.7822
Lb. DRY Air required/ Lb. Fuel	6.7430	5.3944	5.0572	4.7201	4.3829	4.0458	3.7086	3.3715
Lb.Std. Air required / lb. Fuel	6.8306	5.4645	5.1230	4.7814	4.4399	4.0984	3.7568	3.4153
Lb. Excess Air/Lb. Fuel	2.0492	1.6393	1.5369	1.4344	1.3320	1.2295	1.1271	1.0246
Lb. Air Per Hour Req'd.	8.8796	7.1036	6.6599	6.2159	5.7719	5.3279	4.8839	4.4399
Products: Lb. per Lb. Fuel								
Carbon Dioxide (CO2)	1.9287	1.5429	1.4465	1.3501	1.2536	1.1572	1.0608	0.9643
Water Vapor (H2O)(Fuel)	0.6318	0.5054	0.4739	0.4423	0.4107	0.3791	0.3475	0.3159
Water Vapor (H2O)(Free)	0.0000	0.2000	0.2500	0.3000	0.3500	0.4000	0.4500	0.5000
Water Vapor (H2O)(Air)	0.1154	0.0923	0.0866	0.0808	0.0750	0.0693	0.0635	0.0577
Sulfur Dioxide (SO2)	0.0016	0.0013	0.0012	0.0011	0.0010	0.0010	0.0009	0.0008
Hydrogen chloride (HCl)	0.0000	0.0000	0.0000	0.0000	0.0000	0.0000	0.0000	0.0000
Nitrogen (N) (Air)	5.2459	4.1967	3.9344	3.6721	3.4098	3.1475	2.8853	2.6230
Nitrogen (N) (Excess Air)	1.5536	1.2429	1.1652	1.0875	1.0098	0.9321	0.8545	0.7768
Nitrogen (N) (Fuel)	0.0017	0.0014	0.0013	0.0012	0.0011	0.0010	0.0009	0.0009
Excess O2	0.4693	0.3754	0.3520	0.3285	0.3051	0.2816	0.2581	0.2347
Total:Lb.Gas/Lb.Fuel	9.9480	8.1584	7.7110	7.2636	8.8162	6.3689	5.9214	5.4740
Volume of Flue Gas/Lb.Fuel								
Flue Gas Temperature, Deg.F	600	540	520	500	480	460	440	420
Carbon Dioxide (CO2)	33.9031	25.5873	23.5083	21.4933	19.5423	17.6552	15.8321	14.0730
Water Vapor (H2O)(Total)	32.1086	32.3404	32.1957	32.0304	31.8447	31.6385	31.4118	31.1646
Sulfur Dioxide (SO2)	0.0193	0.0146	0.0134	0.0123	0.0111	0.0101	0.0090	0.0080
Hydrogen chloride (HCl)	0.0000	0.0000	0.0000	0.0000	0.0000	0.0000	0.0000	0.0000
Nitrogen (N) (Total)	187.8721	141.7903	130.2696	119.1038	108.2923	97.8353	87.7327	77.9846
Excess O2	11.3434	8.5611	7.8655	7.1913	6.5385	5.9072	5.2972	4.7086
Cu.Ft. Flue Gas/Lb. Fuel	265.2466	208.2936	193.8527	179.8311	166.2290	153.0462	140.2828	127.9389

TABLE 4.2

	A	B
1	Spreadsheet Example:	Spreadsheet software used: Excel ® by Microsoft®
2	FORMULAS	Formulas shown for Column B are same for other columns. Column designation changes.
3		
4	Boiler House Notes:	
5	re: # 51, pg. 3.195	BIOMASS FUEL COMBUSTION
6	Fuel Number:	1
7	Fuel:	Pine Bark
8	Fuel State	solid
9	Fuel Temperature:Deg.F.	80
10	Fuel Moisture Content:	0
11	Water: Lb./Hr. Input	=B10
12	Bone Dry Fuel,Lb./Hr. Input	=(1-D10)
13		
14	Fuel Elements: % mass	
15	Carbon	52.6
16	Hydrogen	7.02
17	Oxygen	40.07
18	Nitrogen	0.17
19	Sulfur	0.08
20	Chlorine	0
21		0
22		0
23	Ash	1.2
24	Fuel Mol Wt.	=(B15/$M$15)+(B16/$M$16)+(B17/$M$17)+(B18/$M$18)+(B19/$M$19)+(B20/$M$20)
25	Gross Heat Value:BTU/lb	=((14544*(B15/100))+(62028*((B16/100)-(B17/100)/8))+(4050*(B19/100)))-(760*(B20/100))
26	Lb. Water per lb of fuel	=B10
27	Btu req'd.: water to steam@1750F.	=(972*B26)*2
28	Net Heat Value:BTU/lb	=(B25*B12)-B27
29	Net BTU/Hr. Input	=B26
30		
31	Theoretical Excess Air:%	30
32		
33	Combustion:	
34	Lb. O2 required / Lb. Fuel	=((((B15/100)*(32/12))+(((B16/100)*(16/2)))+(B19/100))-(B17/100))*B12
35	Lb. DRY Air required/ Lb. Fuel	=(B34*(0.768/0.232))+B34
36	Lb.Std. Air required / lb. Fuel	=B35+(0.013*B35)
37	Lb. Excess Air/Lb. Fuel	=(B31/100)*B36
38	Lb. Air Per Hour Req'd.	=(B36+B37)
39		
40	Products: Lb. per Lb. Fuel	
41	Carbon Dioxide (CO2)	=((B15/100)+((B15/100)*(32/12)))*B12
42	Water Vapor (H2O)(Fuel)	=((B16/100)+((B16/100)*(16/2)))*B12
43	Water Vapor (H2O)(Free)	=(B26)
44	Water Vapor (H2O)(Air)	=B38*0.013
45	Sulfur Dioxide (SO2)	=((B19/100)*2)*B12
46	Hydrogen chloride (HCl)	=((B20/100)*2)*B12
47	Nitrogen (N) (Air)	=B36*0.768
48	Nitrogen (N) (Excess Air)	=B50*(0.768/0.232)
49	Nitrogen (N) (Fuel)	=(B18/100)*B12
50	Excess O2	=(B31/100)*B34
51	Total:Lb.Gas/Lb.Fuel	=SUM(B41:B50)
52		
53	Volume of Flue Gas/Lb.Fuel	
54	Flue Gas Temperature, Deg.F	600
55	Carbon Dioxide (CO2)	=(359/44)*B41*((B54+460)/492)
56	Water Vapor (H2O)(Total)	=(359/18)*(B42+B43+B44)*((B54+460)/492)
57	Sulfur Dioxide (SO2)	=(359/64)*B45*((B54+460)/492)
58	Hydrogen chloride (HCl)	=(359/36.5)*B46*((B54+460)/492)
59	Nitrogen (N) (Total)	=(359/28)*(B47+B48+B49)*((B54+460)/492)
60	Excess O2	=(359/32)*B50*((B54+460)/492)
61	Cu.Ft. Flue Gas/Lb. Fuel	=SUM(B55:B60)

TABLE 4.3

Boiler House Notes:		Page No:
Subject: Typical Biomass fired Boiler Performance		
Ralph L. Vandagriff		Date: 7/24/98
Fuel:		Rice Hulls
Moisture Content:		6.00%
Btu per lb.		6,781
Pounds per hour of fuel		16,522
Predicted Operating Performance Data		
Steam Output	lbs. per hour	85,000
Pressure @ Boiler Outlet	psig	200
Water Temperature entering Boiler	deg. F.	220
Blowdown - pounds water per hour		0
Steam Temperature @	Outlet, deg. F.	saturated
Excess Air in gas leaving Furnace	%	30
Excess Air in Gas leaving stack	%	35
Average Temperature of Gases leaving Furnace	deg. F.	1,850
Gas Temperature leaving Boiler	deg. F.	535
Gas Temperature leaving air preheater	deg. F.	350
Ambient air temperature	deg. F.	80
Air temperature entering air preheater	deg. F.	80
Air temperature leaving air preheater	deg. F.	326
Furnace draft	inches H2O	0.15
Draft loss through Boiler	inches H2O	1.80
Draft loss through air preheater	inches H2O	1.10
Air pressure loss through air preheater	inches H2O	1.50
Air pressure loss through grate	inches H2O	3.00
Pounds Flue Gas per hour		103,490
Pounds Air required per hour for Combustion		86,960
Pounds Air through air preheater		86,960
Heat Release - Btu per cu.ft. per hour		24,000
Heat Release - Btu per sq.ft. per hour	Grate	871,000
Heat Losses - %		
Dry gas		5.97
Hydrogen & moisture in fuel		7.46
Moisture in Air		0.16
Unburned Combustible		1.5
Radiation		0.49
Unaccounted for and/or Manufacturer's Margin		1.5
Total Heat Loss		17.08
Predicted Efficiency		82.92

higher moisture content. With a moisture content of 20%, approximately one-half of the energy goes to raising the temperature of the dry refuse, and one-half goes to evaporating moisture and heating the steam. As the moisture content increases, the percentage of heat required to evaporate the moisture also increases, and at a moisture content of 50%, nearly 80% of the heat is required for evaporation and heating the moisture driven from the refuse.

2. Devolatilizing

The combustible volatiles in refuse should be released between the temperatures of 350°F (450 K) and 980°F (800 K). The devolatilizing starts at the surface and progresses inward as the temperature of the refuse increases. Because of the nonhomogeneous nature of the refuse, some items within the furnace may be completely burned, whereas others are still undergoing this process of thermal decomposition.

3. Ignition

Combustion starts when the volatiles reach ignition temperature. Combustion air must be provided for burning, and this primary air will be provided as underfeed air through the grate system. Both underfeed air and overfeed air are necessary for complete combustion.

See Table 4.4 for information on municipal solid waste combustion.

In very simple terms, the burning of the carbon occurs in this sequence. First, the carbon and oxygen react to form carbon monoxide and some carbon dioxide. Next, the CO burns with oxygen to form carbon dioxide. In the next stage, the carbon dioxide is reduced by the hot carbon to form CO again. The primary air (which contains the oxygen) should be fairly depleted within the refuse bed, and an excess of CO gas develops. The secondary air is then added above the bed to provide the oxygen needed to burn the CO and produce carbon dioxide again. Carbon dioxide is the preferred final gaseous product. Theoretically, the primary air should be sufficient to provide complete combustion of the char, and secondary air should be provided to completely burn the CO. If excess primary air is provided, it then becomes secondary air, injected upward through the refuse bed. This is an undesirable effect because it disturbs the smaller pieces in the refuse bed and imparts an upward velocity to particulates. Some of these particulates may be carried out with the hot gases. Since the secondary air is usually injected horizontally into the combustion chamber at a height of several feet (meters) above the grate, good mixing is promoted. The burning should be uniform across the grate with no tall flames reaching upward to the tube areas.

III. WATER IN THE FUEL

Water will not burn. The heat value per pound of fuel goes down as the moisture content goes up.

One gallon of water at sea level weighs 8.33 lb.

One pound of water is essentially a pint of water.

One pound of water turned into steam at atmospheric pressure (14.696 psia) and at sea level occupies 26.80 ft³ at 212°F.

If your fuel and water in the fuel are at 80°F, then it takes approximately 1100 Btu to turn that pound of water to atmospheric steam at 212°F (1150.4 Btu required at 32°F).

Example	1	2	3	4
Operating (hr/yr)	8,600			
Steam (lb/hr)	100,000			
Steam (psig)	125			
Steam (°F)	Saturated			
Steam (Btu/lb)	1,193			
Fuel:	Southern pine bark			
Moisture content	20%	30%	40%	50%
Fuel temp (°F)	80	80	80	80
Btu/lb; zero moisture	8,900	8,900	8,900	8,900
Available Btu/lb (dry fuel)	7,120	6,230	5,340	4,450
Btu: conversion of water to 0 psig steam at 1750°F	389	583	778	972
Btu/lb of wet bark fuel	6,731	5,647	4,562	3,478
Boiler efficiency	75%			
Steam Btu/hr required	159,066,667			
Pounds fuel per hour	23,632	28,168	34,868	45,735
Fuel: $/ton, delivered	10.00			
Fuel cost: $/yr	$1,016,176.00	$1,211,224.00	$1,499,324.00	$1,966,605.00
TPY: 100% dry fuel	81,294	84,786	89,959	98,331
TPY: Water	20,324	36,337	59,973	98,331

TABLE 4.4

MUNICIPAL SOLID WASTE COMBUSTION

Boiler House Notes: Ralph L. Vandagriff — m. #51, pg 3196
Program: MSWF UEL#1 — Run Date 1-May-00 — Date 8/8/98 3:11 PM — Page:

	Feta	Food Waste	Fruit Waste	Meat Waste	Cardboard	Magazines	Newsprint	Paper-Mixed	Waxed/Carton	Office Sweep	Waste Oil
Fuel Number:	1	2	3	4	5	6	7	8	9	10	11
Fuel State	solid	solid	solid	solid	solid	solid	solid	solid	solid	solid	solid
Fuel Temperature-Deg.F.	70	70	70	70	70	70	70	70	70	70	70
Fuel Moisture Content:	2.0%	70.0%	78.7%	38.8%	5.2%	4.1%	6.0%	10.2%	3.4%	3.2%	0.0%
Water: Lb./Hr. Input	0.020	0.700	0.787	0.398	0.052	0.041	0.060	0.102	0.034	0.032	0.000
Bone Dry Fuel/Lb./Hr. Input	0.980	0.300	0.213	0.612	0.948	0.959	0.940	0.898	0.966	0.968	1.000
Fuel Elements: % mass											
Carbon	73.00	48.00	48.50	59.60	43.00	32.90	49.10	43.40	59.20	24.30	66.90
Hydrogen	11.50	6.40	6.20	9.40	5.90	5.00	6.10	5.60	9.30	3.00	9.60
Oxygen	14.80	37.60	39.50	24.70	44.80	38.60	43.00	44.30	30.10	4.00	5.20
Nitrogen	0.40	2.60	1.40	1.20	0.30	0.10	0.20	0.20	0.10	0.50	2.00
Sulfur	0.10	0.40	0.20	0.20	0.20	0.10	0.20	0.20	0.10	0.20	0.00
Chlorine	0.00	0.00	0.00	0.00	0.00	0.00	0.00	0.00	0.00	0.00	0.00
Ash	0.20	5.00	4.20	4.90	5.80	23.30	1.50	6.30	1.20	68.00	16.30
Fuel Mol Wt	12.2563	8.4487	8.4016	10.4411	7.9211	6.4299	8.4650	7.8993	10.4846	3.6593	10.5514
Gross Heat Value-BTU/lb	16608.8720	8051.7960	7845.0435	12591.8415	6448.1040	4897.5750	7568.9070	6483.0195	12048.8685	5092.9920	15281.4420
Lb. Water per lb of fuel	0.02	0.79	0.79	0.39	0.05	0.04	0.06	0.10	0.03	0.03	0.00
Btu req'd. water to steam @1750F	38.88	1360.80	1538.00	754.27	101.09	79.70	116.64	198.29	66.10	62.21	0.00
Net Heat Value BTU/lb	16,235.85	1,054.74	134.99	6,951.93	6,011.71	4,617.07	7,026.33	5,623.46	11,573.14	4,987.81	15,281.44
Net BTU/Hr. Input	16,236	1,055	135	6,952	6,012	4,617	7,026	5,623	11,573	4,988	15,281
Theoretical Excess Air:%	25.0000	25.0000	25.0000	25.0000	25.0000	25.0000	25.0000	25.0000	25.0000	25.0000	25.0000
Combustion:											
Lb. O2 required / Lb. Fuel	2.6653	0.4260	0.2974	1.2630	1.1117	0.8557	1.2872	1.0599	1.9539	0.8226	2.5000
Lb. DRY Air required/ Lb. Fuel	11.4862	1.8362	1.2820	5.5300	4.7918	3.6866	5.5482	4.5687	8.4220	3.5456	10.7759
Lb. Std. Air required / lb. Fuel	11.6378	1.8601	1.2986	5.6019	4.8541	3.7385	5.6203	4.6281	8.5315	3.5927	10.9159
Lb. Excess Air/Lb. Fuel	2.9094	0.4650	0.3247	1.4005	1.2135	0.9341	1.4051	1.1570	2.1329	0.8992	2.7290
Lb. Air Per Hour Req'd.	14.5470	2.3251	1.6233	7.0023	6.0676	4.6706	7.0254	5.7851	10.6643	4.4908	13.6449
Products: Lb. per Lb. Fuel											
Carbon Dioxide (CO2)	2.6231	0.5280	0.3766	1.3374	1.4947	1.1569	1.6923	1.4390	2.0969	0.8625	2.4530
Water Vapor (H2O)(Fuel)	1.0143	0.1728	0.1169	0.5176	0.5034	0.4316	0.5161	0.4686	0.8065	0.2614	0.8640
Water Vapor (H2O)(Free)	0.0200	0.7000	0.7870	0.3680	0.0520	0.0410	0.0600	0.1020	0.0340	0.0320	0.0000
Water Vapor (H2O)(Air)	0.1891	0.0302	0.0211	0.0910	0.0789	0.0607	0.0913	0.0752	0.1386	0.0584	0.1774
Sulfur Dioxide (SO2)	0.0020	0.0024	0.0009	0.0024	0.0038	0.0019	0.0038	0.0036	0.0019	0.0039	0.0000
Hydrogen chloride (HCl)	0.0000	0.0000	0.0000	0.0000	0.0000	0.0000	0.0000	0.0000	0.0000	0.0000	0.0000
Nitrogen (N)(Air)	8.8377	1.4285	0.9974	4.3022	3.7279	2.8696	4.3164	3.5544	6.5522	2.7592	8.3834
Nitrogen (N)(Excess Air)	2.2057	0.3526	0.2461	1.0618	0.9200	0.7082	1.0652	0.8772	1.6170	0.6809	2.0680
Nitrogen (N)(Fuel)	0.0039	0.0078	0.0030	0.0073	0.0029	0.0010	0.0009	0.0027	0.0010	0.0048	0.0200
Excess O2	0.6653	0.1065	0.0744	0.3207	0.2779	0.2139	0.3216	0.2650	0.4885	0.2057	0.6250
Total-Lb.Gas/Lb.Fuel	15.6922	3.3288	2.6274	8.0287	7.0614	5.4848	8.0678	6.7778	11.7396	4.9987	14.9918
Volume of Flue Gas/Lb.Fuel											
Flue Gas Temperature, Deg F	600	400	380	600	600	600	600	600	600	600	640
Carbon Dioxide (CO2)	46.1108	7.5302	5.2765	21.2920	26.2743	20.3361	29.7484	25.1200	36.2507	15.1613	44.7473
Water Vapor (H2O)(Total)	52.5996	31.4615	31.5643	38.7906	27.2543	22.9144	28.6776	27.7568	43.7519	15.1142	46.4365
Sulfur Dioxide (SO2)	0.0237	0.0235	0.0082	0.0266	0.0458	0.0232	0.0454	0.0434	0.0242	0.0458	0.0000
Hydrogen chloride (HCl)	0.0000	0.0000	0.0000	0.0000	0.0000	0.0000	0.0000	0.0000	0.0000	0.0000	0.0000
Nitrogen (N) (Total)	307.9274	40.0916	27.2858	134.3770	128.4702	98.8589	148.6848	122.4895	204.2035	85.1609	300.2001
Excess O2	16.1052	2.0685	1.4242	7.0211	6.7175	5.1710	7.7779	6.6048	12.2522	4.9719	12.5706
Cu.Ft. Flue Gas/Lb. Fuel	422.7367	81.2134	65.5569	201.5077	188.7621	147.3098	214.9341	181.8145	328.4926	130.4551	407.0906

Boiler House Notes: nr # 51 pg 3.195
Ralph L. Vandagriff
Program: MSWFUEL #2   Date 8/98
Run Date: 1-May-00   3:16 PM   Page:

MUNICIPAL SOLID WASTE COMBUSTION

	12	13	14	15	16	17	18	19	20	21	22
Fuel Number:	12	13	14	15	16	17	18	19	20	21	22
Fuel:	Plastic-Mixed	Polyethylene	Polystyrene	Polyurethane	Polyvinylchloride	Garden Trim	Green Wood	Wood-Mixed	Leather-Mixed	Rubber-Mixed	Textiles-Mixed
Fuel State:	solid	solid	solid	solid	solid	solid	solid	solid	solid	solid	solid
Fuel Temperature Deg.F:	70	70	70	70	70	70	70	70	70	70	70
Fuel Moisture Content:	0.2%	0.2%	0.2%	0.2%	0.2%	60.0%	50.0%	20.0%	10.0%	1.2%	10.0%
Water: Lb./Hr. Input:	0.002	0.002	0.002	0.002	0.002	0.600	0.500	0.200	0.100	0.012	0.100
Bone Dry Fuel, Lb./Hr. Input:	0.998	0.998	0.998	0.998	0.998	0.400	0.500	0.800	0.900	0.988	0.900
Fuel Elements: % mass											
Carbon	60.00	85.20	87.10	63.30	45.20	48.00	50.10	49.50	60.00	69.70	48.00
Hydrogen	7.20	14.20	8.40	6.30	5.60	6.00	6.40	6.00	8.00	8.70	6.40
Oxygen	22.80	0.00	4.00	17.60	1.60	38.00	42.30	42.70	11.60	9.00	40.00
Nitrogen	0.00	0.10	0.20	0.10	0.10	0.30	0.10	0.20	0.40	1.60	2.20
Sulfur	0.00	0.00	0.00	0.10	0.10	0.30	0.10	0.10	0.40	1.60	0.20
Chlorine	0.00	0.00	0.00	2.40	45.40	0.00	0.00	0.00	0.00	0.00	0.00
Ash	10.00	0.40	0.30	4.30	2.00	6.30	1.00	1.50	10.00	20.00	3.20
Fuel Mol Wt.	9.2769	14.1384	11.5489	9.2288	7.8745	9.4008	8.6719	8.4399	9.6625	10.4485	8.5031
Gross Heat Value BTU/lb	11424.6180	21203.5140	17568.0350	11735.3100	9582.4100	7132.7000	7980.6555	7814.2655	12805.4340	14900.5680	7857.6120
Lb. Water per lb of fuel	0.00	0.00	0.00	0.00	0.00	0.80	0.50	0.20	0.10	0.01	0.10
Btu req'd. water to steam @1750F	3.89	3.89	3.89	3.89	3.89	1166.40	972.00	388.80	194.40	23.33	194.40
Net Heat Value BTU/lb	11,397.88	21,157.22	17,529.01	11,707.95	9,559.36	1,686.68	3,018.33	5,702.61	11,330.49	14,696.45	6,877.45
Net BTU/Hr. Input	11,398	21,157	17,529	11,708	9,559	1,687	3,018	5,703	11,330	14,696	6,877
Theoretical Excess Air %	25.0000	25.0000	25.0000	25.0000	25.0000	25.0000	25.0000	25.0000	25.0000	25.0000	25.0000
Combustion:											
Lb. O2 required / Lb. Fuel	1.9441	3.4022	2.9488	2.0130	1.8351	0.5319	0.7130	1.0962	1.9152	2.4509	1.2546
Lb. DRY Air required / Lb. Fuel	8.3798	14.6646	12.7102	8.6766	7.0477	2.2925	3.0733	4.7379	8.2552	10.5642	5.4078
Lb. Std. Air required / Lb. Fuel	8.4867	14.8552	12.8754	8.7894	7.1393	2.3223	3.1132	4.7995	8.3625	10.7016	5.4781
Lb. Excess Air/Lb. Fuel	2.1222	3.7138	3.2188	2.1973	1.7848	0.5806	0.7783	1.1999	2.0906	2.6734	1.3695
Lb. Air Per Hour Req'd.	10.6109	18.5690	16.0942	10.9867	8.9241	2.9029	3.8915	5.9994	10.4531	13.3750	6.8476
Products: Lb. per Lb. Fuel											
Carbon Dioxide (CO2)	2.1956	3.1178	3.1873	2.3164	1.6540	0.6747	0.9185	1.4520	1.9600	2.5250	1.5840
Water Vapor (H2O)(Fuel)	0.6487	1.2754	0.7545	0.5669	0.5030	0.2160	0.2880	0.4320	0.6480	0.7736	0.5184
Water Vapor (H2O)(Free)	0.0000	0.0020	0.0020	0.0020	0.0020	0.6000	0.5000	0.2000	0.1000	0.0120	0.1000
Water Vapor (H2O)(Air)	0.1379	0.2414	0.2092	0.1428	0.1160	0.0377	0.0506	0.0780	0.1359	0.1739	0.0890
Sulfur Dioxide (SO2)	0.0000	0.0000	0.0000	0.0020	0.0020	0.0024	0.0010	0.0016	0.0072	0.0316	0.0039
Hydrogen chloride (HCl)	0.0000	0.0000	0.0000	0.0479	0.9062	0.3852	0.0000	0.0000	0.0000	0.0000	0.0000
Nitrogen (N)(Air)	6.5193	11.4098	9.8883	6.7502	5.4830	1.7636	2.3910	3.6860	6.4224	8.2168	4.2071
Nitrogen (N)(Excess Air)	1.5089	2.6159	2.4404	1.6659	1.3532	0.4402	0.5901	0.9097	1.5850	2.0283	1.0383
Nitrogen (N)(Fuel)	0.0000	0.0010	0.0020	0.0010	0.0010	0.0136	0.0005	0.0016	0.0900	0.0000	0.0198
Excess O2	0.4860	0.8505	0.7372	0.5032	0.4088	0.1330	0.1783	0.2748	0.4788	0.6127	0.3137
Total Lb. Gas/Lb. Fuel	11.5965	19.7143	17.2206	12.0962	10.4291	4.2643	4.9179	7.0957	11.4473	14.3760	7.8739
Volume of Flue Gas/Lb. Fuel											
Flue Gas Temperature, Deg Fees(Flue)	600	600	640	600	560	480	500	520	600	620	540
Carbon Dioxide (CO2)	38.5954	57.9077	59.1419	40.7182	27.9780	10.5170	14.6227	23.5977	34.9056	45.2232	28.2983
Water Vapor (H2O)(Total)	33.6019	68.9594	43.0623	30.5393	25.6775	32.5319	32.6346	28.2057	37.9605	42.0076	28.6770
Sulfur Dioxide (SO2)	0.0000	0.0255	0.0000	0.0241	0.0232	0.0257	0.0109	0.0179	0.0870	0.3693	0.0410
Hydrogen chloride (HCl)	0.0000	0.0000	0.0000	1.0151	18.4779	6.8251	0.0000	0.0000	0.0000	0.0000	0.0000
Nitrogen (N) (Total)	224.5294	415.1967	353.4679	234.1366	181.7374	54.8056	74.5902	117.4091	223.8773	288.4012	137.2112
Excess O2	11.7475	21.7218	18.4906	12.1636	9.5072	2.8500	3.9019	6.1408	11.5726	15.0963	7.1520
Cu.Ft. Flue Gas/Lb. Fuel	308.6743	563.8101	473.1837	318.5899	263.4612	107.5556	125.7603	175.3711	308.1282	391.1105	199.3464

TABLE 4.5

Btu in Wet Biomass Fuel								
Boiler House Notes:							**Page:**	
Btu required to turn 60 deg. F. water to atmospheric steam:								1122.4
Gross Btu per pound of southern pine bark: (zero moisture)								8900
One (1) pound of 100% dry pine bark burned @ 1750 F. produces cu.ft. gas:(zero Excess Air)								641
One(1) pound of zero psig steam @ 1750 deg.F. occupies approximate: cu.ft.:								90
Ralph L. Vandagriff					Actual Firing Conditions - Zero Excess Air			
Date: July 22,1988					One (1) lb. wet fuel			
One Half (1/2) lb. of Water			%	%	Total Btu/lb	Stack Gas	Stack Gas	Total
Btu required to turn free water into steam	Btu required to raise steam to 1750 F.	Total Btu Required	Zero Moisture Fuel/lb	Free Water per lb.	left for Heat Transfer	from dry fuel Cu.Ft.	from free H2O Cu.Ft.	Cu.Ft. @ 1750 F.
			100	0	8,900	641	0	641
11.22	8.22	19.44	99%	1%	8,792	634.59	0.90	635
22.45	16.43	38.88	98%	2%	8,683	628.18	1.80	630
33.67	24.65	58.32	97%	3%	8,575	621.77	2.70	624
44.90	32.86	77.76	96%	4%	8,466	615.36	3.60	619
56.12	41.08	97.20	95%	5%	8,358	608.95	4.50	613
67.34	49.30	116.64	94%	6%	8,249	602.54	5.40	608
78.57	57.51	136.08	93%	7%	8,141	596.13	6.30	602
89.79	65.73	155.52	92%	8%	8,032	589.72	7.20	597
101.02	73.94	174.96	91%	9%	7,924	583.31	8.10	591
112.24	82.16	194.40	90%	10%	7,816	576.90	9.00	586
123.46	90.38	213.84	89%	11%	7,707	570.49	9.90	580
134.69	98.59	233.28	88%	12%	7,599	564.08	10.80	575
145.91	106.81	252.72	87%	13%	7,490	557.67	11.70	569
157.14	115.02	272.16	86%	14%	7,382	551.26	12.60	564
168.36	123.24	291.60	85%	15%	7,273	544.85	13.50	558
179.58	131.46	311.04	84%	16%	7,165	538.44	14.40	553
190.81	139.67	330.48	83%	17%	7,057	532.03	15.30	547
202.03	147.89	349.92	82%	18%	6,948	525.62	16.20	542
213.26	156.10	369.36	81%	19%	6,840	519.21	17.10	536
224.48	164.32	388.80	80%	20%	6,731	512.80	18.00	531
235.70	172.54	408.24	79%	21%	6,623	506.39	18.90	525
246.93	180.75	427.68	78%	22%	6,514	499.98	19.80	520
258.15	188.97	447.12	77%	23%	6,406	493.57	20.70	514
269.38	197.18	466.56	76%	24%	6,297	487.16	21.60	509
280.60	205.40	486.00	75%	25%	6,189	480.75	22.50	503
291.82	213.62	505.44	74%	26%	6,081	474.34	23.40	498
303.05	221.83	524.88	73%	27%	5,972	467.93	24.30	492
314.27	230.05	544.32	72%	28%	5,864	461.52	25.20	487
325.50	238.26	563.76	71%	29%	5,755	455.11	26.10	481
336.72	246.48	583.20	70%	30%	5,647	448.70	27.00	476
347.94	254.70	602.64	69%	31%	5,538	442.29	27.90	470
359.17	262.91	622.08	68%	32%	5,430	435.88	28.80	465
370.39	271.13	641.52	67%	33%	5,321	429.47	29.70	459
381.62	279.34	660.96	66%	34%	5,213	423.06	30.60	454
392.84	287.56	680.40	65%	35%	5,105	416.65	31.50	448
404.06	295.78	699.84	64%	36%	4,996	410.24	32.40	443
415.29	303.99	719.28	63%	37%	4,888	403.83	33.30	437
426.51	312.21	738.72	62%	38%	4,779	397.42	34.20	432
437.74	320.42	758.16	61%	39%	4,671	391.01	35.10	426
448.96	328.64	777.60	60%	40%	4,562	384.60	36.00	421
460.18	336.86	797.04	59%	41%	4,454	378.19	36.90	415
471.41	345.07	816.48	58%	42%	4,346	371.78	37.80	410
482.63	353.29	835.92	57%	43%	4,237	365.37	38.70	404
493.86	361.50	855.36	56%	44%	4,129	358.96	39.60	399
505.08	369.72	874.80	55%	45%	4,020	352.55	40.50	393
516.30	377.94	894.24	54%	46%	3,912	346.14	41.40	388
527.53	386.15	913.68	53%	47%	3,803	339.73	42.30	382
538.75	394.37	933.12	52%	48%	3,695	333.32	43.20	377
549.98	402.58	952.56	51%	49%	3,586	326.91	44.10	371
561.20	410.80	972.00	50%	50%	3,478	320.50	45.00	366

TABLE 4.6

	Boiler House Notes:				Page No:
Date: 11/23/93					
R.L. Vandagriff					
		TABLE OF MOISTURE CONTENT			
	Number of pounds of water for each 100 pounds of dry solids (Wet basis)				
	Moisture Percent	Water: Pounds		Moisture Percent	Water:Pounds
	1	1.0101		51	104.0816
	2	2.0408		52	108.3333
	3	3.0928		53	112.7660
	4	4.1667		54	117.3913
	5	5.2632		55	122.2222
	6	6.3830		56	127.2727
	7	7.5269		57	132.5581
	8	8.6957		58	138.0952
	9	9.8901		59	143.9024
	10	11.1111		60	150.0000
	11	12.3596		61	156.4103
	12	13.6364		62	163.1579
	13	14.9425		63	170.2703
	14	16.2791		64	177.7778
	15	17.6471		65	185.7143
	16	19.0476		66	194.1176
	17	20.4819		67	203.0303
	18	21.9512		68	212.5000
	19	23.4568		69	222.5806
	20	25.0000		70	233.3333
	21	26.5823		71	244.8276
	22	28.2051		72	257.1429
	23	29.8701		73	270.3704
	24	31.5789		74	284.6154
	25	33.3333		75	300.0000
	26	35.1351		76	316.6667
	27	36.9863		77	334.7826
	28	38.8889		78	354.5455
	29	40.8451		79	376.1905
	30	42.8571		80	400.0000
	31	44.9275		81	426.3158
	32	47.0588		82	455.5556
	33	49.2537		83	488.2353
	34	51.5152		84	525.0000
	35	53.8462		85	566.6667
	36	56.2500		86	614.2857
	37	58.7302		87	669.2308
	38	61.2903		88	733.3333
	39	63.9344		89	809.0909
	40	66.6667		90	900.0000
	41	69.4915		91	1,011.1111
	42	72.4138		92	1,150.0000
	43	75.4386		93	1,328.5714
	44	78.5714		94	1,566.6667
	45	81.8182		95	1,900.0000
	46	85.1852		96	2,400.0000
	47	88.6792		97	3,233.3333
	48	92.3077		98	4,900.0000
	49	96.0784		99	9,900.0000
	50	100.0000		100	

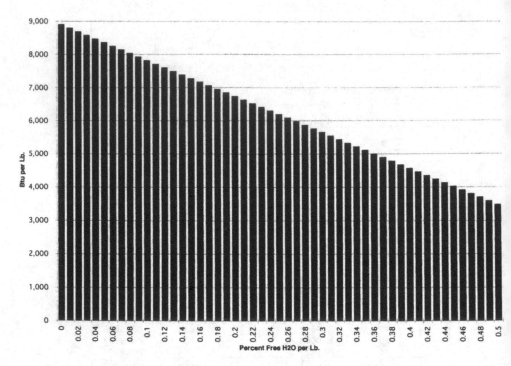

FIGURE 4.1 Southern pine back Btu/lb versus moisture content.

Boiler furnace size required for 3 sec retention time and complete burnout
of fuel.

One (1) pound of 100% dry pine bark when burned at 1750°F, produces
641 ft^3 of gas (zero excess air).

One (1) pound of atmospheric pressure steam at 1750°F occupies approximately 90 ft^3.

Moisture content	20%	30%	40%	50%
Gas from fuel (ft^3)	512.8	448.7	384.6	320.5
Steam (ft^3)	18.0	27.0	36.0	45.0
Gas/lb wet fuel (ft^3)	530.8	475.7	420.6	365.5
Wet fuel (lb/min)	393.87	469.47	581.13	762.25
Combustion gas (ft^3/min)	209,066	223,327	244,423	278,602
Furnace volume required (ft^3)	10,453	11,166	12,221	13,930

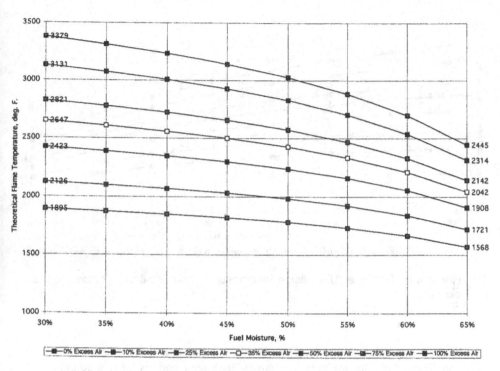

FIGURE 4.2 Southern pine bark flame temperature versus moisture versus excess air.

The larger the combustion gas flow, the larger the ID fan/motor, ducts, etc. required.

See Tables 4.5–4.6 and Figures 4.1–4.5 on the subject of fuel moisture.

IV. SOLID FUELS

A. Amorphous Forms of Carbon [33]

1. Coal

Coal is a form of fossilized wood—wood that has lain buried in the earth for many centuries. There was a time in the history of the earth when climatic conditions promoted a luxuriant growth of trees, ferns, and other plants of all kinds. This period is known as the Carboniferous Age. During that period, in certain regions not far above the level of the sea, vegetable matter accumulated in enormous quantities. Vegetable matter is composed largely of compounds of carbon, hydrogen, and oxygen. This material was later covered by mud, sand, and water. When buried to a considerable depth, it was under great pressure and, at the same

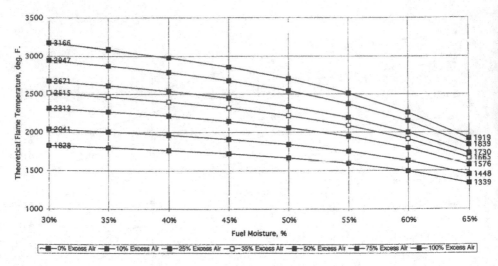

Figure 4.3 Bagasse (U.S.) flame temperature versus moisture content versus excess air.

time, it was subjected to the heat from the interior of the earth. Under these conditions, volatile products, containing hydrogen and oxygen, gradually escaped, and the remainder, composed chiefly of carbon, formed the coal seams of the earth. Coal retains the cellular structures of the plants from which it was derived. These structures can be seen by examining a thin layer of coal with the aid of a microscope. In soft coal the fossil remains of leaves and stems of plants can be seen without a microscope. Trees have been found that are only partially carbonized, one end being coal and the other still wood.

a. Varieties of Coal. There are several kinds of coal, and these differ from each other in the amount of disintegration that has taken place.

Anthracite is a hard dense, shiny coal. Most of the hydrogen and oxygen are driven out of the wood in the process of forming anthracite coal. Since anthracite coal has almost no volatiles, it requires a radiant surface, such as refractory, to sustain combustion. Anthracite coal burns slowly, with practically no flame and without the formation of soot. It is, therefore the most desirable kind of coal to burn in a furnace or stove.

Bituminous, or soft coal, is a coal in which the decomposition has not proceeded as far as in anthracite coal. Some of the carbon is still combined with hydrogen in the form of compounds called "hydrocarbons." These have a high heat content, but burn with a smoky flame, producing soot. Bituminous coal also contains compounds of nitrogen and sulfur.

Lignite contains more hydrogen compounds of carbon than does bitumi-

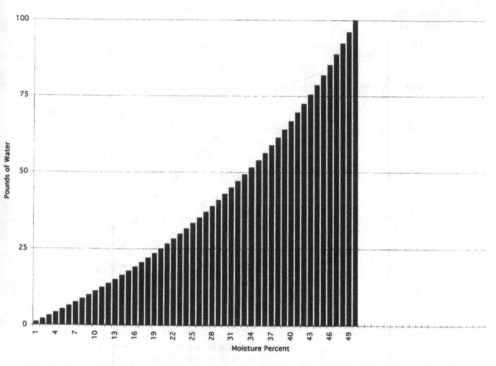

FIGURE 4.4 Moisture content (wet basis, pounds water per 100 lb dry solid).

nous coal. It shows much of the structure of the wood from which it was formed.

Peat is a brown mass of moss and leaves that has undergone, to a slight extent, the same change by which coal is formed. It is usually found in bogs, saturated with water. It must be dried before it can be used as fuel.

2. Wood Charcoal

When wood is heated in the absence of air, gaseous and liquid products distill out, and charcoal remains in the retort. Among the volatile liquids driven out of the wood are methyl alcohol, acetone, and acetic acid. These are three valuable liquids. They are used in enormous quantities in the chemical-manufacturing industries, and before 1925 the distillation of wood for the production of these liquids was a profitable business. All these compounds are now made by chemical processes that are much cheaper than the process of distilling wood. Wood alcohol (methanol), acetic acid, and acetone are not present as such in wood, but are formed from the components of the wood during the heat treatment. The wood is heated in large iron chambers, called retorts. The process of decomposing a substance by heating it in the absence of air is called "destructive distillation." Air is kept out of the apparatus so that the wood will not take fire.

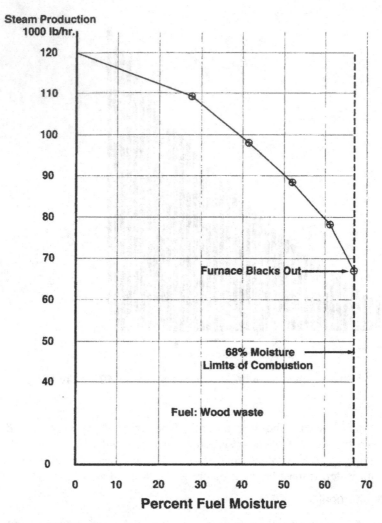

FIGURE 4.5 Effect of fuel moisture on steam production, Moisture combustion test, #10 Boiler, Wood Products Powerhouse, Longview, Washington, 1971.

Charcoal has the property of holding on its surface large quantities of gases. The layer of gas that clings to a solid surface is said to be ''adsorbed.'' Charcoal is very porous, and the surface area presented by a single cubic inch of the substance may amount to hundreds of square feet. Because of its power to adsorb gases, charcoal has been used in gas masks. Charcoal clings also to small particles of solid substances. Water that has been colored by indigo may be decolorized by passing it through a charcoal filter. The dye particles are retained on the surface of the charcoal. Bacteria may be removed from water in the same way, but there is a limit to the capacity of charcoal to hold these things, and the filters become ineffective unless the old carbon is replaced frequently by fresh material.

3. Coke

Bituminous coal is converted into coke by a process quite similar to the conversion of wood into charcoal. When coal is heated in a retort, several different gaseous and liquid products are expelled. Some of the gases that escape are combustible and they are used for fuel. Ammonia is one of the compounds formed in the destructive distillation of coal. Ammonia gas is very soluble in water and can be separated from the fuel gases (carbon monoxide, methane, and hydrogen) by passing the mixed gases through water. Ammonia is more completely removed from the fuel gases by passing the mixture through sulfuric acid. The ammonia combines with the acid, forming ammonium sulfate. The fuel gases escape unchanged, for they do not react with sulfuric acid.

From the liquid and tarry distillation products of coal we obtain benzene, phenol, and many other valuable compounds, also coal tar. The solid residue—coke—is about 90% carbon and 10% mineral matter. The latter appears in the ash when coal or coke is burned. Coke is a porous, gray substance having a high heat value. It is used in enormous quantities in smelting ores. It not only serves to heat the ore, but also acts as a reducing agent, liberating metals from the ores, which usually are oxides. At a high temperature iron oxide is reduced by carbon, as indicated by the equation $2Fe_2O_3 + 3C \rightarrow 3CO_2 + 4Fe$. Coke is superior to coal for this purpose, since it contains much less sulfur to contaminate the metal.

B. Air Distribution: Bituminous Coal [1,15]

In open or archless furnaces burning bituminous coal, with no arches to radiate heat, or to act as baffles for establishing ignition and stabilizing the burning of fuel, wide and rapid changes in rates of burning may not be possible because of unstable ignition, and gas stratification cannot be avoided. However, by proper use of high-pressure overfire air in the furnace, these difficulties can be materially reduced. The combustion characteristics of the particular coal, and its physical size when introduced into the furnace, must be carefully studied to develop the optimum arrangement of overfire air jets. The optimum arrangement speeds up the burning of the gas leaving the fuel bed after ignition is well established. Air introduced into gas traveling at high velocity is not as effective as air introduced into a gas stream traveling at low velocity. If overfire air is introduced too near the fuel bed, burning of the gas may be retarded somewhat. Turbulent mixing of air and gas is desirable, and the pressure and volume of the air used should be sufficient to produce this condition.

In spreader stoker fired units, there is typically 25% excess air at the furnace exit at the designed full-load input. This air is split between the undergrate air, overfire air, and the stoker distribution air. Because of the high degree of suspension burning, air is injected over the fuel bed for mixing to assist the fuel burnout and to minimize smoking. This dictates that 15–20% of the total air be used as overfire air. This air is injected at pressures of 15–30 in. wg through a series of small nozzles arranged along the frontwalls and rearwalls.

Spreader stoker firing, with the air split between undergrate and overfire, is a form of staged combustion and is effective in controlling NO_x. The total airflow is split 65% undergrate and 35% overfire. The 35% includes any air to the coal feeders.

1. Air Preheating

Bituminous coals burn readily on a traveling grate without preheat. However, an air heater may be required for improved efficiency. In these instances the design air temperature should be limited to less than 350°F. The use of preheated air may limit the selection of fuels to the lower-iron, high-fusion coals to prevent undesirable grate-fired bed slagging and agglomerating. The use of preheated air at 350–400°F is necessary for the higher moisture subbituminous coals and lignites. Grate bar design and metallurgy must be taken into account when selecting the air preheat temperature.

C. Slagging, Sooting, and Erosion [1,7,15,26]

The solid portions of the products of combustion (refuse) are a source of operational and maintenance problems. They may stick to the heat-transfer surfaces; they may be deposited in areas of low gas velocity, clogging the gas passages; they may cause corrosion and erosion; or they may help keep the heating surfaces clean by a scrubbing action.

The refuse—which varies according to the type, composition, and temperature of the fuel—may be classified in the following manner.

1. Flue dust: the particles of gas-borne solid matter carried into the products of combustion, including (1) fly ash, the fine particles of ash; (2) cinder, particles of partially burned fuel that are carried from the furnace and from which the volatile gases have been driven off; and (3) sticky ash, ash that is at a temperature between the initial deformation and softening temperatures.

2. Slag: molten or fused refuse, including (1) vitreous slag, a glassy slag; (2) semifused slag, hard slag masses consisting of particles that have partly fused together; (3) plastic slag, slag in a viscous state; and (4) liquid slag, slag in a liquid state.

3. Soot and smoke: unburned combustibles formed from hydrocarbon vapors that have been deprived of oxygen or adequate temperature for ignition.

1. Slagging

Slagging is the formation of molten, partially fused or resolidified deposits on furnace walls and other surfaces exposed to radiant heat. In the furnace the molten fly ash sticks, or plasters itself, as softened slag to the walls. This accumulation reduces the heat transmission of the walls and increases the surface temperature. The slag becomes molten and runs down the walls or drips from the roof. As it runs down the walls (an action called washing), a chemical reaction occurs that causes erosion or slag penetration. This slagging is one of the major causes of high refractory maintenance. Metal walls give the least difficulty from adherence of fly ash, although slag flowing over them will, in time, cause destruction through erosion. If the furnace temperature is not high enough, solidified fly ash may deposit on the walls to a thickness such that the surface temperature equals the ash fusion temperature. Variations in furnace temperature will cause the fly ash to melt or build up until equilibrium is reached. Burning particles of fuel will become embedded in the sticky mass, further increasing the temperature. Around cool openings in the hot zones, the slag may harden and build up into large masses, such as burner "eye brows." Burning particles of fuel may be carried in suspension into the boiler passes. The slagging action may move along with the gas stream as far back into the convection sections as the gas temperature remains above ash-softening temperature. This fusibility (property of the ash to melt, fuse, and coalesce into a homogeneous slag mass) depends on the temperature and ash-softening characteristics of the fuel.

2. Fouling

Fouling is defined as the formation of high-temperature–bonded deposits on convection heat-absorbing surfaces, such as superheaters and reheaters, that are not exposed to radiant heat. In general, fouling is caused by the vaporization of volatile inorganic elements in the fuel during combustion. As heat is absorbed and temperatures are lowered in the convective section of the boiler, compounds formed by these elements condense on ash particles and heating surface, forming a glue that initiates deposition.

3. Clogging

Deposits from burning coal or oil choke the gas passages, reduce the heat transmission rate, and effectively limit the steaming rate. The accumulation may take many forms, including the following:

1. Sponge ash: agglomeration of dry ash particles into structures having a spongy appearance

2. Bridging: agglomeration of refuse and slag that partially or completely blocks the spaces or apertures between heat-absorbing tubes
3. Fouling: agglomeration of refuse in gas passages or on heat-absorbing surfaces that results in undesirable restrictions to the flow of gas or heat
4. Bird-nesting: agglomeration of porous masses of loosely adhering refuse and slag particles in the first tube bank of a watertube boiler
5. Segregation: the tendency of refuse of varying compositions to deposit selectively in different parts of the unit

Beyond the hot zones the ash begins to cool and has a less agglomerate nature. In the rear passes the ash has the flaky, soft characteristics of soot and is easily blown from the tubes.

4. Erosion

Ash erosion usually occurs wherever ash concentrates in streams, such as at the baffle turns of the boiler banks of watertubes and the entrance to firetubes. To prevent this erosion, the gas velocity must be kept low or the gas baffling eliminated as far as possible. Because of the high concentration of fly ash, dry-bottom, pulverized coal furnaces are particularly susceptible to erosion.

5. Corrosion

Deposits tend to set up in the cold-end equipment (air heater, economizer, dust collector) where gas temperatures drop close to, or below, the dew point. Soot deposits have an affinity for absorbing moisture. Coal soot has traces of SO_2 and SO_3; oil soot has sodium and potassium sulfates in addition. These react with moisture to form a dilute, but very corrosive, sulfuric and sulfurous acid, adding to the normal rusting action. Fuel oil slag may contain vanadium pentoxide, which will attack and corrode steels, including those of high chromium content.

6. Effect of Delayed Combustion of Slag Deposits

When active combustion extends into the boiler and superheater, very troublesome deposits of slag on these heating surfaces may result. This slag deposit is caused by higher temperature and partly reducing conditions, both of which make the ash sticky. The delayed combustion is caused by general or local deficiency of air. The local air deficiency is due to insufficient mixing of combustibles with air in the combustion space. The combustibles consist of gas, as well as hot carbon particles. Some of these hot carbon particles are deposited, along with ash, on the boiler tubes and superheater, where they continue to burn, generating heat in the slag deposit. They can be seen as bright red-hot specks that continue to glow for several seconds, until completely burned. New burning particles are continually deposited, and keep the surface of the deposit hot and sticky. The burning particles not deposited can be seen as red streaks in the gas passages of the boiler. The higher the velocity of the gas in these passages, the faster the slag and burning carbon particles are deposited on the surfaces.

7. Furnace Design

Furnaces may be designed to maintain the ash below or above the ash-fusing temperature. When the ash is below the fusing temperature, it is removed in dry or granular form. When low-fusing–ash coal is burned, it is difficult to maintain a furnace temperature low enough to remove the ash in the granular state. Cyclone and some pulverized–coal-fired furnaces are operated at temperatures high enough to maintain ash in a liquid state until it is discharged from the furnace. These units are referred to as "intermittent" or "continuous" slag-tapped furnaces, depending on the procedure used in removing the fluid ash from the furnace. Because this process permits high furnace temperatures, the excess air can be reduced and the efficiency increased.

8. Iron in Coal Ash

Compounds of iron are responsible for much of the misbehavior of coal ash. Therefore, coals with ash high in iron are always under suspicion as causing trouble. If the ash, and particularly the iron, is uniformly distributed through the coal, difficulties are more likely to occur than if the iron compounds are in large pieces, separated from the coal. The large pieces are likely to drop quickly through the reducing zone of the fuel bed, with little reduction of the high oxides to lower oxides of iron. When the coal is pulverized, the larger and heavier pieces of ash are rejected by some pulverizers and do not go through the furnace. In screenings having a high percentage of fines, the ash is uniformly distributed through the coal; therefore, compounds of iron are likely to cause trouble. In some coals, the iron is in the form of pyrite (FeS_2) which, while passing through the furnace, undergoes various changes. Both the iron and sulfur may combine with oxygen; iron forming the lower oxides, and sulfur, SO_2 or SO_3. Sulfur may also combine with the alkaline metals, Na and K, and form sulfur compounds that have a very low fusion temperature.

D. Low-Temperature Deposits [27]

Formation of deposits in the low-temperature zones, such as the economizer and air heater, is usually associated with condensation of acid or water vapor on cooled surfaces. Other types of deposits, especially in the economizer of boilers with bed-type combustion systems, have also been reported while firing coals with relatively small amounts of phosphorus. Phosphatic deposits have been extremely hard, but the problem is restricted to a limited number of boilers, located mostly in Europe.

Condensation of acid or water vapor can be encountered when metal surfaces are allowed to cool below the acid or water dew points. The sulfuric acid dew point depends on the amount of sulfur trioxide present in the flue gases, but it is usually between 250° and 300°F for SO_3 concentrations of 15–30 ppm. The water dew point depends on the coal and air moisture levels, the hydrogen in the coal, the excess air, and the amount of steam used in sootblowing. It is usually in the range of 105°–115°F for coal firing. On air heaters, where metal temperature is a function of both air and flue gas temperature, condensation on low-temperature surfaces of tubular heaters can occur on tubes near the air inlet and flue gas outlet or on cold-end baskets on regenerative heaters as they are being heated by the flue gases on each cycle. Several factors, such as maldistribution of air or flue gases, excessively low exit-gas temperatures and very low air temperatures can aggravate the problem of condensation. Low gas flow during low load, start-up, and other similar periods can also result in condensation of water and acid.

The deposits themselves can be composed of three types of material:

1. The acid attack can produce various amounts of corrosion product next to the metal, depending on the amount of acid available, the temperature, and the type of metal.
2. This wet deposit can trap fly ash which adds to the bulk of the deposit.
3. The acid can react with constituents, such as iron, sodium, and calcium, in the fly ash to form sulfates, which increase the deposit bulk.

The deposits are usually characterized by low pH (highly acidic); many contain hydrated salts and, for most bituminous coals, they are water-soluble. In this case, deposits can sometimes be water-washed from low-temperature surfaces. However, occasionally, when the coal ash contains large amounts of materials, such as calcium, the reaction product ($CaSO_4$) is nearly insoluble, The deposits that form are very hard and difficult to remove by washing. Complete plugging of gas passes also makes removal by water washing more difficult, even when the deposits are water-soluble.

Deposition can be eliminated by operating the metal temperatures well above the acid dew point temperature of the flue gas, but this would result in a significant loss in boiler efficiency. Improvements in design to obtain more uniform air and gas distribution, better materials of construction, and improved cleaning systems have been combined to minimize the low-temperature deposit problem while operating at relatively low exit-gas temperatures.

E. Coal and Solid Fuel Terms

1. Volatile matter: Volatile matter is that portion of the coal driven off as gas or vapor when the coal is heated according to a standardized temperature test (ASTM D3175). It consists of a variety of organic

gases, generally resulting from distillation and decomposition. Volatile products given off by coals when heated have higher hydrogen/carbon ratios than the remaining material.

2. Fixed carbon: Fixed carbon is the combustible residue left after the volatile matter is driven off. In general, the fixed carbon represents that portion of the fuel that must be burned in the solid state.

3. Moisture: Total moisture content of a sample is customarily determined by adding the moisture loss obtained when air-drying the sample and the measured moisture content of the dried sample. Moisture does not represent all the water present in coal; water of decomposition (combined water) and of hydration are not given off under standardized test conditions.

4. Ash: Ash is the noncombustible residue remaining after complete coal combustion and is the final form of the mineral matter present in coal. Ash in coal is determined by ASTM D3174.

5. Sulfur: Sulfur is found in coal as iron pyrites, sulfates, and in organic compounds. It is undesirable because the sulfur oxides formed when it burns contribute to air pollution, and sulfur compounds contribute to combustion system corrosion and deposits. Sulfur in coal is determined by ASTM D3177.

6. Nitrogen: Nitrogen is found in coal molecules and is also an undesirable coal constituent, because the nitrogen oxides (NO_x) that are formed when coal burns contribute to air pollution. Nitrogen in coal is determined by ASTM D3179.

7. Ash—fusion temperatures: Ash-fusion temperatures are a set of temperatures that characterize the behavior of ash as it is heated. These temperatures are determined by heating cones of ground, pressed ash in both oxidizing and reducing atmospheres according to ASTM D1857. When coal ash is heated to high temperatures it becomes soft and sticky, and finally, fluid. These temperatures give an indication of where slagging may occur within a boiler.

The *initial deformation* temperature is related to the temperature at which coal begins to fuse and become soft.

The *softening* temperature is related to the temperature at which the coal ash shows an accelerated tendency to fuse and agglomerate in large masses.

The *fluid* temperature is related to the temperature at which the coal ash becomes fluid and flows in streams.

These temperatures are also affected by the furnace atmosphere; a reducing atmosphere generally gives lower ash-fusion temperatures than does an oxidizing atmosphere. The ash-fusion temperature deter-

TABLE 4.7 Types of Pulverizers for Various Materials

Type of material	Balltube	Impact and attrition	Ballrace	Ringroll
Low-volatile anthracite	×	—	—	—
High-volatile anthracite	×	—	×	×
Coke-breeze	×	—	—	—
Petroleum coke (fluid)	×	—	×	×
Petroleum coke (delayed)	×	×	×	×
Graphite	×	—	×	×
Low-volatile bituminous coal	×	×	×	×
Medium-volatile bituminous coal	×	×	×	×
High-volatile A bituminous coal	×	×	×	×
High-volatile B bituminous coal	×	×	×	×
High-volatile C bituminous coal	×	—	×	×
Subbituminous A coal	×	—	×	×
Subbituminous B coal	×	—	×	×
Subbituminous C coal	—	—	×	×
Lignite	—	—	×	×
Lignite and coal char	×	—	×	×
Brown coal	—	×	—	—
Furfural residue	—	×	—	×
Sulfur	—	×	—	×
Gypsum	—	×	×	×
Phosphate rock	×	—	×	×
Limestone	×	—	—	×
Rice hulls	—	×	—	—
Grains	—	×	—	—
Ores—hard	×	—	—	—
Ores—soft	×	—	×	×

Source: Ref. 6.

mination is an empirical laboratory procedure, and thus it is some-times difficult to make general statements on its appropriateness for all types of applications.

8. Grindability index: The grindability index indicates the ease of pul-verizing a coal in comparison with a reference coal. This index is helpful in estimating mill capacity. The two most common methods for determining this index are the Hardgrove Grindability Method; ASTM D409) and Ball Mill Grindability Method. Coals with a low index are more difficult to pulverize.

TABLE 4.8

Boiler House Notes:														Page No:
Ralph L. Vandagriff					Thermochemical Properties of Biomass Fuels									
Date: 2/26/93														
File: BioAnalysis														
Description:	Weight	Heating Value			Proximate Analysis			Ultimate Analysis							Source
	Lb./Cu.Ft.	BTU/Lb.	Qvh (MJ/Kg)	Qvl (MJ/Kg)	VCM	Ash	FC	C	H	O	N	S	Cl	Residue	
					(% by Weight,DB)										
Field Crops:															
Alfalfa seed straw		7,932	18.45	17.36	72.60	7.25	20.15	46.76	5.40	40.72	1.00	0.02	0.03	6.07	Calif.Ag.85
Barley straw		7,442	17.31	16.24	68.80	10.30	20.90	39.92	5.27	43.81	1.25	0.00	0.00	9.75	Calif.Ag.85
Bean straw		7,506	17.46	16.32	75.30	5.93	18.77	42.97	5.59	44.93	0.83	0.01	0.13	5.54	Calif.Ag.85
Corncobs	12.70	8,070	18.77	17.58	80.10	1.36	18.54	46.58	5.87	45.46	0.47	0.01	0.21	1.40	Calif.Ag.85
Corn stover		7,588	17.65	16.52	75.17	5.58	19.25	43.65	5.56	43.31	0.61	0.01	0.60	6.26	Calif.Ag.85
Cotton stalks		6,806	15.83	14.79	65.40	17.30	17.30	39.47	5.07	39.14	1.20	0.02	0.00	15.10	Calif.Ag.85
Rice straw (fall)		6,999	16.28	15.34	69.33	13.42	17.25	41.78	4.63	36.57	0.70	0.08	0.34	15.90	Calif.Ag.85
Rice straw (old)		6,260	14.56	13.76	62.31	24.36	13.33	34.60	3.93	35.38	0.93	0.16	0.00	25.00	Calif.Ag.85
Safflower straw		8,267	19.23	18.10	77.05	4.65	18.30	41.71	5.54	46.58	0.62	0.00	0.00	5.55	Calif.Ag.85
Wheat straw		7,528	17.51	16.49	71.30	8.90	19.80	43.20	5.00	39.40	0.61	0.11	0.28	11.40	Calif.Ag.85
Wheat straw (5.7%H2O)		6,350				8.90		41.30	5.20	44.40	0.20	0.00		8.90	FPRS1982
Feed and Fiber Processing Waste															
Almond Hulls		7,833	18.22	17.13	71.33	5.78	22.89	45.79	5.36	40.60	0.96	0.01	0.08	7.20	Calif.Ag.85
Almond shells		8,332	19.38	18.17	73.45	4.81	21.74	44.98	5.97	42.27	1.16	0.02	0.00	5.60	Calif.Ag.85
Babassu husks		8,564	19.92	18.83	79.71	1.59	18.70	50.31	5.37	42.29	0.26	0.04	0.00	1.73	Calif.Ag.85
Bagasse, Cuba		7,985						43.15	6.00	47.95	0.00			2.90	CE,1947
Bagasse, Hawaii		8,160						46.20	6.40	45.90	0.00			1.50	CE,1947
Bagasse, Java		8,681						46.03	6.56	45.55	0.18			1.68	CE,1947
Bagasse, Mexico		9,140						47.30	6.08	35.30	0.00			11.32	CE,1947
Bagasse, Peru		8,380						49.00	5.89	43.36	0.00			1.75	CE,1947
Bagasse, Porto Rico		8,368						44.21	6.31	47.72	0.41			1.35	CE,1947
Bagasse, U.S.A.		7,451	17.33	16.24	73.78	11.27	14.95	44.80	5.35	39.55	0.38	0.01	0.12	9.79	Calif.Ag.85
Coconut fiber dust		8,620	20.05	19.02	66.58	3.72	29.70	50.29	5.05	39.63	0.45	0.16	0.28	4.14	Calif.Ag.85
Cocoa hulls		8,186	19.04	17.97	67.95	8.25	23.80	48.23	5.23	33.19	2.98	0.12	0.00	10.25	Calif.Ag.85
Cotton gin trash		7,059	16.42	15.35	67.30	17.60	15.10	39.59	5.26	36.38	2.09	0.00	0.00	16.68	Calif.Ag.85
Grape pomace		8,745	20.34	19.14	68.54	9.48	21.98	52.91	5.93	30.41	1.86	0.03	0.05	8.81	Calif.Ag.85
Macadamia shells		9,093	21.01	20.00	75.92	0.40	23.68	54.41	4.99	39.69	0.36	0.01	0.00	0.56	Calif.Ag.85
Olive pits		9,196	21.39	20.12	78.65	3.16	18.19	48.81	6.23	43.48	0.36	0.02	0.00	1.10	Calif.Ag.85
Peach pits		8,951	20.82	19.62	79.12	1.03	19.85	53.00	5.90	39.14	0.32	0.05	0.00	1.59	Calif.Ag.85
Peanut hulls		8,014	18.64	17.53	73.02	5.89	21.09	45.77	5.46	39.56	1.63	0.12	0.00	7.46	Calif.Ag.85
Pistachio shells		8,280	19.26	18.06	82.03	1.13	16.84	48.79	5.91	43.41	0.56	0.01	0.04	1.28	Calif.Ag.85
Prune pits		10,009	23.28	22.08	76.99	0.50	22.51	49.73	5.90	43.57	0.32	0.00	0.00	0.48	Calif.Ag.85
Rice hulls		6,939	16.14	15.27	65.47	17.86	16.67	40.96	4.30	35.86	0.40	0.02	0.12	18.34	Calif.Ag.85
Rice hulls, Arkansas	9.20	6,781			58.84	18.94	14.74	39.47	4.01	35.71	0.30	0.04		20.47	U.S.Testing
Sugarcane pith (bonedry)		9,100														FPRS
Walnut shells		8,676	20.18	19.02	78.28	0.56	21.16	49.98	5.71	43.35	0.21	0.01	0.03	0.71	Calif.Ag.85
Wheat dust		6,965	16.20	15.16	69.85	13.68	16.47	41.38	5.10	35.19	3.04	0.19	0.00	15.10	Calif.Ag.85
Prunings:															
Almond prunings		8,603	20.01	18.93	76.83	1.63	21.54	51.30	5.29	40.90	0.66	0.01	0.04	1.80	Calif.Ag.85
Black walnut prunings		8,525	19.83	18.65	80.69	0.78	18.53	49.80	5.82	43.25	0.22	0.01	0.05	0.85	Calif.Ag.85
English walnut prunings		8,439	19.63	18.49	80.82	1.08	18.10	49.72	5.63	43.14	0.37	0.01	0.06	1.07	Calif.Ag.85
Cabernet Sauvignon vines		8,181	19.03	17.84	78.63	2.17	19.20	46.59	5.85	43.90	0.83	0.04	0.08	2.71	Calif.Ag.85
Cardinal grape vines		8,259	19.21		78.17	2.22	19.61								Calif.Ag.85
Chenin blanc grape vines		8,224	19.13	17.94	77.28	2.51	20.21	48.02	5.89	41.93	0.86	0.07	0.10	3.13	Calif.Ag.85
Gewurztraminer grape vines		8,237	19.16		77.27	2.47	20.26								Calif.Ag.85
Merlot grape vines		8,100	18.84		77.47	3.04	19.49								Calif.Ag.85
Pinot noire grape vines		8,190	19.05	17.86	76.83	2.71	20.46	47.14	5.82	43.03	0.86	0.01	0.13	3.01	Calif.Ag.85
Ribier grape vines		8,220	19.12		76.97	3.03	20.00								Calif.Ag.85
Thompson seedless vines		8,319	19.35	18.18	77.39	2.25	20.36	47.35	5.77	43.32	0.77	0.01	0.07	2.71	Calif.Ag.85
Tokay grape vines		8,302	19.31	18.12	76.53	2.45	21.02	47.77	5.82	42.63	0.75	0.03	0.07	2.93	Calif.Ag.85
Zinfandel grape vines		8,194	19.06		76.99	3.04	19.49								Calif.Ag.85

9. Free-swelling index: the free-swelling index (ASTM D720) gives a measure of the extent of swelling of a coal and its tendency to agglomerate when heated rapidly. Coals with a high free-swelling index are referred to as caking coals, whereas those with a low index are referred to as free-burning coals.

10. Burning profile: The burning profile of a coal, obtained by thermal

TABLE 4.8 Continued

| Boiler House Notes: | | | | | | | | | | | | | | | Page No: |

Thermochemical Properties of Biomass Fuels

Description:	Weight Lb./Cu.Ft.	BTU/Lb.	Qvh (MJ/Kg)	Qvl (MJ/Kg)	VCM	Ash	FC	C	H	O	N	S	Cl	Residue	Source
		Heating Value			Proximate Analysis (% by Weight,DB)					Ultimate Analysis					
Energy Crops:															
Eucalyptus camaldulensis		8,349	19.42	18.23	81.42	0.76	17.82	49.00	5.87	43.97	0.30	0.01	0.13	0.72	Calif.Ag.85
Eucalyptus globulus		8,267	19.23	18.03	81.60	1.10	17.30	48.18	5.92	44.18	0.39	0.01	0.20	1.12	Calif.Ag.85
Eucalyptus grandis		8,319	19.35	18.15	82.55	0.52	16.93	48.33	5.89	45.13	0.15	0.01	0.08	0.41	Calif.Ag.85
Casuarina		8,358	19.44	18.26	78.94	1.40	19.66	48.61	5.83	43.36	0.59	0.02	0.16	1.43	Calif.Ag.85
Cattails		7,657	17.81	16.31	71.57	7.90	20.53	42.99	5.25	42.47	0.74	0.04	0.38	8.13	Calif.Ag.85
Poplar		8,332	19.38	18.19	82.32	1.33	16.35	48.45	5.85	43.68	0.47	0.01	0.10	1.43	Calif.Ag.85
Sudan grass		7,476	17.39	16.31	72.75	8.65	18.60	44.58	5.35	39.18	1.21	0.08	0.13	9.47	Calif.Ag.85
Wood:															
Alder, Eastern Black		8,054													Kans.St.80
Alder, Red	23.10	8,000													Kans.St.80
Ash, White	42.00	8,920			0.30			49.73	6.93	43.04	0.00	0.00	0.00	0.30	CE.1947
Aspen, Quaking	19.50	8,433													Kans.St.80
Beech	45.00	8,760			0.65			51.64	6.26	41.45	0.00	0.00	0.00	0.65	CE.1947
Birch, White	22.10	8,650			0.29			49.77	6.49	43.45	0.00	0.00	0.00	0.29	CE.1947
Boxelder		8,102													Kans.St.80
Cedar, White	27.50	8,400			0.37			48.80	6.37	44.46	0.00	0.00	0.00	0.37	CE.1947
Chaparral		8,001	18.61	17.58	75.19	6.13	18.68	46.90	5.08	40.17	0.54	0.03	0.02	7.26	Calif.Ag.85
Cottonwood, shavings(Ark.)	5.40	7,920			87.60	0.94		46.70	6.00	44.90	0.00	0.08	0.00		Ri.Furn.87
Cottonwood, Black		8,800													Kans.St.80
Cottonwood, Common	28.00	8,217													Kans.St.80
Cottonwood, Siouxland	23.10	7,876													Kans.St.80
Cypress	32.00	9,324			0.40			54.98	6.54	38.08	0.00	0.00	0.00	0.41	CE.1947
Cypress - HHV		9,870													FPRS
Elm, American	35.00	7,414			0.74			50.35	6.57	42.34	0.00	0.00	0.00	0.74	Kans.St.80
Elm, American - HHV		8,810													FPRS
Fir, Douglas	34.00	8,438			0.80			52.30	6.30	40.50	0.10	0.00	0.00	0.80	CE.1947
Fir, White		8,577	19.95	18.74	83.17	0.25	16.58	49.00	5.98	44.75	0.05	0.01	0.01	0.20	Calif.Ag.85
Hemlock, Western	29.00	8,056			2.20			50.40	5.80	41.40	0.10	0.10	0.00	2.20	CE.1947
Hickory	39.90	8,039			0.73			49.67	6.49	43.11	0.00	0.00	0.00	0.73	CE.1947
Hickory, Shagbark	50.00	7,252													Kans.St.80
Locust, Black	41.20	8,474	19.71	18.55	80.94	0.80	18.26	50.73	5.71	41.93	0.57	0.01	0.08	0.97	Calif.Ag.85
Madrone		8,345	19.41	18.20	82.99	0.57	16.44	48.00	5.96	44.95	0.06	0.02	0.01	1.00	Calif.Ag.85
Manzanita		8,298	19.30	18.09	81.29	0.82	17.89	46.18	5.94	44.68	0.17	0.02	0.01	1.00	Calif.Ag.85
Maple - HHV		8,580													FPRS,1986
Maple, Red	22.60	7,265			1.35			50.64	6.02	41.74	0.25	0.00	0.00	1.35	CE.1947
Maple, Silver		8,054													Kans.St.80
Maple, Sugar	44.00	7,235													Kans.St.80
Oak, Black		8,180			0.15			48.78	6.09	44.98	0.00	0.00	0.00	0.15	CE.1947
Oak, Red	44.00	8,690			0.02			49.49	6.62	43.74	0.00	0.00	0.00	0.02	CE.1947
Oak, Red, shavings(Ark.)	11.00	7,810			84.20	0.35		46.70	6.37	44.80	0.05	0.05	0.00		Ri.Furn.87
Oak, White	48.00	8,810			0.24			50.44	6.59	42.73	0.00	0.00	0.00	0.24	CE.1947
Pine, Pitch		11,320			1.13			59.00	7.19	32.68	0.00	0.00	0.00	1.13	CE.1947
Pine, Ponderosa	28.00	8,607	20.02	28.80	82.54	0.29	17.17	49.25	5.99	44.36	0.06	0.03	0.01	0.30	Calif.Ag.85
Pine, Southern		9,000													FPRS 1982
Pine, Southern, roots		8,605													FPRS 1982
Pine, White, western	27.00	8,900			0.12			52.55	6.08	41.25	0.00	0.00	0.00	0.12	CE.1947
Pine, Yellow		9,610			1.31			52.60	7.02	40.07	0.00	0.00	0.00	1.31	CE.1947
Poplar	29.00	8,920			0.65			51.64	6.26	41.45	0.00	0.00	0.00	0.65	CE.1947
Poplar, shavings (Ark.)	4.90	7,200			85.30	0.10		46.60	6.20	45.30	0.09	0.05	0.00		Ri.Furn.87
Redwood		8,908	20.72	19.51	79.72	0.36	19.92	50.64	5.98	42.88	0.05	0.03	0.02	0.40	Calif.Ag.85
Redwood, sapwood		8,732	20.31		80.12	0.67	19.21								Calif.Ag.85
Redwood, heartwood	28.00	9,089	21.14		80.28	0.17	19.55								Calif.Ag.85
Redwood, mill waste		9,020	20.98		81.19	0.18	18.63								Calif.Ag.85
Sweetgum		7,450													Kan.ST.80
Sycamore	28.70	8,230													Kan.St.80
Tanoak		8,138	18.93	17.73	80.93	1.67	17.40	47.81	5.93	44.12	0.12	0.01	0.01	2.00	Calif.Ag.85
Tanoak, sapwood		8,199	19.07		83.61	1.03	15.36								Calif.Ag.85
Willow, Black		7,198													Kan.St.80
Willow, Sandbar	22.50	7,982													Kan.St.80

TABLE 4.8 Continued

Boiler House Notes:													Page No:
				Thermochemical Properties of Biomass Fuels									
Description:	Weight	Heating Value			Proximate Analysis				Ultimate Analysis					Source
	Lb./Cu.Ft.	BTU/Lb.	Qvh (MJ/Kg)	Qvl (MJ/Kg)	VCM	Ash	FC (% by Weight,DB)	C	H	O	N	S	Cl	Residue	
Tree Bark														
Alder, Red		8,406												FPRS 1976
Aspen, quaking		8,496												FPRS 1976
Beech		7,640												FPRS 1976
Birch, paper		9,486												FPRS 1976
Birch, yellow		9,144												FPRS 1976
Birch, white		10,310												FPRS 1976
Blackgum		7,936												FPRS 1976
Cedar, western red		8,694												FPRS 1976
Cottonwood, black		9,000												FPRS 1976
Elm, American		6,921												FPRS 1976
Elm, soft		7,600												FPRS 1976
Fir, balsam		8,946												FPRS 1976
Fir, Douglas		9,792												FPRS 1976
Hemlock, eastern		8,874												FPRS 1976
Hemlock, western		9,396												FPRS 1976
Larch, western		8,280												FPRS 1976
Maple, hard		8,230												FPRS 1976
Maple, soft		8,100												FPRS 1976
Maple, sugar		7,373												FPRS 1976
Oak, red		8,082												FPRS 1976
Oak, white		7,074												FPRS 1976
Pine, jack		8,930												FPRS 1976
Pine, lodgepole		10,260												FPRS 1976
Pine, ponderosa		9,100												FPRS 1976
Pine, slash		9,002												FPRS 1976
Pine, southern (FPRS 1976)		9,360												FPRS 1976
Pine, southern (FPRS 1982)		9,000												FPRS 1982
Pine, western white		9,090												FPRS 1976
Poplar		8,810												FPRS 1976
Redwood (FPRS 1976)		8,350			72.60	0.40	27.00	51.90	5.10	42.40	0.10	0.10	0.00	0.40	FPRS 1976
Redwood (Calif.Ag. 1985)		8,418												Calif.Ag. 85
Spruce		8,740			69.60	3.80	26.60	51.80	5.70	38.40	0.20	0.10	0.00	3.80	FPRS 1976
Spruce, black		8,610												FPRS 1976
Spruce, Engelmann		8,424												FPRS 1976
Spruce, pine		8,825												FPRS 1976
Spruce, red		8,630												FPRS 1976
Spruce, White		8,530												FPRS 1976
Sweetgum		7,526												FPRS 1976
Sycamore		7,403												FPRS 1976
Tamarack		9,010												FPRS 1976
Tanoak		7,911												Calif.Ag. 85
Willow, black		7,249												FPRS 1976

gravimetric analysis, is a plot of the rate of weight loss when a coal sample is heated at a fixed rate. The burning profile of a solid fuel offers an indication of ignition and burning characteristics by comparing it with other burning profiles of fuels with known performance.

11. Ash analysis Ash analyses give percentages of inorganic oxides present in an ash sample. Ash analyses are used for evaluation of the corrosion, slagging, and fouling potential of coal ash. The ash constituents of interest are silica (SiO_2), alumina (Al_2O_3), titania (TiO_2), ferric oxide (Fe_2O_3), lime (CaO), magnesia (MgO), potassium oxide (K_2O), sodium oxide (Na_2O), and sulfur trioxide (SO_3). An indication of ash behavior can be estimated from the relative percentages of each constituent.

See Tables 4.7–4.13 for further details.

TABLE 4.9

Boiler House Notes					Page No.
Ralph L. Vandagriff					
Data: Southern Hardwoods – 6" Stemwood					
Species	Pounds/cu.ft.	Btu per Pound, Ovendry, HHV			Ash,percent
	green stemwood	Stemwood	Branchwood	Stembark	Branchbark	(average)
Ash, green	51.6	7,695	7,727	7,472	7,606	
Ash, white	53.6	8,033	8,013	7,695	7,816	0.30
Elm, American	58.7	7,770	7,857	6,840	6,904	0.50
Elm, winged	64.4	7,917	7,869	7,019	6,889	
Hackberry	56.5	7,882	7,867	7,147	7,141	
Hickory, true	60.8	8,163	7,931	7,586	7,259	1.00
Maple, red	52.6	7,846	7,829	7,595	7,384	0.30
Oak, black	65.5	7,680	7,692	7,642	7,847	0.50
Oak, blackjack	69.4	7,739	7,739	7,766	7,907	1.30
Oak, cherrybark	65.8	7,848	7,737	7,582	7,655	
Oak, chestnut						0.40
Oak, laurel	63.3	7,828	7,653	7,897	7,806	
Oak, northern red	64.1	7,791	7,776	7,879	7,926	0.20
Oak, post	68.1	7,889	7,845	7,191	7,728	0.50
Oak, scarlet	65.7	7,798	7,673	8,041	7,894	0.10
Oak, Shumard	65.9	7,789	7,745	7,970	7,913	
Oak, southern red	64.6	7,919	7,839	7,983	7,798	0.40
Oak, water	63.6	7,876	7,833	7,930	7,918	0.30
Oak, white	67.2	7,676	7,507	7,328	7,574	0.65
Sweetbay	54.8	7,736	7,802	7,822	7,886	0.20
Sweetgum	62.3	7,667	7,690	7,200	7,214	0.30
Tupelo, black	59.3	7,867	7,814	7,788	8,176	0.50
Yellow Poplar	52.2	7,774	7,811	7,696	7,666	0.30
re: #55						

TABLE 4.10

Boiler House Notes:									Page No:	
Date: 11/29/93										
N.Little Rock, Arkansas USA		**Thermochemical Analysis; Miscellaneous Fuels**								
By: R.L.Vandagriff										
File: Misc.SolidsAnalysis										
Description:	Moisture	Net Heat Value			Ultimate	Analysis				
	Content	BTU/Lb.	C	H	O	N	S	Cl	Ash
	Percent	(Approximate)								
FATS	2.00%	16,236	73.00	11.50	14.80	0.40	0.10	0.00	0.20
FOOD WASTE	70.00%	1,055	48.00	6.40	37.60	2.60	0.40	0.00	5.00
FRUIT WASTE	78.70%	135	48.50	6.20	39.50	1.40	0.20	0.00	4.20
MEAT WASTE	38.80%	6,952	59.60	9.40	24.70	1.20	0.20	0.00	4.90
CARDBOARD	5.20%	6,012	43.00	5.90	44.80	0.30	0.20	0.00	5.00
MAGAZINES	4.10%	4,617	32.90	5.00	38.60	0.10	0.10	0.00	23.30
NEWSPRINT	6.00%	7,026	49.10	6.10	43.00	0.10	0.20	0.00	23.30
PAPER-MIXED	10.20%	5,623	43.40	5.80	44.30	0.30	0.20	0.00	6.00
WAXED CARTON	3.40%	11,573	59.20	9.30	30.10	0.10	0.10	0.00	1.20
OFFICE SWEEP	3.20%	4,868	24.30	3.00	4.00	0.50	0.20	0.00	68.00
WASTE OIL	0.00%	15,281	66.90	9.60	5.20	2.00	0.00	0.00	16.30
PLASTIC - MIXED	0.20%	11,398	60.00	7.20	22.80	0.00	0.00	0.00	10.00
POLYETHYLENE	0.20%	21,157	85.20	14.20	0.00	0.10	0.10	0.00	0.40
POLYSTYRENE	0.20%	17,529	87.10	8.40	4.00	0.20	0.00	0.00	0.30
POLYURETHANE	0.20%	11,708	63.30	6.30	17.60	6.00	0.10	2.40	4.30
POLYVINYLCHLORIDE	0.20%	9,559	45.20	5.60	1.60	0.10	0.10	45.40	2.00
GARDEN TRIM	60.00%	1,687	46.00	6.00	38.00	3.40	0.30	0.00	6.30
GREEN WOOD	50.00%	3,018	50.10	6.40	42.30	0.10	0.10	0.00	1.00
LEATHER - MIXED	10.00%	11,330	60.00	8.00	11.60	10.00	0.40	0.00	10.00
TEXTILES - MIXED	10.00%	6,877	48.00	6.40	40.00	2.20	0.20	0.00	3.20
RUBBER - MIXED	1.20%	14,698	69.70	8.70	9.00	0.00	1.60	0.00	20.00

TABLE 4.11

E.F.W., Inc.				Boiler House Notes:					Page No:
Date: 11/19/93									
N.Little Rock, Arkansas USA				**Thermochemical Analysis of Rubber Tires**					
By: R.L.Vandagriff									
File: TireAnalysis									
Description:	HHV			Ultimate	Analysis				Source
	BTU/Lb.	C	H	O	N	S	Cl	Ash	
Rubber Tire Chips	13,138	65.02	5.96	7.27	0.25	1.19	0.00	14.76	Comb.Systems
1/2" Rubber Fuzz	14,036	69.74	0.63	0.34	0.05	1.30	0.07	16.48	Babcock&Wilcox
2" Chips, No metal	14,652	72.15	6.74	9.67	0.36	1.23	0.09	8.74	Babcock&Wilcox
2" Chips, with Metal	13,340	67.00	5.81	1.64	0.25	1.33	0.03	23.19	Babcock&Wilcox
Shredded	16,569	83.87	7.09	2.17	0.24	1.23	0.00	4.78	Pyropower
Tire Derived Fuel (TDF)	15,469	77.37	6.81	5.12	0.37	1.62	0.00	8.32	FosterWheeler
Steel Belted	12,422	64.20	5.00	4.40	0.10	0.91	0.00	25.20	EPI,Inc.
Fiberglass	15,112	75.80	6.62	4.39	0.20	1.29	0.00	11.70	EPI,Inc.
Polyester	16,508	83.50	7.08	1.72	0.01	1.20	0.00	6.50	EPI,Inc.
Nylon	15,785	78.90	6.97	5.42	0.01	1.51	0.00	7.20	EPI,Inc.
Kevlar Belted	17,121	86.50	7.35	2.11	0.01	1.49	0.00	2.50	EPI,Inc.
Average:	14,923	74.91	6.01	4.02	0.17	1.30	0.02	11.76	

TABLE 4.12

Boiler House Notes	Date: 7/9/98										Page No.
Ralph L. Vandagriff											

Progressive Stages of Transformation of Vegetal Matter Into Coal

Fuel Classification by Rank	Locality	Moisture Content As received	Proximate Analysis			Analysis on dry basis					Btu Heating Value (dry basis)
			VM (volatile matter)	FC (fixed carbon)	Ash	S	H2	C	N2	O2	
								Ultimate Analysis			
Wood		46.9	78.1	20.4	1.5		6.0	51.4	0.1	41.0	8,835
Peat	Minnesota	64.3	67.3	22.7	10.0	0.4	5.3	52.2	1.8	30.3	9,057
Lignite	North Dakota	36.0	49.8	38.1	12.1	1.8	4.0	64.7	1.9	15.5	11,038
Lignite	Texas	33.7	44.1	44.9	11.0	0.8	4.6	64.1	1.2	18.3	11,084
Sub-bituminous C	Wyoming	22.3	40.4	44.7	14.9	3.4	4.1	61.7	1.3	14.6	10,598
Sub-bituminous B	Wyoming	15.3	39.7	53.6	6.7	2.7	5.2	67.3	1.9	16.2	12,096
Sub-bituminous A	Wyoming	12.8	39.0	55.2	5.8	0.4	5.2	73.1	0.9	14.6	12,902
Bituminous High Volatile C	Colorado	12.0	38.9	53.9	7.2	0.6	5.0	73.1	1.5	12.6	13,063
Bituminous High Volatile B	Illinois	8.6	35.4	56.2	8.4	1.8	4.8	74.6	1.5	8.9	13,388
Bituminous High Volatile A	Pennsylvania	1.4	34.3	59.2	6.5	1.3	5.2	79.5	1.4	6.1	14,396
Bituminous Medium Volatile	West Virginia	3.4	22.2	74.9	2.9	0.6	4.9	86.4	1.6	3.6	15,178
Bituminous Low Volatile	West Virginia	3.6	16.0	79.1	4.9	0.8	4.8	85.4	1.5	2.6	15,000
Semi-anthracite	Arkansas	5.2	11.0	74.2	14.8	2.2	3.4	76.4	0.5	2.7	13,142
Anthracite	Pennsylvania	5.4	7.4	75.9	16.7	0.8	2.6	76.8	0.8	2.3	12,737
Meta-anthracite	Rhode Island	4.5	3.2	82.4	14.4	0.9	0.5	82.4	0.1	1.7	11,624

re: # 14

TABLE 4.13

Boiler House Notes:												Page No:	
Subject: Coal Properties													
Ralph L. Vandagriff		Date: 7/16/98											
			Typical Properties of U.S. Coals										
Source	Coal Name	ASTM Classification	Btu/Lb. as rec'd.	**Proximate Analysis**					**Ultimate Analysis**				
				Moisture	Volatile Matter Dry	Ash	Fixed Carbon		C	H	O	N	S
Ohio or Pennsylvania	Pittsburgh #8	Bituminous(HV)	12,540	5.2%	40.2%	9.1%	50.7%		74.0%	5.10%	7.90%	1.80%	2.30%
Illinois	Illinois #6	Bituminous(HV)	10,300	17.6%	44.2%	10.8%	45.0%		69.0%	4.90%	10.00%	1.00%	4.30%
Pennsylvania	Upper Freeport	Bituminous(MV)	12,970	2.2%	28.1%	13.4%	56.5%		74.9%	4.70%	4.97%	1.27%	0.76%
Wyoming	Spring Creek	Subbituminous	9,190	24.1%	43.1%	5.7%	51.2%		70.3%	5.00%	17.69%	0.96%	0.35%
Montana	Decker	Subbituminous	9,540	23.4%	40.8%	5.2%	54.0%		72.0%	5.00%	16.41%	0.95%	0.44%
North Dakota		Lignite	7,090	33.3%	43.6%	11.1%	45.3%		63.3%	4.50%	19.00%	1.00%	1.10%
Texas	S.Hallsville	Lignite	7,080	37.7%	45.2%	10.4%	44.4%		66.3%	4.90%	18.20%	1.00%	1.20%
Texas	Bryan	Lignite	3,930	34.1%	31.5%	50.4%	18.1%		33.6%	3.30%	11.10%	0.40%	1.00%
Texas	San Miguel	Lignite	2,740	14.2%	21.2%	66.8%	10.0%		18.4%	2.30%	9.01%	0.29%	1.20%
re: #1													

			Low-Rank World Coals										
Source	Coal Name	ASTM Classification	Btu/Lb. as rec'd. HHV	**Proximate Analysis**					**Ultimate Analysis**				
				Moisture	Volatile Matter Dry	Ash			C	H	O	N	S
Belgium	Florennes	Lignite (B)	11,910	63.3%	53.6%	2.2%			70.1%	5.00%	23.50%	0.70%	0.70%
Canada	Alberta	Subbituminous	13,150	24.2%	39.2%	7.3%			76.6%	5.50%	15.80%	1.60%	0.50%
Czechoslovakia	Northern Bohemia	Subbituminous(A)	13,690	20.2%	49.0%	2.7%			77.6%	5.80%	14.70%	1.00%	0.90%
		Subbituminous(C)	13,560	31.2%	50.5%	6.4%			76.0%	6.00%	15.70%	1.10%	1.20%
	Central Bohemia	Subbituminous(B)	13,770	21.1%	38.1%	5.9%			79.5%	5.00%	13.10%	1.20%	1.20%
		Subbituminous(B)	13,640	23.2%	43.6%	15.9%			77.1%	5.20%	11.10%	1.50%	5.10%
	Southern Bohemia	Lignite (A)	12,240	44.8%	56.5%	7.4%			70.2%	5.40%	19.50%	0.90%	4.00%
	Western Bohemia	Lignite (A)	13,680	39.7%	51.7%	8.0%			76.6%	6.10%	15.20%	1.20%	0.90%
	Slovakia	Lignite(A)	11,870	36.1%	51.3%	12.2%			69.6%	4.90%	19.50%	1.50%	4.50%
Federal Republic of Germany	Rhine Region	Lignite(B)	11,580	62.2%	55.3%	3.3%			69.1%	4.90%	24.10%	0.90%	1.00%
	Bavaria,Peissenberg	Subbituminous(A)	13,200	9.7%	52.8%	11.9%			72.6%	5.40%	12.40%	1.60%	7.80%
Greece													
Athens Area	Peristeri	Subbituminous(C)	12,140	22.4%	46.9%	6.5%			71.8%	4.70%	19.80%	1.90%	1.80%
Island of Euboea	Aliveri	Lignite(A)	11,600	33.3%	56.1%	11.3%			66.0%	5.20%	23.90%	0.90%	2.00%
West Macedonia	Ptolemais	Lignite(B)	10,340	59.0%	57.9%	12.2%			65.4%	4.30%	27.30%	1.90%	1.10%
Italy													
Lucania	Mercure	Lignite(B)	11,020	58.6%	58.6%	10.1%			65.4%	5.20%	25.30%	1.80%	2.30%
Toscana	Pietrafitta	Lignite(B)	11,640	60.6%	57.3%	9.9%			70.5%	4.80%	19.90%	1.90%	2.90%
Toscana	Valdarno	Lignite(B)	11,350	51.9%	56.8%	5.3%			66.5%	5.20%	26.30%	1.00%	1.00%
Umbria	Spoleto	Lignite(A)	11,140	31.8%	60.4%	7.8%			64.6%	5.30%	27.30%	1.10%	1.50%
Federation of Maylaya													
Selangor	Batu Arang	Subbituminous(A)	13,640	17.7%	48.7%	7.5%			77.1%	5.80%	15.10%	1.80%	0.40%
Poland		Lignite(B)	12,510	53.9%	60.8%	7.3%			71.4%	5.8%	20.20%	0.60%	2.00%
		Subbituminous(A)	13,540	17.6%	44.0%	6.7%			77.7%	5.10%	14.30%	0.90%	2.00%
U.S.S.R.													
Chelyabinsk Basin		Subbituminous(B)	12,910	17.6%	43.6%	9.9%			76.2%	5.20%	16.20%	1.90%	0.50%
Rhiczechinsk		Lignite(A)	11,910	37.0%	43.7%	6.9%			72.2%	4.30%	22.10%	1.10%	0.30%
Borneo-Serawak													
Upper Rajang Valley		Subbituminous(C)	12,390	23.3%	48.1%	1.3%			72.4%	4.90%	21.20%	1.40%	0.10%
East Germany													
Niederlausitz		Lignite(B)	11,720	59.6%	55.6%	2.5%			68.1%	4.70%	25.80%	0.70%	0.70%
Bautzen		Lignite(B)	12,100	63.4%	57.6%	3.9%			70.1%	5.30%	21.90%	0.70%	2.00%
Oberlausitz		Lignite(B)	12,020	58.3%	54.4%	3.2%			67.7%	5.20%	26.00%	0.60%	0.50%
re: #13													

V. MISCELLANEOUS FUEL

A. Petroleum Coke [1,13]

Petroleum coke is a by-product of a process in which residual hydrocarbons are converted to lighter distillates. Two processes are employed: delayed coking and fluid coking.

1. Delayed Coking

In the delayed coking process, reduced crude oil is heated rapidly and sent to coking drums. The delayed coke resembles run-of-mine coal, except that it is dull black. Proximate analysis varies with the feed crude stock. The components range as follows:

Moisture	3–12%
Volatile matter	9.0–15.0%
Fixed carbon	71–88%
Ash	0.2–3.0%
Sulfur	2.9–9.0%
Btu/lb, dry	14,100–15,600

Some delayed cokes are easy to pulverize and burn whereas others are difficult.

2. Fluid Coking

The fluid coking process uses two vessels, a reactor and a burner. The coke formed in this process is a hard, dry, spherical solid resembling black sand. Again the proximate analysis varies with the crude feed stock. The range of analysis is

Fixed carbon	90–95%
Volatile matter	3–6.5%
Ash	0.2–0.5%
Sulfur	4.0–7.5%
Btu/lb, dry (HHV)	14,100–14,600

Fluid coke can be pulverized and burned, or it can be burned in a cyclone furnace or a fluidized bed. All three types of firing require supplemental fuel to aid ignition.

Note: Waterwall boilers that burn fuel on grates need a fuel with volatile content of 18% and higher, to sustain combustion. To burn petroleum coke as the only fuel, would be extremely difficult if not impossible.

5

Steam Boiler Feedwater

Boiler Feedwater; Oxygen in Boiler Feedwater; Deaerators.

I. BOILER FEEDWATER [1,12]

An ample supply of boiler feedwater of good quality is a necessity for economic and efficient operation of a steam plant. Among the numerous ill effects arising from the use of unsuitable feedwater the following may be mentioned:

1. Tube failures
2. Crystallization or embrittlement and corrosion of boiler steel
3. Loss of heat due to the deposit of scale, dirt, or oil on the heating surfaces
4. Length of time apparatus must be out of service for cleaning, inspection, and repairs
5. Investment in spare equipment
6. Loss of heat owing to blowing down boilers, heaters, and such
7. Increased steam consumption of prime movers because of accumulation of scale or dirt in valves, nozzles, and buckets
8. Foaming and priming

The organic constituents of the foreign matter in raw water are of vegetable and animal origin and are taken up by the water in flowing over the ground or by direct contamination with sewage and industrial refuse. Feedwater containing organic matter may cause foaming, because the suspended particles collect on the surface of the water in the boiler and impede the liberation of the steam bubbles arising to the surface. The inorganic impurities in suspension or in colloidal solution consist of clay, silica, iron, alumina, and the like. The more common soluble inorganic impurities are calcium, magnesium, potassium, and sodium in the form of carbonates, sulfates, chlorides, and nitrates. Raw water also contains a certain quantity of gases in solution, such as air, CO_2, and occasionally, hydrogen sulfide.

The most widely known evidence of the presence of scale-forming ingredients in feedwater is known as hardness. If the water contains only such ingredients as the bicarbonates of lime, magnesia, and iron, which may be precipitated as normal carbonates by boiling at 212°F, it is said to have temporary hardness. Permanent hardness is due to the presence of sulfates, chlorides, and nitrates of lime, magnesia, and iron that are not completely precipitated at a temperature of 212°F.

A. Scale

Mud or suspended mineral matter, if introduced into the boiler with the feedwater, will eventually form a deposit on the heating surfaces. Iron, aluminum, and silicon in colloidal solution will also tend to produce scale, but by far the greater part of the objectionable scale deposit results from the salts of calcium and magnesium. The salts are in solution in the cold raw water and constitute "hardness."

Raw water from surface or subsurface sources invariably contains in solution some degree of troublesome scale-forming materials, free oxygen, and sometimes acids. Because good water conditioning is essential in the operation of any steam cycle, these impurities must be removed.

Dissolved oxygen will attack steel and the rate of attack increases sharply with a rise in temperature. High chemical concentrations in the boiler water and feedwater cause furnace tube deposition and allow solids carryover into the superheater and turbine, resulting in tube failures and turbine blade deposition or erosion.

As steam plant operating pressures have increased, the water treatment system has become more critical. This has led to the installation of more complete and refined water treatment facilities.

The temperature of the feedwater entering an economizer should be high enough to prevent condensation and acid attack on the gas side of the tubes. Dew point and rate of corrosion vary with the sulfur content of the fuel and with the type of firing equipment.

II. OXYGEN IN BOILER FEEDWATER

A. Oxygen

The presence of oxygen accelerates the combining of iron and water. Oxygen can react with iron hydroxide to form a hydrated ferric oxide or magnetite. Generally localized, this reaction forms a pit in the metal. Severe attack can occur if the pit becomes progressively anodic in operation. Oxygen reacts with hydrogen at the cathodic surface and depolarizes the surface locally. This permits more iron to dissolve, gradually creating a pit.

The most severe corrosion action occurs when a deposit covers a small area. The creation of a differential aeration cell about the deposit can lead to a severe local action. The metal beneath the deposit is lower in oxygen than areas surrounding it, becomes anodic, and is attacked. *Pitting is most prevalent in stressed sections of boiler tubing, such as at welds and cold-worked sections, and at surface discontinuities in the metal.*

Efficient operation of boilers requires the exclusion of oxygen from the feedwater. The normal guaranteed value of oxygen leaving the deaerating heater or a deaerating condenser is less than 0.005 ppm. Some problem areas are air leakage during boiler start-up, the addition of nondeaerated water to the feedwater, the aerated heater drips into the condensate, and feedwater exposure to air during short outages. During an outage, auxiliary steam should be admitted to the deaerator to maintain a pressure of 3–5 psig and prevent oxygen contamination of the feedwater. During an outage, low-pressure feedwater heaters and related extraction piping are often under negative pressure, and any leading valves, pumps, or flanges will provide a path for oxygen introduction into the system.

Acceptable feedwater oxygen levels during steady-state operation do not necessarily mean that oxygen concentration is within safe limits. And do not be lulled into a false sense of security if oxygen levels are excessive only for a short time. Considerable damage can still occur. Thus, use of dissolved-oxygen monitors is important, particularly during load swings and start-up operations.

B. pH of Boiler Feedwater [1,13]

Of equal importance with oxygen is the control of boiler-water pH. Small deviations from the recommended boiler water limits will result in tube corrosion. Large deviations can lead to the destruction of all furnace wall tubes in a matter of minutes.

The primary cause of acidic and caustic boiler-water conditions is raw water leakage into the boiler feedwater system. The water source determines whether the in-leakage is either acid-producing or caustic-producing. Freshwater from lakes and rivers, for example, usually provides dissolved solids that hydrolyze in the boiler water environment to form a caustic, such as sodium hydroxide. By contrast, seawater and water from recirculation cooling water systems with cooling towers contain dissolved solids that hydrolyze to form acidic compounds.

Strict tolerance levels on leakage should be established for all high-pressure boilers. Set a limit of 0.5-ppm dissolved solids in the feedwater for normal operation; allow 0.5–2 ppm for short periods only.

Another potential source of acidic and caustic contaminants is the makeup demineralizer, where regenerant chemicals such as sulfuric acid and caustic may inadvertently enter the feedwater system. Chemicals incorrectly applied during boiler water treatment also can be corrosive, as for example, sodium hydroxide used in conjunction with sodium phosphate compounds to treat boiler water. Corrosion can occur if the sodium hydroxide and sodium phosphate are not added to the water in the proper proportion.

III.   DEAERATORS [6,14,16,45,56]

Deaerators or deaerating heaters serve to degasify feedwater and thus reduce equipment corrosion, also to heat feedwater and improve thermodynamic efficiency, and to provide storage, positive submergence, and surge protection on the boiler feed pump suction.

The predominant gases in feedwater are oxygen, carbon dioxide, hydrogen sulfide, and ammonia. These dissolved gases in a power plant feedwater supply can produce corrosion and pitting, and they must be removed to protect boiler tubes, steam drums, piping, pumps, and condensate systems.

Removal of oxygen and carbon dioxide from boiler feedwater and process water at elevated temperature is essential for adequate condition. Water in the deaerator must be heated to and kept at saturation temperature, because the gas solubility is zero at the boiling point of the liquid, and be mechanically agitated by spraying or cascading over trays for effective scrubbing, release, and removal of gases. Gases must be swept away by an adequate supply of steam. Because the water is heated to saturation conditions, the terminal temperature difference is zero, with maximum improvement in associated turbine heat rate. Extremely low partial gas pressures, dictated by Henry's law, call for large volumes of scrubbing steam.

Steam deaerators break up water into a spray or film, then sweep the steam across and through them to force out dissolved gases. In the steam deaerator, there is a heating section and a deaerating section, plus storage for hot deaerated water. The entering steam meets the hottest water first, to thoroughly scrub out the last remaining fraction of dissolved gas. The latter is carried along with steam as it flows through the deaerator. The direction of steam flow may be across, down, or counter to the current. The steam picks up more noncondensable gas as it goes through the unit, condensing as it heats the water. The bulk of the steam condenses in the first section of the deaerator, when it contacts entering cold water. The remaining mixture of noncondensable gases is discharged to atmosphere through a vent condenser.

The normal guaranteed value of oxygen leaving the deaerating heater is less than 0.005 ppm, which is near the limit of chemical detectability. To achieve this low residual, it is necessary to exclude air leakage into the condenser, to prevent the addition of nondeaerated water to the condensate or feedwater, to prevent the addition of aerated heater drips into the condensate, and to assure the exclusion of air into the feedwater cycle during short outages of the boiler.

The *tray-type* deaerator is prevalent. Although it has some tendency to scale, it will operate at wide load conditions and is practically independent of water inlet temperature. Trays can be loaded to some 10,000 lb/ft² hr⁻¹, and the deaerator seldom exceeds 8 ft in height.

The *spray type* uses a high-velocity steam jet to atomize and scrub the preheated water. In industrial plants, the operating pressures must be stable for this type of deaerator to be applied satisfactorily. It requires a temperature gradient (e.g., 50°F minimum) to produce the fine sprays and vacuums with the cold water required.

A. Estimating Deaerator Steam Requirements

Because the deaerator is a direct-contact heat exchanger, the calculations are the same.

1. Assumptions:
   1. There is no loss of Btu through the shell; perfect insulation.
   2. Loss of Btu through the vent to atmosphere is negligible.
   3. The process is a constant and steady flow.
   4. Energy entering = energy leaving.

2. S = steam flow (lb/hr)
   W_1 = water entering (lb/hr)
   W_2 = water leaving (lb/hr)

   h_s = steam enthalpy (Btu/lb)
   h_1 = enthalpy of entering water (Btu/lb)
   h_2 = enthalpy of water leaving (Btu/lb)

3. Estimated steam required (lb/hr) $S = \dfrac{W_2(h_2 - h_1)}{hs - h_1}$

4. Estimated deaerator water required (lb/hr) $W_2 - S$

Example

Boiler (lb/hr)		100,000
Steam pressure (psig)		125
Steam temperature		Saturated
Steam enthalpy (Btu/lb)	(h_s)	1,194.0
Blowdown (%)		3
Total boiler feedwater required (lb/hr)	(W_2)	103,000
Water temperature to deaerator (°F)		70
Water entering deaerator; enthalpy (Btu/lb)	(h_1)	$70 - 32 = 38$
Deaerator operating pressure (psig)		15

Water temperature leaving deaerator (°F)   250
Water leaving deaerator; enthalpy (Btu/lb)   (h_2)   250 − 32 = 218

Deaerating steam required (lb/hr) $\dfrac{103{,}000\ (218 - 38)}{1{,}194 - 38}$ = 16,038 lb/hr (S)

Water to deaerator required (lb/hr) 103,000 − 16,038 = 86,962 lb/hr (W_1)

B. Comparison of Deaerator Types [56]

A temperature rise of at least 50°F (30°C) over the temperature of the incoming water is required to make spray-type deaerators perform effectively at full load. Furthermore, the effectiveness of deaeration in a spray-type unit seriously decreases as loads decrease, for at operating loads of less than 25% of design rating, the heating steam requirements are not sufficient to maintain a high steam flow.

Tray-type deaerators, on the other hand, use a different principle for the release of gases. It is still necessary to provide a large water surface area to give adequate opportunity for the release of the dissolved gases. This is accomplished by the water cascading from one tray tier to another, exposing as much surface as possible to the scrubbing action of the steam.

Tray-type deaerators are designed with as many as 24 tray tiers to permit an adequate surface exposure. With this design method, the same amount of water surface is exposed to the steam for gas release, regardless of inlet water temperature. Also, as operating loads decrease under the design maximum, the ratio of surface to throughput water increases, and thinner films for release of gases are provided. This ensures effective deaeration under all inlet water temperatures and flow conditions.

Therefore, when a deaerator is operated under varying load conditions or inlet water temperature, the tray-type deaerator gives the most satisfactory results over the entire operating range.

Translating the foregoing points into terms of application, it follows that industrial plants that may operate under low-load conditions (at night, on weekend, or during the summer) will find that a tray-type deaerator gives better assurance of satisfactory oxygen removal than a spray type. In addition, central station deaerators serving turbines that also operate over a wide range of load conditions can get better deaeration with a tray-type unit.

This is not an indictment of spray-type deaerators. They do have definite places of application, and, when operated within their specified limits, will do an excellent job of gas removal. In marine service, where it is not possible to maintain the trays in a level state because of the toss and roll of a ship, spray-type units are always used. Waters that are seriously scaling will cause less difficulty in a spray-type than a tray-type unit.

See Tables 5.1 and 5.2 for data on properties of water.

Boiler House Notes: Page:

By: Ralph L. Vandagriff Date: 5/15/96

PROPERTIES OF WATER

Temperatures 32°F – 210°F

Temp Deg. F	Temp Deg. C	Enthalpy Btu/lb	Saturation Pressure InHg (Abs.)	Specific Volume Cu Ft/lb	Specific Volume Gallon/lb	Density Pound/Gal	Density Lb./Cu Ft	Density Gr./Cu.Cm.	Kinematic Viscosity Centistokes	Kinematic Viscosity Centipoise
32	0.00	0.00	0.1803	0.016022	0.1199	8.3403	62.414	0.9997	1.79	1.7895
33	0.56	1.01	0.1878	0.016021	0.1199	8.3472	62.418	0.9998	1.75	1.7497
34	1.11	2.02	0.1955	0.016021	0.1198	8.3472	62.418	0.9998	1.72	1.7197
35	1.67	3.02	0.2035	0.016021	0.1198	8.3472	62.420	0.9998	1.68	1.6797
36	2.22	4.03	0.2118	0.016021	0.1198	8.3472	62.420	0.9998	1.66	1.6597
37	2.78	5.04	0.2203	0.016021	0.1198	8.3472	62.420	0.9998	1.63	1.6297
38	3.33	6.04	0.2292	0.016019	0.1198	8.3472	62.425	0.9999	1.6	1.5999
39	3.89	7.04	0.2383	0.016019	0.1198	8.3472	62.425	0.9999	1.56	1.5599
40	4.44	8.05	0.2478	0.016019	0.1198	8.3472	62.425	0.9999	1.54	1.5399
41	5.00	9.05	0.2576	0.016019	0.1198	8.3472	62.426	0.9999	1.52	1.5199
42	5.56	10.05	0.2677	0.016019	0.1198	8.3472	62.426	0.9999	1.49	1.4899
43	6.11	11.06	0.2782	0.016019	0.1198	8.3472	62.426	0.9999	1.47	1.4699
44	6.67	12.06	0.2891	0.016019	0.1198	8.3472	62.426	0.9999	1.44	1.4399
45	7.22	13.04	0.3004	0.016021	0.1198	8.3472	62.420	0.9998	1.42	1.4198
46	7.78	14.06	0.3120	0.016021	0.1198	8.3472	62.420	0.9998	1.39	1.3898
47	8.33	15.07	0.3240	0.016021	0.1198	8.3472	62.420	0.9998	1.37	1.3698
48	8.89	16.07	0.3364	0.016021	0.1198	8.3472	62.420	0.9998	1.35	1.3498
49	9.44	17.07	0.3493	0.016023	0.1198	8.3472	62.410	0.9997	1.33	1.3296
50	10.00	18.05	0.3626	0.016023	0.1199	8.3403	62.410	0.9997	1.31	1.3096
51	10.56	19.07	0.3764	0.016023	0.1199	8.3403	62.410	0.9997	1.28	1.2796
52	11.11	20.07	0.3906	0.016023	0.1199	8.3403	62.410	0.9997	1.26	1.2596
53	11.67	21.07	0.4052	0.016026	0.1199	8.3403	62.400	0.9995	1.24	1.2394
54	12.22	22.07	0.4203	0.016026	0.1199	8.3403	62.400	0.9995	1.22	1.2194
55	12.78	23.07	0.4359	0.016028	0.1199	8.3403	62.390	0.9994	1.20	1.1992
56	13.33	24.06	0.4520	0.016028	0.1199	8.3403	62.390	0.9994	1.19	1.1892
57	13.89	25.06	0.4686	0.016028	0.1199	8.3403	62.390	0.9994	1.17	1.1693
58	14.44	26.06	0.4858	0.016031	0.1199	8.3403	62.380	0.9992	1.16	1.1591
59	15.00	27.06	0.5035	0.016031	0.1199	8.3403	62.380	0.9992	1.14	1.1391
60	15.56	28.06	0.5218	0.016033	0.1199	8.3403	62.370	0.9990	1.12	1.1189
62	16.67	30.05	0.5601	0.016036	0.12	8.3333	62.360	0.9989	1.09	1.0888
64	17.78	32.05	0.6009	0.016038	0.12	8.3333	62.350	0.9987	1.06	1.0586
66	18.89	34.05	0.6442	0.016044	0.12	8.3333	62.330	0.9984	1.03	1.0284
68	20.00	36.04	0.6903	0.016046	0.12	8.3333	62.320	0.9982	1	0.9982
70	21.11	38.05	0.7392	0.016049	0.1201	8.3264	62.310	0.9981	0.98	0.9781
75	23.89	43.03	0.8750	0.016059	0.1201	8.3264	62.270	0.9974	0.9	0.8977
80	26.67	48.04	1.0321	0.016072	0.1202	8.3195	62.220	0.9966	0.85	0.8471
85	29.44	53.00	1.2133	0.016085	0.1203	8.3126	62.170	0.9958	0.81	0.8066
90	32.22	58.02	1.4215	0.016098	0.1204	8.3056	62.120	0.9950	0.76	0.7562
95	35.00	62.98	1.6600	0.016113	0.1205	8.2988	62.060	0.9941	0.72	0.7157
100	37.78	68.00	1.9325	0.016129	0.1207	8.2850	62.000	0.9931	0.69	0.6853
110	43.33	77.98	2.5955	0.016134	0.1209	8.2713	61.980	0.9928	0.61	0.6056
120	48.89	87.97	3.4458	0.016205	0.1212	8.2508	61.710	0.9885	0.57	0.5634
130	54.44	97.96	4.5251	0.016244	0.1215	8.2305	61.560	0.9861	0.51	0.5029
140	60.00	107.95	5.8812	0.016292	0.1219	8.2034	61.380	0.9832	0.47	0.4621
150	65.56	117.95	7.5690	0.016343	0.1223	8.1766	61.190	0.9801	0.44	0.4313
160	71.11	127.96	9.6520	0.016395	0.1226	8.1566	60.990	0.9769	0.41	0.4005
170	76.67	137.97	12.1990	0.016450	0.1231	8.1235	60.790	0.9737	0.38	0.3700
180	82.22	148.00	15.2910	0.016510	0.1235	8.0972	60.570	0.9702	0.36	0.3493
190	87.78	158.04	19.014	0.016573	0.124	8.0645	60.340	0.9665	0.33	0.3190
200	93.33	168.09	23.467	0.016636	0.1245	8.0321	60.110	0.9628	0.31	0.2985
210	98.89	178.15	28.755	0.016706	0.125	8.0000	59.860	0.9588	0.29	0.2781

Temperatures 212°F – 400°F (Flc. Prop WaterRt)

Temp Deg. F	Temp Deg. C	Enthalpy Btu/lb	Saturation Pressure Psig	Specific Volume Cu Ft/lb	Specific Volume Gallon/lb	Density Pound/Gal	Density Lb./Cu Ft	Density Gr./Cu.Cm.	Kinematic Viscosity cS	Kinematic Viscosity CP
212	100.00	180.170	0.000	0.016720	0.1251	7.9936	59.810	0.9580		
214	101.11		0.593					0.958		
216	102.22		1.205					0.957		
218	103.33		1.837					0.956		
220	104.44	188.230	2.490	0.016776	0.1255	7.9681	59.610	0.9548		
222	105.56		3.165					0.954		
224	106.67		3.861					0.953		
226	107.78		4.579					0.952		
228	108.89		5.320					0.952		
230	110.00	198.330	6.095	0.016849	0.126	7.9365	59.350	0.9507		
232	111.11		6.871					0.95		
234	112.22		7.683					0.949		
236	113.33		8.521					0.948		
238	114.44		9.384					0.947		
240	115.56	208.450	10.272	0.016926	0.1266	7.8989	59.080	0.9463		
242	116.67		11.188					0.946		
244	117.78		12.131					0.945		
246	118.89		13.102					0.943		
248	120.00		14.101					0.943		
250	121.11	218.590	15.144	0.017007	0.1272	7.8616	58.800	0.9419	0.24	0.2260
252	122.22		16.188							
254	123.33		17.277							
256	124.44		18.397							
258	125.56		19.549							
260	126.67	228.760	20.731	0.017088	0.1278	7.8247	58.520	0.9374		
262	127.78		21.950							
264	128.89		23.201							
266	130.00		24.486							
268	131.11		25.806							
270	132.22	238.950	27.162	0.017176	0.1285	7.7821	58.220	0.9326		
272	133.33		28.556							
274	134.44		29.986							
276	135.56		31.454							
278	136.67		32.961							
280	137.78	249.170	34.504	0.017265	0.1291	7.7459	57.920	0.9278		
290	143.33	259.400	42.860	0.017361	0.1299	7.6982	57.600	0.9226		
300	148.89	269.700	52.309	0.017449	0.1305	7.6628	57.310	0.9180	0.2	0.1836
310	154.44	280.000	62.984	0.017550	0.1313	7.6161	56.980	0.9127		
320	160.00	290.400	74.947	0.017658	0.1321	7.5700	56.630	0.9071		
330	165.56		88.364	0.017759	0.1329	7.5245	56.310	0.9020		
340	171.11	311.30	103.296	0.017870	0.1337	7.4794	55.960	0.8964		
350	176.67			0.017989	0.1346	7.4294	55.590	0.8904	0.17	0.1514
360	182.22	332.30	138.31	0.018109	0.1355	7.3801	55.220	0.8845		
370	187.78			0.018235	0.1364	7.3314	54.840	0.8784		
380	193.33	353.60	181.03	0.018359	0.1374	7.2780	54.470	0.8725		
390	198.89			0.018501	0.1384	7.2254	54.050	0.8658		
400	204.44	375.10	232.56	0.018639	0.1394	7.1736	53.650	0.8594	0.15	0.1269

re: # 11

TABLE 5.2

BOILER HOUSE NOTES:			Page No:
Subject: Boiler Feedwater Btu			
Ralph L. Vandagriff			
Date: 6/10/98	Feedwater Temperature	Feedwater Sensible Heat (hf)	Feedwater Density Specific Weight
Deaerator Pressure, psig.	deg. F.	Btu/lb.	Pound per Gallon
0	212.0	180.2	7.9936
1	215.4	183.6	7.9828
2	218.5	186.8	7.9730
3	221.5	189.8	7.9635
4	224.5	192.7	7.9540
5	227.4	195.5	7.9448
6	230.0	198.1	7.9365
7	232.4	200.6	7.9272
8	234.8	203.1	7.9179
9	237.1	205.5	7.9090
10	239.4	207.9	7.9001
11	241.6	210.1	7.8923
12	243.7	212.3	7.8847
13	245.8	214.4	7.8772
14	247.9	216.4	7.8697
15	249.8	218.4	7.8629
16	251.7	220.3	7.8556
17	253.6	222.2	7.8483
18	255.4	224.0	7.8414
19	257.2	225.8	7.8345
20	258.8	227.5	7.8284

C. Specifications: Tray-Type Deaerator [*Source*: Kansas City Deaerator Company]

1. The deaerator shall be a vertical direct contact, tray-type unit designed for two-stage, counterflow operation welded to a horizontal storage such that deaeration is guaranteed over the full load range from 0 to 100% of specified capacity.

2. Design conditions are as follows:

   a. Total outlet capacity (lb/hr) _____

   b. Makeup water (lb/hr, min) _____
      Makeup water (lb/hr, max) _____
      Makeup water temperature (°F, min) _____
      Makeup water temperature (°F, max) _____

   c. Condensate return (lb/hr, min) _____
      Condensate return (lb/hr, max) _____
      Condensate return temp (°F, min) _____
      Condensate return temp (°F, max) _____

   d. High-temperature flashing drains

      lb/hr min _____
      lb/hr max _____
      Temperature (°F, min) _____
      Temperature (°F, max) _____

   e. Deaerator design pressure (psig) 30

   f. Deaerator operating pressure (psig) _____
      Steam will be supplied by user to maintain a saturation temperature corresponding to positive design operating pressure. _____

   g. Storage volume (gal) _____
      as measured below overflow level. Overflow level is not to exceed 90% of the storage volume.

   h. Minutes of operation w/ storage, approx. _____
      at rated capacity.

3. The first stage shall consist of a 304 stainless steel water box assembly containing spring-loaded, variable orifice spray valves mounted to spray into a stainless steel vent condensing chamber and containing a stainless steel vent pipe positioned to permit efficient venting to the atmosphere. Spray valves shall be cast from type 316 stainless steel and shall produce a hollow cone, thin-filmed spray pattern over the range of 5%–200% of rated valve capacity to assure rapid heating and stable venting. Valves shall operate with the use of Teflon guides to assure quiet operation.

4.   The second stage shall consist of tray assemblies housed in a tray enclosure constructed of carbon steel (*304 stainless steel optional*). The tray enclosure shall be closed on five sides to eliminate oxygen coming in contact with the carbon steel head/shell. A parallel flow is acceptable if the vessel is lined with stainless steel to protect the shell from traces of oxygen. Counterflow movement of water and steam shall be provided such that the water leaving the bottom layer of trays will be "stripped" by pure steam entering the heater, and such that corrosive gases do not contact the vessel heads or shell. Trays shall be type 430 stainless steel, not less than 16 gauge, and shall be assembled with stainless steel rivets, not stamped or welded.

5.   The deaerator shall be constructed in accord with the ASME Code, Section VIII, Division 1, for unfired pressure vessels and designed for 30-psig design pressure (*50 psig optional; full vacuum optional*). All steel plate shall be ASTM grade SA-516-70. A corrosion allowance of at least 1/16 in. (*1/8 in. optional*) shall be included over the ASME calculated thicknesses.

6.   Access to deaerator internals and storage shall be through an 18-in.–diameter access door, and an adequately sized manhole shall be provided for storage vessel access.

7.   The deaerator shall include heavy-duty saddles that provide continuous support over a contact arc of not less than 120 degrees of storage vessel circumference.

8.   All connections necessary to accommodate piping and specified accessories shall be included so as to comprise a complete, working unit, NPT couplings shall be provided for connections 2 1/2 in. and smaller. Connections above 3-in. connections will be pad flange, flanged pipe, or NPT coupling. All interconnecting piping and bridles will be furnished by purchaser. The deaerator shall be shipped as complete as possible. Final piping assembly shall be furnished by the purchaser.

9.   The stainless steel vent condensing system shall provide for the efficient release of oxygen and carbon dioxide with minimum loss of steam, and shall be complete with stainless steel vent pipe.

10.   The deaerator shall be guaranteed.

a.   To deliver specific capacity at the saturated steam temperature corresponding to the steam pressure within the heater.

b.   To deliver specified capacity with an oxygen content not to exceed 0.005 cc/L as determined by the Winkler test, or equal, and with a free carbon dioxide content of zero as determined by the APHA test, one time through without recirculating pumps.

c.   To be free from defects in material and workmanship for a period of 1 year from date of initial operating or 18 months from date of shipment, whichever occurs first.

11. *Optional*: The deaerator shall be in accordance with the *Heat Exchange Institute* 1992, 5th ed., and shall include the following to assure a safe, reliable vessel.

- 1/8-in. corrosion allowance
- Full penetration welds
- Post weld heat treatment
- Wet fluorescent magnetic particle inspection
- Stainless steel inlet and water box
- Allowable nozzle velocity

12. *Optional*: The following accessories shall be provided;

One (1) high-capacity pressure relief valve (full relief by others). Set at vessel design pressure.

One (1) vacuum breaker valve.

One (1) water inlet regulating valve to pass___lb/hr at inlet pressure from___psi to___psi (*optional by-pass*).

One (1) internal level controller w/pneumatic actuator. (*Optional—two [2] mechanical float switches for low [and high] water level. SPDT Nema 4 rated.*)

One (1) mechanical overflow tray pr valve with float control.

One (1) steam pressure reducing valve (PRV). To reduce available steam pressure from___psig to deaerator operating pressure. (*optional by-pass*).

One (1) dial pressure gauge complete with a siphon and cock.

One (1) stainless steel trim vent valve.

Two (2) separable socket thermometers.

13. *Optional*: When a *Package unit* is requested, the following additional components shall be included. The package assemble shall be factory preassembled and shall be broken down for shipment due to clearances.

a. Structural stand build per OSHA. Standard is prefit up to match up with deaerator. The stand will have proper anchor bolt holes for attachment to site foundation.

b. Boiler feed pump is sized to accommodate 100% of rated capacity with low NPSH required. The pump will be attached to the pump skid that is welded to the structural stand. A quantity of___pump(s) is required for a capacity of___GPM each (0.002099 × pph = GPM). Discharge pressure to boiler is___psig. The pump is a centrifugal type and have seals and materials suitable for continuous operation up to 250°F (*300°F optional*). Each pump is driven by a___HP, 3600 RPM,___volt, 3-phase, open drip proof (ODP) motor (*optional:*

TEFC). Pumps shall have low NPSH requirements which shall not be exceeded at any time during operation.

c. Boiler feed pump suction piping is included. The suction piping will consist of a gate valve, Y strainer, and expansion joint. The piping will be fit up between the deaerator and the boiler feed pump(s), if shipment clearances allow. Boiler feed pump recirculation piping is to be supplied with isolation valves and orifice plates.

d. A control panel is to be furnished completely wired, tested, and mounted on the stand. The enclosure is Nema 4 (*Nema 12 optional*). Internals included flange mounted main breaker, fused control power transformer, motor fuse protection, motor starters with overload heaters, hand-off autoselection switch and "on" pilot lights. Panel is capable of housing up to four pump starters.

e. *Optional*: A bridle piping assembly is supplied prefit up to the deaerator tanks bridle connections. The bridle can be tapped for any customer usage.

14. Any deviations from, or exceptions to, the above specifications must be clearly stated in bid; otherwise, bidder will be expected to deliver equipment exactly as specified.

D. Specifications: Spray-Type Deaerator [*Source*: Kansas City Deaerator Company]

1. The deaerator shall be a horizontal, direct contact, spray-type unit with integral storage, and shall be designed for two-stage operation such that deaeration is guaranteed over the full load range without recirculation pumps. The deaerator shall have a stainless steel liner in the spary area to reduce oxygen pitting of the unit in the area of concentrated oxygen.

2. Design conditions are as follows:

a. Total outlet capacity (lb/hr)   _____

b. Makeup water (lb/hr, min)   _____
   Makeup water (lb/hr, max)   _____
   Makeup water temperature (°F, min)   _____
   Makeup water temperature (°F, max)   _____

c. Condensate return (lb/hr, min)   _____
   Condensate return (lb/hr, max)   _____
   Condensate return temp (°F, min)   _____
   Condensate return temp (°F, max)   _____

d. High-temperature flashing drains

   lb/hr min   _____
   lb/hr max   _____

Temperature (°F, min)   _____

Temperature (°F, max)   _____

e. Deaerator design pressure (psig)   30

f. Deaerator operating pressure (psig.)   2–5 psig
  Steam will be supplied by user so as to maintain a saturation temperature corresponding to positive design
  operating pressure.

g. Storage volume, gallons: _____
  as measured below overflow level. Overflow level not
  to exceed 75% of the tank diameter.

h. Minutes of operation w/storage, approx. 10.0 (15.0
  min optional) at rated capacity.

3. The first stage shall consist of a 304 stainless steel water box assembly containing spring-loaded, variable orifice spray valves mounted to spray into the vent condensing chamber and containing a stainless steel vent pipe positioned to permit efficient venting to the atmosphere. Spray valves shall be cast from type 316 stainless steel and shall produce a hollow cone, thin-filmed spray pattern over the range of 5–200% of rated valve capacity to assure rapid heating and stable venting. Valves shall operate with the use of Teflon guides to assure quiet operation. A heavy-gauge collection basin shall be provided to accumulate pre-heated water for entry into the scrubber.

4. The second stage shall consist of a deaerating steam scrubber, receiving preheated water from a collection basin downcomer, and receiving pure steam from a variable area steam distributor, such that a vigorous scrubbing action occurs to strip the water of its final traces of oxygen and carbon dioxide. Scrubber outlet shall be positioned to permit exiting steam to flow into the vent condensing chamber for stage 1 heating, deaerating, and venting.

5. The deaerator shall be constructed in accord with the ASME Code, Section VIII, Division 1, for unfired pressure vessels and designed for 30-psig–design pressure (*50 psig optional; full vacuum optional*). All steel plate shall be ASTM grade SA-516-70. A corrosion allowance of at least 1/16 in. (1/8 in. optional) shall be included over the ASME calculated thicknesses.

6. Access to deaerator internals and storage shall be provided through a 12″ × 16″ manhole.

7. The deaerator shall include heavy-duty saddles that provide continuous support over a contact arc of not less than 120 degrees of storage vessel circumference.

8. All connections necessary to accommodate piping and specified accessories shall be included such as to comprise a complete, working unit, NPT couplings shall be provided for connections 2 1/2 in. and smaller. Connections above 3 in. connections will be pad flange, flanged pipe, or NPT coupling. All interconnecting piping and bridles will be furnished by purchaser. The deaerator shall be

shipped as complete as possible. Final piping assembly shall be furnished by the contractor.

9. The vent condensing system shall provide for the efficient release of oxygen and carbon dioxide with minimum loss of steam, and shall be complete with stainless steel vent pipe.

10. The deaerator shall be guaranteed

    a. To deliver specified capacity at the saturated steam temperature corresponding to the steam pressure within the heater.

    b. To deliver specified capacity with an oxygen content not to exceed 0.005 cc/L as determined by the Winkler test, or equal, and with a free carbon dioxide content of zero as determined by the APHA test, one time through without recirculating pumps.

    c. To be free from defects in material and workmanship for a period of 1 year from date of initial operating or 18 months from date of shipment, whichever occurs first.

11. *Optional*: The deaerator shall be in accordance with the *Heat Exchange Institute 1992*, 5th Ed. and shall include the following to assure a safe, reliable vessel.

- 1/8 in. corrosion allowance
- Full penetration welds
- Post weld heat treatment
- Wet fluorescent magnetic particle inspection
- Stainless steel inlet and water box
- Allowable nozzle velocity

12. *Optional*: The following accessories shall be provided;

One (1) high-capacity pressure relief valve (full relief by others). Set at vessel design pressure.

One (1) vacuum breaker valve.

One (1) water inlet regulating valve to pass_____lb/hr at inlet pressure from_____psi to_____psi (*optional by-pass*).

One (1) internal level controller w/pneumatic actuator. (*Optional: two [2] mechanical float switches for low [and high] water level. SPDT Nema 4 rated.*)

One (1) mechanical overflow tray pr valve with float control.

One (1) steam pressure reducing valve (PRV). To reduce available steam pressure from—psig to deaerator operating pressure (*Optional by-pass*).

One (1) dial pressure gauge complete with a siphon and cock.

One (1) stainless steel trim vent valve.

One (1) separable socket thermometer.

13. *OPTIONAL*: When a *Package unit* is requested, the following additional components shall be included. The package assemble shall be factory pre-assembled and shall be broken down for shipment due to clearances.

    a. Structural stand build per OSHA. Stand is prefit up to match up with deaerator. The stand will have proper anchor bolt holes for attachment to site foundation.

    b. Boiler feed pump is sized to accommodate 100% of rated capacity with low NPSH required. The pump will be attached to the pump skid which is welded to the structural stand. A quantity of___pump(s) is required for a capacity of___GPM each ($0.002099 \times$ pph = GPM). Discharge pressure to boiler is___psig. The pump is a centrifugal type and have seals and materials suitable for continuous operation up to 250°F (*300°F optional*). Each pump is driven by a___HP, 3600 RPM,___volt, 3-phase open drip proof (ODP) motor (*optional: TEFC*). Pumps shall have low NPSH requirements that shall not be exceeded at any time during operation.

    c. Boiler feed pump suction piping is included. The suction piping will consist of a gate valve, Y-strainer and expansion joint. The piping will be fit up between the deaerator and the boiler feed pump(s), if shipment clearances allow. Boiler feed pump recirculation piping is to be supplied with isolation valves and orifice plates.

    d. A control panel is to be furnished completely wired, tested, and mounted on the stand. The enclosure is Nema 4 (*Nema 12 optional*). Internals included flange mounted main breaker, fused control power transformer, motor fuse protection, motor starters with overload heaters, hand-off autoselection switch and "on" pilot lights. Panel is capable of housing up to four pump starters.

    e. *Optional*: A bridle piping assembly is supplied prefit up to the deaerator tanks bridle connections. The bridle can be tapped for any customer usage.

14. Any deviations from, or exceptions to, the above specifications must be clearly stated in bid; otherwise, bidder will be expected to deliver equipment exactly as specified.

6

Boiler Feedwater Pumps

Questions and Answers; Selection of Boiler Feed Pumps; Boiler Water Level in
Steel Drum; Steel Drum Water Level Control.

I. QUESTIONS AND ANSWERS

A. Boiler Feed Pumps

Question. What is a boiler feed pump?

Answer. It is used in steam-generating power plants to deliver feedwater
to the boiler. Depending on the feed cycle, the boiler feed pump may take its
suction from a condensate pump discharge, a deaerating heater, or, in small
plants, directly from the makeup source external to the feed cycle. A boiler feed
pump generally handles water at a temperature of 212°F or higher.

B. Suction Conditions

Question. How do suction conditions affect pump design?

Answer. To cause liquid flow into the impeller of a centrifugal pump, an
outside source of pressure must be provided. This may be static head when the
suction source level is above the pump center line, atmospheric pressure when
the suction source is below the pump center line, or both. When the net absolute
pressure above the vapor pressure of the liquid is small, the pump suction water
passages must be comparatively large to keep the velocities in them down to a
low value. If this is not done, the absolute pressure at the suction eye of the
impeller may drop below the vapor pressure and the pumped liquid will flash
into vapor, preventing further pumping. When abnormal suction conditions exist,
an oversized or special pump is generally required, which costs more and has a
lower speed and efficiency than a pump of equal capacity and head for normal
suction conditions.

C. Vapor Pressure of Water

Question.   What is the vapor pressure of water at 212°F, expressed in feet of water?

Answer.   Because the vapor pressure of water at 212°F is 14.7 psia (standard barometric pressure at sea level), the equivalent head in feet of 62°F water is 14.7 × 2.31 = 33.9 ft. The specific gravity of water at 212°F is 0.959. Then water at this temperature has a vapor pressure of 33.9 ÷ 0.959 = 35.4 ft. of 212°F water.

When figuring pump heads, care must be taken to convert pressures to feet of liquid at the pumping temperature and not to use conversion factors applying to other temperatures.

D. Effect of Water Temperature on Pump Brake Horsepower

Question.   I have been told that the power consumption of a boiler feed pump increases as the feedwater temperature increases. How can this be, since the gravity of water decreases with higher temperature and it should take less power to handle a lighter fluid?

Answer.   Your informants are quite correct, and the power consumption of a feed pump does increase exactly in an inverse ratio of the specific gravity (sp gr) of the feedwater it handles. The apparent paradox arises because a boiler feed pump must be selected to handle a given weight of feedwater of so many pounds per hour and to develop a certain pressure in pounds per square inch rather than a total head of so many feet. The formula for brake horsepower (bhp) is

$$bhp = \frac{gpm \times head~in~feet \times specific~gravity}{3960 \times efficiency}$$

Thus, as long as a fixed volume in gallons per minute is pumped against a fixed head in feet, the bhp will decrease with specific gravity. However, if we convert pounds per hour into gallons per minute and pounds per square inch pressure into feet of head, we find that the relation with specific gravity changes.

$$gpm = \frac{lb/hr}{500 \times sp~gr} \quad and \quad head~in~feet = \frac{psi \times 2.31}{sp~gr}$$

and, therefore,

$$bhp = \frac{lb/hr/(500 \times sp~gr) \times (psi \times 2.31)/sp~gr \times sp~gr}{3960 \times efficiency}$$

By simplifying this relation, we obtain

$$bhp = \frac{lb/hr \times psi}{857,000 \times efficiency} \times \frac{1}{sp\ gr}$$

And we see that, as the temperature increases and the specific gravity decreases, there is an increase in power consumption.

E. Effect of Changing Heater Pressure on NPSH

Question. Our boiler feed pumps take their suction from a deaerating heater operating at 5 psig. We would like to revamp this installation and operate the heater at 15 psig so we could tie it into our process steam line. The heater is suitable for this pressure, but we are concerned over the boiler feed pump requirements. Will the heater have to be raised to compensate for the higher temperature water that the pump will handle?

Answer. Assuming that the boiler feed pumps are now operating satisfactorily and the existing submergence is sufficient, there will be no need to alter the installation. A centrifugal pump requires a certain amount of energy in excess of the vapor pressure of the liquid pumped to cause flow into the impeller. By definition, the net positive suction head (NPSH) represents this net energy referred to the pump center line, *over and above* the vapor pressure of the liquid. This definition makes the NPSH automatically independent of any variations in temperature and vapor pressure of the feedwater.

The feedwater in the storage space of a direct-contact heater from which the boiler feed pump takes its suction is under a pressure corresponding to its temperature. Therefore, the energy available at the first-stage impeller, over and above the vapor pressure, is the static submergence between the water level in the storage space and the pump center line less the friction losses in the suction piping.

The method of determining the available NPSH as well as the total suction pressure illustrates that whether the heater pressure is 5 or 15 psig, the available NPSH will not change. The total suction pressure, on the other hand, will be increased by the change in heater pressure.

F. Cleaning Boiler Feed Pump Suction Lines

Question. We have experienced seizures of boiler feed pumps shortly after the initial start-up and traced these difficulties to the presence of foreign matter in the lines. This foreign matter apparently gets into the clearance spaces in the pump and damages the pump to a considerable extent. What special precautions are recommended to avoid such difficulties?

Answer. Boiler feed pumps have internal running clearances from 0.020 in. to as low as 0.012 in. on the diameter (that is, from 0.010 to 0.006 in. radially), and it is obvious that small particles of foreign matter, such as mill scale, left in the piping or brittle oxides can cause severe damage should they get into these clearances. Incidentally, it has been the general experience that an actual seizure does not occur while the pump is running, but rather as it is brought down to rest. But since boiler feed pumps are frequently started and stopped during the initial plant start-up period, seizures are very likely to occur if foreign matter is present.

The actual method used in cleaning the condensate lines and the boiler feed-pump suction piping varies considerable in different installations. But the essential ingredient is always the use of a temporary strainer located at a strategic point. Generally, the cleaning out starts with a very thorough flushing of the condenser and deaerating heater, if such is used in the feedwater cycle. It is preferable to flush all the piping to waste before finally connecting the boiler feed pumps. If possible, hot water should be used in the latter flushing operation, as additional dirt and mill scale can be loosened at higher temperatures. Some installations use a hot phosphate and caustic solution for this purpose.

Temporary screens or strainers of appropriate size must be installed in the suction line as close to the pump as possible. It is difficult to decide what constitutes sound practice in choosing the size of the openings. If 8-mesh screening is used and assuming that 0.025-in. wire is used, the openings are 0.100 in., and that is too coarse to remove particles large enough to cause difficulties at the pump clearances, which may be from 0.006 to 0.01 in. radially. If there is an appreciable quantity of finely divided solids present, and if the pump is stationary during flushing, some solids would be likely to pack into the clearances and cause damage when the pump is started.

The safest solution consists of using a strainer with 40–60 mesh and flushing with the pump stationary until the strainer remains essentially clean for a half day or longer. After that, a somewhat coarser mesh can be used if it is necessary to permit circulation at a higher rate. But it is very important that the pumps be turned by hand both before and after flushing to check whether any foreign matter has washed into the clearances. If the pump "drags" after flushing, it must be cleared before it is operated.

Unless the system is thoroughly flushed before the pump is started, the use of a fine-mesh screen may cause trouble. For instance, 40-mesh screening with 0.015-in. wire leaves only 0.010-in. openings, and these would clog up instantly unless a very thorough cleaning job was done initially. Pressure gauges must be installed both upstream and downstream from the screen, and the pressure drop across it watched most carefully. As soon as dirt begins to build up on the screen and the pressure drop starts to climb, the pump should be stopped and the screen cleaned out.

If the size of the free-straining area is properly selected to minimize the pressure drop, the start-up strainer may be left in the line for a considerable period before the internal screen is removed. Alternatively, the entire unit may be removed and replaced with a spool piece. The strainer is then available for the next start-up.

G. Common Recirculation Line for Several Boiler Feed Pumps

Question. We are planning an installation of three boiler feed pumps of which two are intended to run at full load and the third is a spare. Must each pump be provided with its own by-pass recirculation line, pressure-reducing orifice, and by-pass control valve, or can a single common line, orifice, and valve be used for the protection of the boiler feed pumps?

Answer. Decidedly, each pump requires its own recirculation line and controls. There are several reasons for this:

1. If a common recirculation system is used, the danger arises that, when the flow is reduced to nearly the minimum, one of the two pumps operating may develop a slight excess of discharge pressure and shut the check valve of the other pump, allowing it to run against a fully closed discharge with no by-pass.
2. Although two pumps normally operate to carry full load, there will be times when a single pump will be running at loads below 50%. At other times, when pumps are being switched, all three pumps may be running for short periods. If the orifice capacity was selected to pass a flow equal to twice the minimum flow of a single pump, the recirculation would be twice that which is necessary whenever one pump was running alone and only two-thirds of that required whenever all three pumps were on the line. The first is wasteful, and the third does not afford the necessary protection.

Separate by-pass recirculation lines should be provided, origination between the pumps and their check valves. Each line must have its own orifice and its own control valve. These individual lines can be manifolded into a single return header to the deaerator on the downstream side of the pressure-reducing orifices and control valves.

H. Use of Cast-Steel Casings for Boiler Feed Pumps

Question. We have recently issued specifications for three boiler feed pumps designed for 750-psig–discharge pressure. In our desire to purchase equipment with something better than cast iron for the casing material, we specified that pump casings were to be made of cast carbon steel. We were much surprised when several bidders refused to quote cast-steel casings and suggested that we choose between cast iron and a 5% chrome steel. We had heard that chrome steels are used for the higher-pressure range in boiler feed pumps. Can you tell us why cast steel may be unsuitable? Can you also suggest what clues we may look for in our existing installation to determine whether we need to go to more expensive materials than cast iron casings and standard bronze fittings?

Answer. The mechanics of corrosion–erosion attack in boiler feed pumps first became the subject of considerable attention in the early 1940s, when it became desirable to use feedwater of a scale-free character. This desire led to the use of lower pH values and to the elimination of various mineral salts that, theretofore, had acted as buffering agents. As you state, the practice was instituted to use chrome steels throughout the construction of high-pressure boiler feed pumps. But this does not mean that limiting this practice to high-pressure pumps only is justifiable. It is true that the feedwaters used in the lower-pressure plants may not necessarily undergo the same degree of purification. Nevertheless, cases have been brought to my attention in which evidence of corrosion–erosion occurred in feed pumps operating at pressures as low as 325 psi.

The exact nature of the attack and the causes leading to it are not, in my opinion, fully understood, as minor variations in the character of the feedwater or in its pH seem to produce major variations in the results. At the same time, certain very definite facts have become established:

1. If not necessarily the cause, at least it is recognized that low pH is an indicator of potential corrosion–erosion phenomena.
2. Feedwaters that coat the interior of the pump with red or brownish oxides (Fe_2O_3) generally do not lead to such trouble.
3. If interior parts are coated with black oxide (Fe_3O_4), severe corrosion–erosion may well be expected.
4. When feedwaters are corrosive, cast iron seems to withstand the corrosion infinitely better than plain carbon steel. Chrome steels, however, with a chromium content of 5% or higher, withstand the action of any feedwater condition so far encountered. Some manufacturers have a preference for 13% chrome steels for the impellers, wearing rings, and other pump parts other than the casing.

If you install boiler feed pumps that are not fully stainless steel fitted, it is important to carry out frequent tests of the pump performance. This step will help you avoid sudden interruption of service. Corrosion–erosion attack comes on rather unexpectedly, and its deteriorating effects are very rapid once the attack has started. If protective scale formation is absent, the products of corrosion are washed away very rapidly, constantly exposing virgin metal to the attack from the feedwater.

Thus, if the original pump capacity is liberally selected, there may be no indication that anything is wrong until such time that deterioration of metal has progressed to the point that the original margin has been "eaten up." The resulting breakdown immediately assumes the proportions of an emergency, because the net available capacity is no longer sufficient to feed the boiler. Unless additional spare equipment is available, the power plant operator may find himself in an unenviable spot.

To avoid such an unforseen emergency, it is recommended that complete tests of the pump performance be carried out at, say, not less than 3-month intervals if the pump is built of materials that may be subject to corrosion–erosion attack. From a more constructive point of view, it is wise to investigate the effect of the feedwater used on the materials in the pump in question. If any indication exists that these materials may be inadequate, replacement parts of stainless steel should be ordered. If it is intended to replace only the internal parts by stainless steel and to retain the original cast iron casing, the replacement program should be carried out at the first opportunity, rather than waiting to the end of the useful life of the original parts. Otherwise, the deterioration of parts that form a fit with the casing may lead to internal leakage which, in turn, will cause the destruction of casing fits and will make ultimate repairs extremely costly.

But whatever your decision is on the internal parts of the pumps you contemplate to purchase, I earnestly recommend that you *do not* specify cast steel casings. Too many sad experiences have been traced to their use.

I. Cavitation

Question. What is cavitation?

Answer. "Cavitation" describes a cycle of phenomena that occur in flowing liquid because the pressure falls below the vapor pressure of the liquid. When this occurs, liquid vapors are released in the low-pressure area and a bubble or bubbles form. If this happens at the inlet to a centrifugal pump, the bubbles are carried into the impeller to a region of high pressure, where they suddenly collapse. The formation of these bubbles in a low-pressure area and their sudden collapse later in a high-pressure region is called cavitation. Erroneously, the word is frequently used to designate the effects of cavitation, rather than the phenomenon itself.

Question. In what form does cavitation manifest itself in a centrifugal pump?

Answer. The usual symptoms are noise and vibration in the pump, a drop in head and capacity with a decrease in efficiency, accompanied by pitting and corrosion of the impeller vanes. The pitting is physical effect that is produced by the tremendous localized compression stresses caused by the collapse of the bubbles. Corrosion follows the liberation of oxygen and other gases originally in solution in the liquid.

J. Standby Pump and Problem Symptoms

Question. How is a standby boiler feed pump held ready for operation?

Answer. It is held ready with suction and discharge gate valves open and the discharge check valve closed to prevent reverse flow through the pump. To maintain the idle pump at near operating temperature, feed water flows through it from the open suction through the warm-up valve between pump and discharge check valve. To avoid wasting feedwater, the warm-up valve drains return to the feed cycle at a lower pressure point than the feed pump suction. In an emergency, a cold pump may be put in operation without warming up. However, as a rule, it should by heated for about 30 min before starting.

Question. Is there a quick rule for determining minimum permissible capacity of a boiler feed pump and, thus, the necessary bypass?

Answer. To limit the temperature rise in a boiler feed pump to 15°F, do not reduce its capacity below 30 gpm for each 100-hp input to the pump at shutoff.

Question. What are the most common symptoms of troubles and what do they indicate?

Answer. Symptoms may be hydraulic or mechanical. In the hydraulic group, a pump may fail to discharge, or it may develop insufficient capacity or pressure, lose its prime after starting, or take excessive power. Mechanical symptoms may show up at the stuffing boxes and bearings or in pump vibration, noise, or overheating. See the following chart for symptoms of pump troubles and possible causes.

Symptoms	Possible cause of trouble[a]
Pump does not deliver liquid	1,2,3,4,6,11,14,16,17,22,23
Insufficient capacity delivered	2,3,4,5,6,7,8,9,10,11,14,17,20,22,23,29,30,31
Insufficient pressure developed	5,14,16,17,20,22,29,30,31
Pump loses prime after starting	2,3,5,6,7,8,11,12,13
Pump requires excessive power	15,16,17,18,19,20,23,24,26,27,29,33,34,37
Stuffing box leaks excessively	13,24,26,32,33,34,35,36,38,39,40
Packing has short life	12,13,24,26,28,32,33,34,35,36,37,38,39,40
Pump vibrates or is noisy	2,3,4,9,10,11,21,23,24,25,26,27,28,30,35,36, 41,42,43,44,45,46,47
Bearings have short life	24,26,27,28,35,36,41,42,43,44,45,46,47
Pump overheats and seizes	1,4,21,22,24,27,28,35,36,41

[a] See following numerical list of reasons.

Forty-Seven Possible Causes of Trouble

Suction Troubles
1. Pump not primed
2. Pump or suction pipe not completely filled with liquid
3. Suction lift too high
4. Insufficient margin between suction pressure and vapor pressure
5. Excessive amount of air or gas in liquid
6. Air pocket in suction line
7. Air leaks into suction line
8. Air leaks into pump through stuffing boxes
9. Foot valve too small
10. Foot valve partially clogged
11. Inlet of suction pipe insufficiently submerged
12. Water-seal pipe plugged
13. Seal cage improperly located in stuffing box, preventing sealing fluid from entering space to form the seal

System Troubles
14. Speed too low
15. Speed too high
16. Wrong direction of rotation
17. Total head of system higher than design head of pump
18. Total head of system lower than pump design head
19. Specific gravity of liquid different from design
20. Viscosity of liquid differs from that for which designed
21. Operation at very low capacity
22. Parallel operation of pumps unsuitable for such operation

Mechanical Troubles
23. Foreign matter in impeller
24. Misalignment
25. Foundations not rigid
26. Shaft bent
27. Rotating part rubbing on stationary part
28. Bearings worn
29. Wearing rings worn
30. Impeller damaged
31. Casing gasket defective, permitting internal leakage
32. Shaft or shaft sleeves worn or scored at the packing
33. Packing improperly installed
34. Incorrect type of packing for operating conditions
35. Shaft running off-center because of worn bearings or misalignment
36. Rotor out of balance, resulting in vibration
37. Gland too tight, resulting in no flow of liquid to lubricate packing
38. Failure to provide cooling liquid to water-cooled stuffing boxes
39. Excessive clearance at bottom of stuffing box between shaft and casing, causing packing to be forced into pump interior

40. Dirt or grit in sealing liquid, leading to scoring of shaft or shaft sleeve
41. Excessive thrust caused by a mechanical failure inside the pump or by the failure of the hydraulic balancing device, if any
42. Excessive grease or oil in antifriction bearing housing or lack of cooling, causing excessive bearing temperature
43. Lack of lubrication
44. Improper installation of antifriction bearings (damage during assembly, incorrect assembly of stacked bearings, use of unmatched bearings as a pair, etc.)
45. Dirt getting into bearings
46. Rusting of bearings due to water getting into housing
47. Excessive cooling of water-cooled bearing, resulting in condensation in the bearing housing of moisture from the atmosphere.

Source: Refs 21 and 22.

II. SELECTION OF BOILER FEED PUMPS [23,24]

How do you go about selecting a boiler feed pump? There is no easy answer. There is no simple way to "play it safe." In fact, playing it safe and arbitrary use of safety factors can make a bad problem worse.

A boiler feed pump cannot be properly selected and applied without an intelligent and informed analysis of the boiler feedwater system. Feed pumps are a relatively small part of a boiler package. They can easily be taken for granted, even by a person who has been designing or operating boilers for years. The boiler feed pump must fit in with the complete boiler system. In the great majority of cases, any changes in the design or operation of the system are repaid in the price of a less-expensive pump alone, to say nothing of operating and maintenance savings.

All centrifugal pumps are designed to operate on liquids. Whenever they are used on mixtures of liquid and vapor or air, shortened rotating element life can be expected. If the liquid is high temperature or if boiler feedwater with vapor (steam) is present, rapid destruction of the casing can also occur. This casing damage is commonly called *wire drawing* and is identified by worm-like holes in the casing at the parting which allow liquid to bypass behind the diaphragms or casing wearing rings. Whenever wire drawing is detected, an immediate check of the entire suction system must be made to eliminate the source of vapor. Vapor may be present in high-temperature water for several reasons. The available net positive suction head (NPSH) may be inadequate, resulting in partial or serious cavitation at the first-stage impeller and formation of some free vapor. The pump may be required to operate with no flow, resulting in a rapid temperature rise within the pump to above the flash point of the liquid, unless a proper bypass line with orifice is connected and is open. (This can also cause seizure of the rotating element.) The submergence over the entrance into the suction line may be inadequate, resulting in vortex formation and entrainment of vapor of air. When a pump becomes vapor-bound or loses its prime, a multistage pump becomes unbalanced and exerts a maximum thrust load on the thrust bearing. This frequently results in bearing failure; if it is not detected immediately, it may ruin the entire rotating element because of the metal-to-metal contact when the rotor shifts and probable seizure in at least one place of the pump.

Boiler feed is a demanding pump service, and careful pump selection is required. Some of the factors that must be carefully considered follow.

A. Rating

Boiler feed pumps are commonly selected to handle the flow required by the boiler under maximum firing rate, and to this, a generous factor of safety is added. The maximum firing rate is needed for a few days in the dead of winter, when the plant is running full blast. On weekends and during the summer, the pump may be running at one-fourth, or less, of capacity. Sizing for a future boiler does not help this situation.

Boilers are rated for both steam flow and pressure—for example, 60,000 lb/hr at 150 psig saturated steam production, or 200,000 lb/hr at 600 psig and 750°F superheat. The steam flow can be converted to equivalent water requirement as follows:

$$\text{gpm} = \frac{0.002 \times \text{steam flow (lb/hr)}}{\text{sp gr of boiler feedwater}}$$

Example. A 60,000 lb/hr boiler being fed 220°F feedwater (0.955 sp gr) would require:

$$\text{gpm} = \frac{0.002 \times 60,000}{0.955} = 126 \text{ gal/min (gpm) boiler feedwater}$$

The rated steam flow tells how much steam the boiler can make when it is being fired to full capacity. The boiler feed pump flow is then commonly sized to pump an equivalent amount of feedwater into the boiler. A safety factor of 10–20% is added to take care of fluctuations in the boiler water level, pump wear, and such. This flow is the absolute maximum that can possibly occur.

But, boilers are rarely operated at full load. A boiler sized for heating a building on a cold winter day may operate at 50% capacity during the spring and fall. Similarly, a process steam boiler may operate at part load on a night shift, and at essentially no load on a weekend. In multiple boiler installations, common practice is to bring a second boiler on line when the load on one boiler reaches 80% of rating.

Part-load boiler operation must be considered if a proper fit between boiler and boiler feed pump is to be obtained. The least pump cost is obtained by sizing the pump for the normal steam flow, not the boiler rating. This might require, for example, one pump rated at 75% of boiler capacity, and one rated at 25%. The largest pump would handle normal loads. The smaller pump would handle light weekend and summer loads; and in parallel with the larger pump, handle the peak loads occurring a few times a year. The 75:25% split is obviously only an example—each application must be individually studied. For an application to be considered satisfactory, the pump must operate within the efficiency lines on the CDS when the boiler is operating at *normal* (not maximum) steam flow.

The head that is to be developed by the feed pump must also be checked. The head must be high enough to pump against the boiler operating pressure as well as the friction in the boiler feedwater control (or controls—there may be two in series). For instance, a 300-psig (725 ft of head) discharge pressure would be required for a boiler operating at 200 psig and had two feedwater controls, each with a pressure drop of 50 psig. However, consider that the boiler in the foregoing example might be rated at 200 psig, but would always be operated at 150 psig; and the actual pressure drop through each feedwater control might be only 25 psig. The pump would only need to develop 200 psig (485 ft of head).

B. NPSH

A boiler feed pump normally takes suction from a deaerator or deaerating heater. A deaerator is a closed vessel in which the feedwater is heated by direct contact with steam to remove air that could cause corrosion in the boiler. In a properly operated deaerator, the water is heated to boiling. Vapor pressure (VP) is equal to deaerator pressure (P), and the two cancel. $NPSH_A$ is then equal to the static level (L_H) minus friction (H_f). Unfortunately, the steam pressure in the deaerator may fluctuate. If it drops 1 psi or about 2.5 ft of water, there are immediately 2.5 ft less NPSH available. These fluctuations must be considered in computing the NPSH available to the pump.

If the NPSH available in the system is less than the NPSH required by the pump, the pump will cavitate. The effects of cavitation can range from a rumbling noise in mild cases, to impeller damage, thrust bearing failure, and internal seizures at the wearing rings. The consequences of cavitation are so well known that there is a tendency to use excessive safety factors. This will often take the form both of understating the available NPSH, and stating that this NPSH be met at higher rates of flow that would be encountered. These requirements can be met only with an oversized pump. The excessive initial, operating and maintenance costs of the oversized pumps are not as dramatic as cavitation and tend to be overlooked. Either understanding or overstating the available NPSH can cause serious problems. Accurate computation of available NPSH is essential to proper pump selection.

NPSH problems are becoming more frequent by a trend toward setting the deaerator as low as possible, especially on newer installations and on package units. The more compact unit obtained may be somewhat less expensive to build and install. However, this approach also decreases the NPSH available. It becomes particularly serious for larger units, beyond about 300 gpm, where NPSH required values at the BEP of 15–20 ft are required. The savings from making the deaerator package compact are small, are offset by the costs of a larger, oversized pump with low efficiency and higher power costs.

The more compact the deaerator, the less NPSH will be available.

C. Parallel Pump and/or Boiler Operation

Boilers and pumps are often interconnected so that one or more pumps can feed one or more boilers. This allows greater flexibility of operation and reduces the need for installed spare equipment. However, one of two things often happen. The largest pump, selected to feed two or three boilers, is operating to feed one boiler. Once again, a pump is being run continuously at half or quarter flow. Other times two pumps with different characteristics are operated in parallel, with the result that one pump is backed off the line and runs for long periods at shutoff.

D. Bypass Orifices

Boiler controls can throttle a pump so that it will run at shutoff for short periods. To prevent overheating, an orifice is installed to continuously bypass a small amount of water. However, consider a boiler feed pump sold for 100-gpm, 650-ft discharge head with 5-ft NPSH available. The pump needs 5-ft NPSH at all points from shutoff to 100 gpm. The water in the pump is just at the flash point. *Any* temperature rise will allow the water to flash, and put the pump into cavitation. The pump may seize or destroy a bearing. To protect against this, pump manufacturers required the NPSH available be at least 1 ft greater than that required by the pump at shutoff.

III. BOILER WATER LEVEL IN STEAM DRUM [25]

A. Shrink and Swell and Boiler Water Circulation

When the steam load on a drum-type boiler is increased, steam bubbles rise through the riser tubes of the boiler at a faster rate. The circulation of the water is from the steam drum to the mud drum in the "downcomer" tubes and then as a mixture of steam and water up through the "riser" tubes to the steam drum. The application of more heat to the riser or waterwall tubes generates more steam, reducing the heat applied to the riser or waterwall tubes reduces steam bubble formation when less steam is required.

There are two different areas in the steam drum. The area above the water level is where the steam scrubbers and separators are located. This area receives the steam–water mixture from the waterwall or riser tubes and separates the water from the steam, returning the water to the water space. This water space, in the lower part of the steam drum, is relatively quiet and is where the feedwater is admitted. Also this area is where the drum water level is measured.

If steam bubbles rise in the boiler tubes, these tubes are acting as risers. If they are connected into the steam drum water space, the rising steam bubbles may cause the measured water level to appear unstable. Whether a boiler tube acts as a riser or a downcomer is dependent on the amount of heat received by the tube. The amount of heat received is dependent on the temperature of the flue gases that pass around the tube and the circulation through the tube.

When the boiler is being operated in a steady-state condition the steam drum contains a certain mass of water and steam bubbles below the surface of the water. In this steady-state condition there is an average mixture density. As long as the boiler steaming rate is constant, the steam–water mixture has the same volumetric proportions, and the average mixture density is constant.

When the steam demand increases, the concentration of steam bubbles under the water must increase. There is a temporary lowering of steam drum pressure and, as a result, the volumetric proportions in the water–steam mixture change and the average density of the mixture, with some adjustment for the temporary change in boiler pressure, decreases. The immediate result is an increase in the volume of the steam–water mixture caused by the average density decreasing. Because the only place to expand is upward in the steam drum, this causes an immediate increase in the drum water level, even though additional water has not been added. This water level increase is known as *swell*.

When the steam demand decreases, there is a slight temporary increase in steam drum pressure, there are fewer steam bubbles in the mixture, the average density increases and the volume of the steam–water mass decreases. This causes an immediate drop in the steam drum water level, although the mass of water and steam has not changed. This sudden drop caused by a decrease in steam demand is called *shrink*.

Thus, under steady-steaming conditions there is less water in the steam drum when steaming rate is high and more water at a low-steaming rate with the drum water level at the normal set point. This supports the fact that energy storage in the boiler water is higher at lower loads and lower at higher loads.

If the water has "swelled" owing to increased steam demand, the boiler water inventory must be reduced to bring the drum water level down to the normal water level (NWL). If the water has "shrunk" owing to a decrease in steam demand, the water inventory must be increased to return the water level to the NWL.

If steam demand increases and boiler feedwater flow is immediately increased by the same amount, the water inventory would remain constant. In this case, the steam drum level would be forced to remain in the swell condition. To counteract this, the water flow change is delayed so some of the excess inventory can be converted to steam and the drum water level returned to the set point. The reverse is true when steam demand is reduced; this requires an addition to water inventory, by delaying the reduction in the water flow rate.

One of the key factors that can affect the magnitude of the swell or shrink is the size of the steam drum. With greater steam drum volume, the swell or shrink will be less and no apparent change in the other factors. Another factor is higher steam pressure, thus steam density is greater. This reduces the shrink and swell, because of the effect on mixture density. This all translates into

Use single-element feedwater controllers with small scotch marine boilers.
Use two- or three-element feedwater controllers with large boilers and process steam boilers.
Use three-element feedwater controllers with large process boilers.

V. STEAM DRUM WATER LEVEL CONTROL [14]

When a boiler has a drum it is necessary to regulate the flow of feedwater and steam in such a manner that the level of water in the drum is held at a constant level. Water level is affected by the pressure in the drum, by the temperature of the water, and by the rate at which heat is being added.

To get a picture of what happens in the drum with variation in load, assume a state of equilibrium with water at the desired level. If load is increased, causing a temporary drop of pressure in the drum, the steam bubbles and the water will increase, tending to make the water swell and raising the water level. At the same time the increase in load requires increased flow of feedwater to the drum, and this feedwater is comparatively cool by comparison with the near saturation temperature of water already in the drum. This increase in feedwater flow cools the water in the drum and causes the level to shrink or fall.

A water level control is designed to maintain the required amount of water in the steam generator over the operating conditions. In its simplest form, a valve controls feedwater flow so that the water level is maintained constant. The basic process that is being controlled here is an integrating one. The level is the integral of the inlet water flow and the outlet steam flow. This basic system is shown schematically in Figure 6.1a. The simplified block diagram with the appropriate transfer functions is shown in Figure 6.1b.

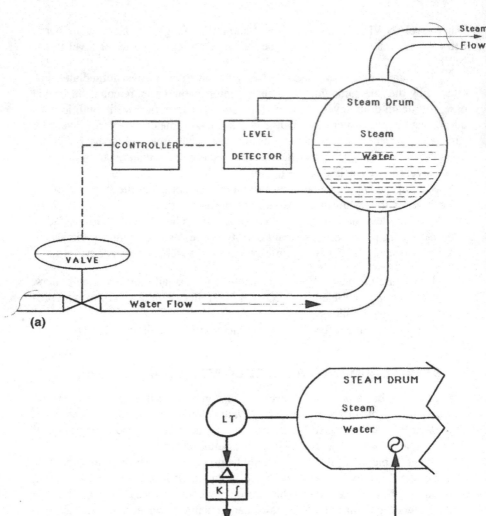

FIGURE 6.1   (a) Schematic of a water level controller with one detector element.
(b) Block diagram of a single-element simple feedback feedwater control

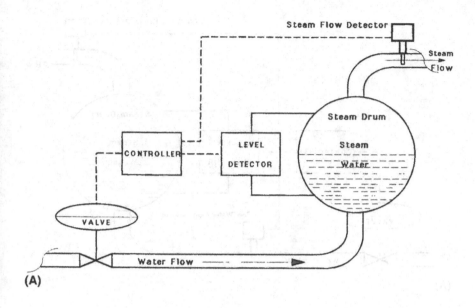

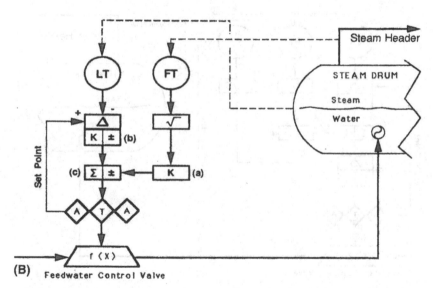

FIGURE 6.2 (A) Schematic of a two-detector elements water level controller. (B) Two-element feedwater control system (feedforward plus feedback).

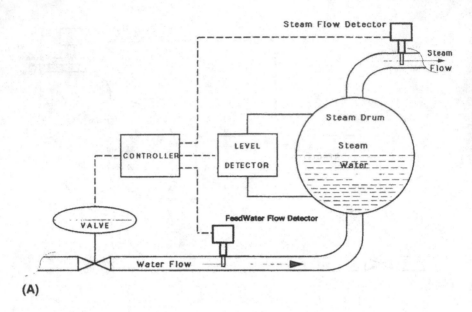

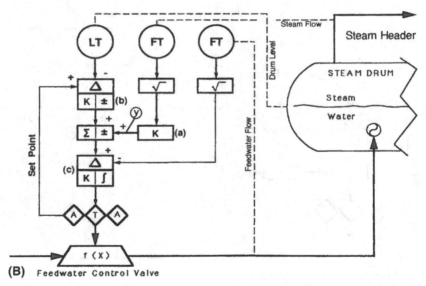

FIGURE 6.3 (A) Schematic of a water level controller with three detector elements. (B) Three-element feedwater control system (feedforward, feedback, plus cascade).

Although this arrangement has certain advantages, it has two severe limitations. The first is that the drum area tends to be large compared with the amounts of water stored so that the level changes slowly. The second is that changes in water level owing to changes in the density of the water with pressure changes will put severe transients on a system that is relatively slow acting. These density changes (swell) are usually a function of steam flow.

To anticipate the effect of rapid changes in steam flow, a steam flow signal is used to control the feedwater and the level is used as an adjustment on the water flow. The level is an integral of the error between the steam flow and the water flow and serves to correct any errors in the balance between these two flows. This system is shown schematically in Figure 6.2A, followed by a related block diagram in Figure 6.2B.

In practice, another detector is commonly added to anticipate level changes and take care of transients in the feedwater system. This measures the feedwater flow and is shown schematically in Figure 6.3a and block diagram form in Figure 6.3b. Although this is the most complicated of the three control systems illustrated, it provides more versatility in response characteristics. This has advantages in terms of speed of response and disadvantages because of opportunities for unstable actions. A combination of experience and study is required to determine which of the three systems of drum level control should be applied in specific cases.

7

Stack Gases

Dry Gas Loss; Oxides of Nitrogen; Smoke from Combustion; Combustion Control; Stack Gas Sampling and Analysis.

I. DRY GAS LOSS

There are two variables in dry gas loss: stack temperature of the gas and the weight of the gas leaving the unit. Stack temperature varies with the degree of deposition on the heat-absorbing surfaces throughout the unit and with the amount of excess combustion air. The effect of excess air increases the gas weight and raises the exit gas temperature; both effects reduce efficiency. A 40°F (22°C) increase in stack gas temperature on coal-fired installations gives an approximate 1% reduction in efficiency.

When increased losses of dry gas are indicated, the source of the increase must be determined and appropriate actions can be planned. It must be determined if the dry gas loss is the result of low absorption or increased gas weight. Lower absorption is usually the result of slagging or fouling. This could be the result of inadequate soot-blowing which could be operational-, maintenance-, or design-related. It could also be the result of improper combustion owing to the condition of the fuel supply equipment, fuel-burning equipment, or fuel–air control system. It is also possible that the problem is the result of a change in the fuel.

If the problem is excessive gas weight owing to excess air, investigation should focus on the fuel burning or the fuel–air control system. Once the causes of the loss of efficiency have been narrowed to a relative few, it will take additional off-line and on-line investigation to pinpoint the actual sources.

High exit gas temperatures and high draft losses with normal excess air indicate dirty heat-absorbing surfaces and the need for soot-blowing or soot-blower maintenance. Generally, high excess air increases exit gas draft losses and indicates the need to adjust the fuel/air ratio or perform maintenance on the control devices as just discussed. Even though this may be the most prevalent cause, high excess air can also be caused by excessive casing leaks, excessive cooling or sealing air, or high air heater leakage [1].

Usually the largest factor affecting boiler efficiency, dry gas loss increases with higher exit gas temperatures or excess air values. Every 35°–40°F increment in exit gas temperature will lower boiler efficiency by 1%. A 1% increase in excess air by itself decreases boiler efficiency by only 0.05%. On most boilers, however, increased excess air leads to higher exit gas temperature. Consequently, increases in excess air can have a twofold effect on unit efficiency.

Usually, coal-fired units are designed to operate with 20–30% excess air. To operate a boiler most efficiently, therefore, an operator must have a reliable means of assessing the quantity of excess air leaving the boiler. In situ oxygen recorders that measure the oxygen at the boiler or economizer outlet, are the best information source. They must, however, be checked daily for proper calibration and maintained as necessary. The operator should sustain the required excess air by making sure the controls are in the correct mode or by manual bias of the fuel/air ratio [13].

II. OXIDES OF NITROGEN [1,13]

Unlike sulfur oxides, which are formed only from sulfur in the fuel, nitrogen oxides (NO_x) are formed from both fuel-bound nitrogen and nitrogen contained in the combustion air introduced into the furnace.

Nitrogen oxides in the form of NO and NO_2 are formed during combustion by two primary mechanisms: thermal NO_x and fuel NO_x. Importantly, even though NO_x consists usually of 95% NO and only 5% NO_2, the normal practice is to calculate concentrations of NO_x as 100% NO_2. Thermal NO_x results from the dissociation and oxidation of nitrogen in the combustion air. The rate and degree of thermal NO_x formation is dependent on oxygen availability during the combustion process and is exponentially dependent on combustion temperature. Thermal NO_x reactions occur rapidly at combustion temperatures in excess of 2800°F (1538°C). Thermal NO_x can be reduced by operating at low excess air as well as by minimizing gas temperatures throughout the furnace by the use of low-turbulence diffusion flames and large water-cooled furnaces. Thermal NO_x is the primary source of NO_x formation from natural gas and distillate oils because these fuels are generally low or devoid of nitrogen. Fuel NO_x, on the other hand, results from oxidation of nitrogen organically bound in the fuel. Fuel-bound nitrogen in the form of volatile compounds is intimately tied to the fuel hydrocarbon chains. For this reason, the formation of fuel NO_x is linked to both fuel nitrogen content and fuel volatility. Inhibiting oxygen availability during the early stages of combustion, during fuel devolatilization, is the most effective means of controlling fuel NO_x formation.

Numerous combustion process NO_x control techniques are commonly used. These vary in effectiveness and cost. In all cases, control methods are aimed at reducing either thermal NO_x, fuel NO_x, or a combination of both.

A. Low Excess Air

Low excess air effectively reduces NO_x emissions with little, if any, capital expenditure. Low excess air is a desirable method of increasing thermal efficiency and has the added benefit of inhibiting thermal NO_x. If burner stability and combustion efficiency are maintained at acceptable levels, lowering the excess air may reduce NO_x by as much as 10–20%. The success of this method largely depends on fuel properties and the ability to carefully control fuel and air distribution to the burners. Operation may require more sophisticated methods of measuring and regulating fuel and airflow to the burners and modifications to the air delivery system to ensure equal distribution of combustion air to all burners.

B. Two-Stage Combustion

Two-stage combustion is a relatively long-standing and accepted method of achieving significant NO_x reduction. Combustion air is directed to the burner zone in quantities less than is theoretically required to burn the fuel, with the remainder of the air introduced through overfire air ports. By diverting combustion air away from the burners, oxygen concentration in the lower furnace is reduced, thereby limiting the oxidation of chemically bound nitrogen in the fuel. By introducing the total combustion air over a larger portion of the furnace, peak flame temperatures are also lowered. Appropriate design of a two-stage combustion system can reduce NO_x emissions by as much as 50% and simultaneously maintain acceptable combustion performance. The following factors must be considered in the overall design of the system.

1. Burner Zone Stoichiometry

The fraction of theoretical air directed to the burners is predetermined to allow proper sizing of the burners and overfire air ports. Normally, a burner zone stoichiometry in the range of 0.85–0.90 will result in desired levels of NO_x reduction without notable adverse effects on combustion stability and turndown.

2. Overfire Air Port's Design

Overfire air ports must be designed for thorough mixing of air and combustion gases in the second stage of combustion. Ports must have the flexibility to regulate flow and air penetration to promote mixing both near the furnace walls and toward the center of the furnace. Mixing efficiency must be maintained over the anticipated boiler load range and the range in burner zone stoichiometries.

3. Burner Design

Burners must be able to operate at lower air flow rates and velocities without detriment to combustion stability. In a two-stage combustion system, burner zone stoichiometry is typically increased with decreasing load to ensure that burner air velocities are maintained above minimum limits. This further ensures positive windbox to furnace differential pressures at reduced loads.

4. Overfire Air Port Location

Sufficient residence time from the burner zone to the overfire air ports and from the ports to the furnace exit is critical to proper system design. Overfire air ports must be located to optimize NO_x reduction and combustion efficiency and to limit change to furnace exit gas temperatures.

5. Furnace Geometry

Furnace geometry influences burner arrangement and flame patterns, residence time, and thermal environment during the first and second stages of combustion. Liberal furnace sizing is generally favorable for lower NO_x, as combustion temperatures are lower and residence times are increased.

6. Airflow Control

Ideally, overfire air ports are housed in a dedicated windbox compartment. In this manner, air to the NO_x ports can be metered and controlled separately from air to the burners. This permits operation at desired stoichiometric levels in the lower furnace and allows for compensation to the flow split as a result of airflow adjustments to individual burners or NO_x ports.

Additional flexibility in controlling burner fuel and airflow characteristics is required to optimize combustion under a two-stage system. Consequently, improved burner designs are emerging to address the demand for tighter control of airflow and fuel-firing patterns to individual burners.

In the reducing gas of the lower furnace, sulfur in fuel forms hydrogen sulfide (H_2S), rather than sulfur dioxide (SO_2) and sulfur trioxide (SO_3). The corrosiveness of reducing gas and the potential for increased corrosion of lower furnace wall tubes is highly dependent on H_2S concentration. Two-stage combustion, therefore is not recommended when firing high-sulfur residual fuel oils.

C. Flue Gas Recirculation

Flue gas recirculation (FGR) to the burners is instrumental in reducing NO_x emissions when the contribution of fuel nitrogen to total NO_x formation is small. Accordingly, the use of gas recirculation is generally limited to the combustion of natural gas and fuel oils. By introducing flue gas from the economizer outlet into the combustion airstream, burner peak flame temperatures are lowered, and

NO_x emissions are significantly reduced. Air foils are commonly used to mix recirculated flue gas with the combustion air. Flue gas is introduced in the sides of the secondary air-measuring foils and exits through slots downstream from the air measurement taps. This method ensures thorough mixing of flue gas and combustion air before reaching the burners and does not affect the airflow-metering capability of the foils. In general, increasing the rate of flue gas recirculation to the burners results in an increasingly significant NO_x reduction. Target NO_x emission levels and limitations on equipment size and boiler components dictate the practical limit of recirculated flue gas for NO_x control. Other limiting factors include burner stability and oxygen concentration of the combustion air. Oxygen content must be maintained at or above 17% on a dry basis for safe and reliable operation of the combustion equipment.

III. SMOKE FROM COMBUSTION [34]

A. Year 1989

Smoke is a suspension of solid or liquid particulate in a gaseous discharge that produces a visible effect. Generally, the particles range from fractions of a micron to over 50 μm in diameter. The visibility of smoke is a function of the quantity of particles present, rather than the weight of particulate matter. The weight of particulate emission, therefore, is not necessarily indicative of the optical density of the emission. For instance, a weight density of so many grains of emissions per cubic foot of gas is not directly related to the opacity of the discharge. Neither is the color of a discharge related to opacity, or smoke density. Smoke can be either black or nonblack (white smoke).

1. White Smoke

The formation of white or other opaque, nonblack smoke is usually due to insufficient furnace temperatures when burning carbonaceous materials. Hydrocarbons heated to a level at which evaporation or cracking occurs within the furnace produce white smoke. The temperatures are not high enough to produce complete combustion of these hydrocarbons. With stack temperatures in the range of 300°–500°F, many hydrocarbons will condense to liquid aerosols and, with the solid particulate present, will appear as nonblack smoke.

Increased furnace or stack temperatures and increased turbulence are two methods of controlling white smoke. Turbulence helps ensure thermal uniformity within the off-gas flow.

Excessive airflow may provide excessive cooling, and an evaluation of reducing white smoke discharges includes investigating the air quantity introduced into the furnace. Inorganic chemicals in the exit gas may also produce nonblack smoke. For instance, sulfur and sulfur compounds appear yellow in a discharge; calcium and silicon oxides appear light to dark brown.

2. Black Smoke

Black smoke is formed when hydrocarbons are burned in an oxygen-deficient atmosphere. Carbon particles are found in the off-gas. Causes of oxygen deficiency are poor atomization, inadequate turbulence (or mixing), and poor air distribution within a furnace chamber. These factors will each generate carbon particulates that, in the off-gas, produce dark, black smoke.

Pyrolysis reactions occur within an oxygen-starved atmosphere. This generates stable, less complex hydrocarbon compounds that form dark, minute particulate, generating black smoke.

A common method of reducing or eliminating black smoke has been steam injection into the furnace. The carbon present is converted to methane and carbon monoxide as follows:

$$3C \text{ (smoke)} + 2H_2O \rightarrow CH_4 + 2CO$$

Similar reactions occur with other hydrocarbons present, and the methane and carbon monoxide produced burn clean in the heat of the furnace, eliminating the black carbonaceous smoke that would have been produced without steam injection:

$$CH_4 + 2O_2 \rightarrow CO_2 + 2H_2O \quad \text{(smokeless)}$$

$$2CO + O_2 \rightarrow 2CO_2 \quad \text{(smokeless)}$$

Steam injection normally requires from 20 to 80 lb of steam per pound of flue gas, or 0.15–0.50 lb of steam per pound of hydrocarbon in the gas stream. (It should be noted that there is some controversy over the effect of steam injection on carbonaceous discharges. Some argue that the steam primarily produces good mixing, and that the turbulence, or effective mixing of air, eliminates the smoke discharge as opposed to methane generation.)

B. Year 1928 [12]

1. Loss from Visible Smoke

Soot is formed by the incomplete combustion of the hydrocarbon constituents of a fuel. All hydrocarbons are unstable at furnace temperatures, and unless air to ensure complete combustion is mixed with them at the time they are distilled, they are quickly decomposed, the ultimate product consisting mostly of soot, H_2, and CO. Soot is formed at the surface of the fuel bed by heating the hydrocarbons in the absence of air; it is not formed by the hydrocarbons striking the comparatively cool heating surface of the boiler. As a matter of fact, only a small trace of hydrocarbon gases reaches the boiler heating surface, provided there is a supply of air above the fire; hydrocarbons that do so are prevented from decomposition by the reduction in temperature by contact. Once formed, it is difficult to burn it in the atmosphere of the furnace, because the oxygen is greatly rarefied, the gases containing only a few percent of free oxygen.

Experience with burning soft coal shows that, if soot is once formed, a large percentage remains floating in the gases after all the other gaseous combustibles have been completely burned. Part of the soot is deposited on the tubes and throughout the boiler setting, whereas the rest is discharged through the stack with the gaseous products of combustion. A smoky chimney does not necessarily indicate an inefficient furnace, because the fuel loss due to visible smoke seldom exceeds 1%. As a matter of fact, a smoky chimney may be much more economical than one that is smokeless. Thus, a furnace operating with very small air excess may cause considerable visible smoke and still give a higher evaporation than one made smokeless by a very large air excess. There will be some loss from CO, unburned hydrocarbons, and soot in the former case, but in the latter this may be offset by the excessive loss caused by the heat carried away in the chimney gases. In general, however, smoky chimneys indicate serious losses, not because of the soot, but because of the unburned, invisible combustible gases. The loss under this paragraph heading refers strictly to the visible combustible discharged up the stack and not that deposited on the tubes and in various parts of the setting. With natural draft the latter seldom exceeds a fraction of 1% of the heat value of the fuel.

With a very high rate of combustion under forced draft, the loss from combustible in the cinders may range as high as 10% or more. A well designed furnaces, properly operated, will burn many coals without smoke up to a certain rate of combustion. Further increase in the amount burned will result in smoke and lower efficiency due to deficient furnace capacity. Small sizes of coal ordinarily burn with less smoke than larger sizes, but develop lower capacities. In the average hand-fired furnace, washed coal burns with lower efficiency and makes more smoke than raw coal. Most coals that do not clinker excessively can be burned with a smaller percentage of black smoke than those that clinker badly.

C. Year 1923

The question of smoke and smokelessness in burning fuels has recently become a very important factor of the problem of combustion. Cities and communities throughout the country have passed ordinances relative to the quantities of smoke that may be emitted from a stack, and the failure of operators to live up to the requirements of such ordinances, resulting as it does in fines and annoyance, has brought their attention forcibly to the matter.

The whole question of smoke and smokelessness is largely a comparative one. There are any number of plants burning a wide variety of fuels in ordinary hand-fired furnaces, in extension furnaces, and on automatic stokers that are operating under service conditions practically without smoke. It is safe to say, however, that no plant will operate smokelessly under any and all conditions of service, nor is there a plant in which the degree of smokelessness does not largely depend on the intelligence of the operating force.

When a condition arises in a boiler room requiring the fires to be brought up quickly, the operators in handling certain types of stokers will use their slice bars freely to break up the green portion of the fire over the bed of partially burned coal. In fact, when a load is suddenly thrown on a station, the steam pressure can often be maintained only in this way, and such use of the slice bar will cause smoke with the very best type of stoker. In a certain plant using a highly volatile coal and operating boilers equipped with ordinary hand-fired furnaces, extension hand-fired furnaces, and stokers, in which the boilers with the different types of furnaces were on separate stacks, a difference in smoke from the different types of furnaces was apparent at light loads, but when a heavy load was thrown on the plant, all three stacks would smoke to the same extent and it was impossible to judge which type of furnace was on one or the other of the stacks.

In hand-fired furnaces, much can be accomplished by proper firing. A combination of the alternate and spreading methods should be used, the coal being fired evenly, quickly, lightly, and often, and the fires worked as little as possible. *Smoke can be diminished by giving the gases a long travel under the action of heated brickwork before they strike the boiler heating surfaces.* Air introduced over the fires and the use of heated arches, for mingling the air with the gases distilled from the coal will also diminish smoke. Extension furnaces will undoubtedly lessen smoke where hand firing is used, owing to the increase in length of gas travel and because this travel is partially under heated brickwork. *Where hand-fired grates are immediately under the boiler tubes and a highly volatile coal is used, if sufficient combustion space is not provided, the volatile gases, distilled as soon as the coal is thrown on the fire, strike the tube surfaces and are cooled below the burning point before they are wholly consumed, and pass through the boiler as smoke.* With an extension furnace, these volatile gases are acted on by the radiant heat from extension furnace arch, and this heat, with the added length of travel, causes their more complete combustion before striking the heating surfaces than in the former case.

Smoke may be diminished by employing a baffle arrangement that gives the gases a fairly long travel under heated brickwork and by introducing air above the fire. In many cases, however, special furnaces for smoke reduction are installed at the expense of capacity and economy.

From the standpoint of smokelessness, undoubtedly the best results are obtained with a good stoker, properly operated. As already stated, the best stoker will cause smoke under certain conditions. Intelligently handled, however, under ordinary operating conditions, stoker-fired furnaces are much more nearly smoke-less than those that are hand-fired, and are, to all intents and purposes, smokeless. In practically all stoker installations, there enters the element of time for combustion, the volatile gases as they are distilled being acted on by ignition or other arches before they strike the heating surfaces. In many instances, too, stokers are installed with an extension beyond the boiler front, which gives an added length of travel, during which the gases are acted on by the radiant heat from the ignition or supplementary arches, *and here again, we see the long travel giving time for the volatile gases to be properly consumed.*

To repeat, it must be clearly borne in mind that the question of smoke-lessness is largely one of degree, and dependent to an extent much greater than is ordinarily appreciated upon the handling of the fuel and the furnaces by the operators, be these furnaces hand-fired or automatically fired.

D. Smoke and Efficiency of Combustion [31]

Although there is perhaps no phase of combustion that has been so fully discussed as that which results in the production of smoke, the common understanding of the loss from this source is at best vague and is, at least partly, based on misconception. For this reason a brief consideration of smoke is included here, regardless of the amount of data on the subject available elsewhere.

Of the numerous and frequently unsatisfactory definitions of smoke that have been offered, that of the Chicago Association of Commerce Committee in its report *Smoke Abatement and the Electrification of Railway Terminals in Chicago* is perhaps the best. This report defines smoke as "the gaseous and solid products of combustion, visible and invisible, including . . . mineral and other substances carried into the atmosphere with the products of combustion."

From the standpoint of combustion loss, it is necessary to lay stress on the term "visible and invisible." The common conception of the extent of loss is based on the visible smoke, and such conception is so general that practically all, if not all, smoke ordinances are based on the visibility, density, or color of escaping stack gases. As a matter of fact, the color of the smoke, which is imparted to the gases by particles of carbon, cannot be taken as an indication of the stack loss. The invisible or practically colorless gases issuing from a stack may represent a combustion loss many times as great as that due to the actual carbon present in the gases, and but a small amount of such carbon is sufficient to give color to large volumes of invisible gases that may or may not represent direct combustion losses. A certain amount of color may also be given to the gases by particles of flocculent ash and mineral matter, neither of which represents a combustion loss. The amount of such material in the escaping gases may be considerable where stokers of the forced draft type are used and heavy overloads are carried.

The carbon or soot particles in smoke from solid fuels are not due to the incomplete combustion of the fixed carbon content of the fuel. Rather, they result from the noncombustion or incomplete combustion of the volatile and heavy hydrocarbon constituents, and it is the wholly or partially incomplete combustion of these constituents that causes smoke from all fuels—solid, liquid, or gaseous.

If the volatile hydrocarbons are not consumed in the furnace and there is no secondary combustion, there will be a direct loss resulting from the noncombustion of these constituents. Although certain of these unconsumed gases may appear as visible smoke, the loss from this source cannot be measured with the ordinary flue-gas analysis apparatus and must of necessity be included with the unaccounted losses.

Where the combustion of the hydrocarbon constituents is incomplete, a portion of the carbon component ordinarily appears as soot particles in the smoke. In the burning of hydrocarbons, the hydrogen constituent unites with oxygen before the carbon; for example, for ethylene (C_2H_4).

$$C_2H_4 + 2O = 2H_2O + 2C$$

If, after the hydrogen is "satisfied," there is sufficient oxygen present with which that carbon component may unite, and temperature conditions are right, such combination will take place and combustion will be complete. If, on the other hand, sufficient oxygen is not present, or if the temperature is reduced below the combining temperature of carbon and oxygen, the carbon will pass off unconsumed as soot, regardless of the amount of oxygen present.

The direct loss from unconsumed carbon passing off in this manner is probably rarely in excess of 1% of the total fuel burned, even for the densest smoke. The loss from unconsumed or partially consumed volatile hydrocarbons, on the other hand, although not indicated by the appearance of the gases issuing from a stack, may represent an appreciable percentage of the total fuel fired.

While the loss represented by the visible constituents of smoke leaving a chimney may ordinarily be considered negligible, there is a loss owing to the presence of unconsumed carbon and tarry hydrocarbons in the products of combustion which, although not a direct combustion loss, may result in a much greater loss in efficiency than that from visible smoke. These constituents adhere to the boiler-heating surfaces and, acting as an insulating layer, greatly reduce the heat-absorbing ability of such surfaces. From the foregoing it is evident that the stack losses indicated by smoke, whether visible or invisible, result almost entirely from improper combustion. Assuming a furnace of proper design and fuel-burning apparatus of the best, there will be no objectionable smoke where there is good combustion. On the other hand, a smokeless chimney is not necessarily indicative of proper or even of good combustion. Large quantities of excess air in diluting the products of combustion naturally tend toward a smokeless stack, but the possible combustion losses corresponding to such an excess air supply have been shown.

E. Year 1915 [35]

1. Conditions for Complete and Smokeless Combustion

If air is passed upward through a deep bed of ignited carbon devoid of volatile matter, there is a tendency for any CO_2 that is formed in lower layers to be reduced to CO when coming into contact with the carbon above. If this CO is not subsequently supplied with a proper *amount of air* while still at a high temperature, it will pass off unoxidized and this will result in a loss of heat that would otherwise be made available. It is, therefore, important that an adequate air supply and a suitable *temperature* be maintained in the upper part of, and just above, the bed of fuel. This air may either pass through the bed or be supplied from above.

The foregoing applies to the combustion of coke and charcoal as well as to carbon. Anthracite coal, which is mostly fixed carbon, behaves similarly, but in this instance there is also a small amount of volatile matter that must be properly burned. These fuels, which have little or no volatile matter, give *short flames* above the fuel bed, the flames being due to the combustion of CO and the small quantity of volatile matter present.

When coal possessing a considerable amount of *volatile matter* is placed on a hot bed of fuel, the greater part of the volatile portion distills off as the temperature rises, and the residue, which is coke, burns in the manner just described. The more serious problem that confronts the engineer in this case is the complete oxidation of the combustible part of this volatile matter. Evidently in the ordinary up-draft furnaces that are fired from above the combustion of this part of the fuel must occur above the fuel bed, just as with CO; and so that the combustible gases may be completely burned, the following four conditions must exist: (1) There must be *sufficient air* just above the fuel bed, supplied either from above or through the fuel bed itself; (2) this air must be properly distributed and intimately *mixed* with the combustible gases; (3) the mixture must have a *temperature* sufficiently high to cause ignition (some of the combustible gases, when mixed with the burned gases present above the fuel, have an ignition temperature of approximately 1450°F); and (4) there must be sufficient *time* for the completion of combustion; that is, the combustion must be complete before the gases become cooled by contact with the relative cold walls of the boiler (which are at a temperature of about 350°F) or with other cooling surface.

To prevent the *stratification* of the air and gases, special means are sometimes adopted, such as employing steam jets above the fire and using baffle walls, arches, and piers in the passage of the flame, to bring about an intimate mixture.

So that the air used above the fuel bed will not chill and extinguish the flame, it should be *heated* either by passing it through the fuel bed, or through passages in the hotter parts of the furnace setting, or in some other way before mingling with the gases; or also the mixture of gases and air should be made to pass over or through hot portions of the fuel bed, or should be brought into contact with furnace walls, or other brickwork, which is at a temperature sufficiently high to support the combustion.

So that the flame will not be chilled and extinguished by coming in contact with cold objects, it should be protected by the hot furnace walls until combustion is complete. The *furnace* should have proper volume to accommodate the burning gases, and, when the conditions are such that the flame is long, the distance from the fuel bed to the relatively cold boiler surfaces with which the gases first come in contact, should be at least as great as the length that the flame attains when the fire is being forced. The *length of flame* depends on the amount and character of the volatile matter in the fuel, on the rapidity of combustion, and on strength of draft. It varies from a few inches, with coke and anthracite coal, to 8 ft or even more with highly volatile coals—even 20 ft has been reached with some western coals.

To have complete combustion of all the fuel in a furnace it is necessary that uniform conditions prevail throughout the fuel bed; and to bring this about it is essential that the fuel itself be uniform in character. Therefore, the best results are obtained with coal that has been graded for *size*. This is especially true with anthracite coal, which ignites slowly and is more difficult to keep burning than volatile coals. This coal requires a rather strong draft, and unless the bed is uniform, the rush of air through the less dense portions tends to deaden the fire in those regions; hence, good results can be obtained with this coal only when it is uniform in size and evenly distributed.

Smoke may be composed of unconsumed, condensible tarry vapors, of unburned carbon freed by the splitting of hydrocarbons, of fine noncombustible matter (dust), or of a combination of these. It is an indication of incomplete combustion, hence, of waste, and in certain communities, is prohibited by ordinance as a public nuisance. Smoke can be avoided by using a smokeless fuel, such as coke or anthracite coal; or, when the more volatile coals are used, by bringing about complete combustion of the volatile matter. In general, the greater the proportion of the *volatile content* of the coal the more difficult it is to avoid smoke, although much depends on the character of the volatile matter. Coals that smoke badly may give from 3–5% lower efficiencies than smokeless varieties.

For each kind of coal and each furnace there is usually a range in the *rate of combustion* within which it is comparatively easy to avoid smoke. At higher rates, owing to the lack of furnace capacity, it becomes increasingly difficult to supply the air, mix it, and bring about complete combustion. Hence, when there is both a high volatile content in the coal and a rapid rate of combustion, it is doubly difficult to obtain complete and smokeless combustion.

However, although smoke is an indication of incomplete and, hence, inefficient combustion, it may sometimes be more profitable, because of lower price or for other reasons, to use a coal with which it is difficult to avoid smoke, provided the latter is not a nuisance or is not prohibited by statute.

F. Year 1914 [36]

Smoke is unburned carbon in a finely divided state. The amount of carbon carried away by the smoke is usually small, not exceeding 1% of the total carbon in the coal. Its presence, however, often indicates improper handling of the boiler, which may result in a much larger waste of fuel. Smoke is produced in a boiler when the incandescent particles of carbon are cooled before coming into contact with sufficient oxygen to unite with them. It is necessary that the carbon be in an incandescent condition before it will combine with the oxygen. Any condition of the furnace that results in carbon being cooled below the point of incandescence before sufficient oxygen has been furnished to combine with it will result in smoke. Smoke once formed is very difficult to ignite, and the boiler furnace must be handled such that it does not produce smoke. Fuels very rich in hydrocarbons are most likely to produce smoke. When the carbon gas liberated from the coal is kept above the temperature of ignition and sufficient oxygen for its combustion is added, it burns with a red, yellow, or white flame. The slower the combustion the larger the flame. When the flame is chilled by the cold heating surfaces near it taking away heat by radiation, combustion may be incomplete, and part of the gas and smoke pass off unburned. If the boiler is raised high enough above the grate to give room for the volatile matter to burn and not strike the tubes at once, the amount of smoke given off and of coal used will both be reduced.

IV. COMBUSTION CONTROL

A. Flue Gas Analysis for Combustion Control

Several flue gas analyses are useful in combustion control trimming loops. Generally the analyses and their combinations are as follows:

1. Analyses

1. Percentage oxygen ($\%O_2$): Excess combustion air is a function of the percentage of oxygen.
2. Percentage opacity: Measuring smoke of particulate matter in the flue gas. Environmental standards come into play here also.
3. Percentage carbon dioxide ($\%CO_2$): When total combustion air is greater than 100% of that theoretically required, excess combustion air is a function of the percentage carbon dioxide.
4. Carbon monoxide (CO) or total combustible in the ppm range: This

measurement is that of unburned gases. Measurement in the ppm range is necessary if desired control precision is to be obtained.

2. Uses

1. The percentage oxygen is an individual control index. [This is the most tested, with control application since the early 1940s.]
2. The ppm CO or total combustible as an individual control index. [Application of this method began in approximately 1973.]
3. The percent oxygen in combination with ppm CO or total combustible. [Applications began in approximately 1977.]
4. The percentage carbon dioxide in combination with ppm CO. [Applications began in approximately 1977.]
5. The percentage oxygen in combination with percentage opacity. [Applications began in approximately 1977.]

B. Pros and Cons of Measurement Methods and Gases Selected for Measurement [48]

This boils down to

1. The selection of the constituent gas or gases for measurement in relation to their intended use
2. The quality of measurement of these gases based on the capability of normally used measurement methods.

C. Selection of the Constituent Gas or Gases

Since the mid-1970s, there has been somewhat of a controversy between the use of percentage oxygen for trimming control as opposed to CO in the ppm range. Another controversy from the 1940s was the use of percentage CO_2 versus percentage oxygen.

Percentage of oxygen won the battle with percentage of CO_2 many years ago owing to its nonambiguity in the low excess air ranges. The same percentage CO_2 reading may mean either an excess or deficiency of air. The percentage oxygen is relatively unaffected by the carbon/hydrogen ratio of the fuel. The percentage of span for the same excess air span is greater for percentage of oxygen, and the measurement accuracy of percentage oxygen analyzers is better than for those measuring percentage of CO_2.

For the use of ppm CO, a theory was advanced that optimum low-cast operation is obtained if the CO can be kept at a constant setpoint, somewhere in the range of 250–400 ppm. The second part of this was that only a constant setpoint ppm CO single-element feedback controller trimming the airflow control was necessary.

It is now generally agreed that CO in the ppm range should not be used alone. This measurement can provide useful information, but if used it should be used in combination with percentage oxygen or CO_2. Both of these are valid, but, for the foregoing reasons, the preponderance of the argument favors the use of percentage oxygen. A constant setpoint ppm CO has been highly touted as the final arbiter for optimizing the combustion process. We now know that if a final control from ppm CO is used, the setpoint should be variable for boiler load, just as the percentage oxygen setpoint should vary as the load changes.

The percentage oxygen can be used alone for control, but should be used with a variable setpoint related to boiler load. The control should include a small margin of excess air. A margin of 3–4% excess air (0.6–0.8% oxygen) is usually suggested.

Because of this margin, the ppm CO control can theoretically be operated closer to the limit. Even the ppm CO control, however, should have some margin. The ppm CO measurement is "noisy" with constant fluctuation above and below the operating setpoint. Because of the nonlinearity of the ppm CO versus excess air curve, the loss for each fluctuation above the operating point is greater than the opposite fluctuations below the operating point. To be most economical, the average of the two should not exceed the loss of the indicated optimum point.

V. STACK GAS SAMPLING AND ANALYSIS [6]

Stack gas analyzers are either *extractive* or *in situ*. The term refers to the way the gas sample is delivered into the analyzer. In extractive sampling, a gas sample is drawn out of the duct or stack, then passed through a sample conditioner, where it is prepared for analysis by a remotely located analyzer. In situ analysis features an analyzer mounted on the stack with its sampling apparatus directly in contact with the stack gases.

Maintaining the sampling equipment has long been as much a problem as performing the analysis. Consequently, an analyzer that does not require a sample gas stream is welcomed by the industrial user. The in situ analyzer, which became available in the 1970s, provides this benefit and has been accepted for industrial and utility applications. Both types of sampling are discussed.

A. Extractive Sampling

In extractive sampling, a gas is drawn out of a duct or stack by an aspirator or a pump, and it then passes through a sample conditioner on its way to the measuring instrument. Analyzers generally operate on a dry basis. Sample conditioning may include (1) filtration to remove particulates, (2) refrigeration to remove water vapor, (3) heating or insulation of lines to maintain proper processing temperatures, and (4) introduction of standard composition gases for calibration, so that the zero and span adjustments for scaling and calibration can be made.

Extractive sampling starts with a probe inserted into the stack or duct. The probe must include filters, a potential source of plugging. An air- or water-operated vacuum educator draws the sample into the probe. Primary filters may pass up to 50-μm particles, and secondary filters may take out all particles but those smaller than 1 μm.

The filtered sample enters a conditioning chamber mounted near the sampling point. Acid mist condensables and entrained liquids are removed by a chiller. The sample is then heated, filtered to remove the last trace of particulates, and dried below its dew point.

Extractive sampling allows location of the analyzer at a remote site, but an interval must necessarily lapse between the time at which the sample is pulled from the gas stream and the time at which analysis takes place. Thus, the reading is always somewhat behind the process. More seriously, sampling equipment is vulnerable to plugging and loss of sample. Good consistent maintenance is required to make it work.

B. In Situ Sampling

When sampling for oxygen is done in place, the measurement element is mounted on a probe directly in the hot gas stream. Also, a sample can be drawn out of the gas stream and passed through an analyzer mounted on the side of the duct or stack.

In situ measurement of CO, CO_2, SO_2, NO_x, and unburned hydrocarbons combines a light source shining across the stack with a receiver–analyzer. It is based on absorption spectroscopy, measuring in the ultraviolet, visible, and near-infrared portions of the optical spectrum. The molecules of each different materials vibrate at specific frequencies, which cancel equivalent light frequencies in the light beam.

Detection of the absorbed frequencies in the spectrum from a narrowband source identifies the components and their concentrations. Maintenance is required to prevent fouling of windows at the transmitting and receiving ends of the light path, but this fouling can normally be minimized by a continuous air purge at both ends of the beam. The preferred location of sampling for O_2 is in the breeching; for other gases, on the stack.

C. Gas Analysis

Regardless of the sampling method, gas analyzer design follows a pattern that makes use of a measuring cell and a reference cell. The material in the reference cell is of standard composition, often air, and the results of the measurement in this cell are predictable. The same is made in both cells, and the results are compared. The concept is no different from that of the Wheatstone bridge, which follows the same pattern in an electrical context. Differences in analyzers occur because of the physical principles invoked to make the measurements. Some are described in the following:

1. Paramagnet Properties

A paramagnetic material is attracted by a magnetic field, whereas a diamagnetic one is repelled. Oxygen is one of the few gases that is paramagnetic (nitric oxide is another). The magnetization produced by a magnetic field of unity strength in a paramagnetic gas varies inversely as the absolute temperature. By properly combining a magnetic field gradient and a thermal gradient, it is possible to induce and sustain a convective gas flow that is dependent on the percentage of oxygen in a gas sample containing no other paramagnetic gas. Changes in gas flow rate are measured by the effect produced on the resistance of a temperature-sensitive element.

2. Thermal Conductivity

The rate at which heat from a heated electrical element will be conducted through a gas mixture is a function of the composition of the gas. The thermal conductivity of the gas mixture is proportional to the product of the mole fraction of each gas in the mixture and its respective thermal conductivity. With a thermistor as the detector in a sample gas and another thermistor in a reference gas, when a constant temperature is maintained at the heat source, the difference in temperature of the two detectors is an indication of relative concentration when the thermal conductivities of the sample and the reference are known. Power plant applications include measurement of CO_2 in stack gas and hydrogen purity in hydrogen-cooled generators.

3. Heat of Combustion

The concentration of combustible gases in a sample stream is converted to an electric signal by oxidation of these gases and measurement of the signal produced by the combustion, which takes place on the surface of a measuring filament. The filament is coated with a catalyst to permit controlled combustion at lower than ignition temperatures. It is one active arm of a Wheatstone bridge circuit; the other arm is an uncoated reference filament. Both filaments are exposed to the same pressure, composition, flow rate, and temperature of sample. Differences in resistance are a function of changes caused by combustion at one filament. The signal is proportional to the concentration of combustible material in the gas stream.

D. Leading Analyzers

Leading analyzer designs today for stack gas analysis follow the same pattern of sample and reference cell measurement, but use photoconductive cells combined with optical filters for spectrographic analysis. Cell design depends on the

TABLE 7.1

Characteristics of Air and Gas Cleaning Devices

Ralph L. Vandagriff

Service Most Suitable For Contaminant:
1 = Melodors, Gases
2 = Lints, Dusts, Pollens & Tobacco Smoke
3 = Dusts, fumes, Smokes & Mists
4 = Gases, Vapors & Melodors

General Class	Specific Type	Description of Device	Service Most Suitable For Contaminant	Optimum Particle Size μm	Optimum Concentration mg/cu.meter	Limits of Gas Temp. deg. F	Usual Water & Power Rates gpm,kw or hp	Usual Face Velocity Ft/Min. through	Usual Air Resistance In. Wg.	Remarks	Usual Efficiency % by Wt.	Increase with	Disposition of Collected Material — Usual State & Location	Automatic Options
Odor Adsorbers	Shallow Bed	Activated charcoal beds in cells or cartridges	1	(Molecular)	<20	0-100	none	50-120 bed	<0.3		<95	declining yet	Discarded as charcoal	Charcoal frequently salvaged
Air Washers	Spray Chamber	One or two coarse spray banks followed by bent plate eliminators	1 & 2	>20	<10	40-700	2-5 gpm/1000 cfm	300-500 chamber	<0.4		<25	increasing yet	Slurry (or solution) in water	
	Wet Cell	Wetted glass or synthetic fiber cells followed by bent-plate eliminators	1 & 2	>5	<10	40-700	2-5 gpm/1000cfm	200-350 cells	<0.7	clean	>25	increasing yet	Slurry (or solution) in water	
	Corrugated Fill	Wetted paper, glass, or plastic fill followed by drift eliminator	1 & 2	>10	<10	40-100	2-5 gpm/1000 cfm	200-500 fill	<0.5		>25	increasing yet	Slurry (or solution) in water	
Electrostatic Precipitator	Two-stage plate	Ionizing (+) wires followed by collecting (-) plates	2	<1	<2	0-260	01-.02 kw/1000 cfm	275-500 plates	<0.3		<90	declining yet	Slurry in oil,washed off plated	Intermittent wash
low voltage	Two-stage filter	Ionizing (+) wires followed by filter (-) cells	2	<1	<2	0-180	01-.02 kw/1000 cfm	200-300 cells	<0.2	initial	>50	declining yet	Discarded with filters	
Air Filters viscous coated	Throwaway	Deep bed of coarse glass, vegetable or synthetic fibers in cells	2	>5	<4	0-180	none	300-500 cells	<0.1	initial	<25	increasing yet	Discarded with filters	
	Washable	Deep bed of metal wires, screens or ribbons in cells	2	>5	<4	0-260	none	300-500 cells	<0.1	initial	<25	increasing yet	Slurry in oil,washed off filter	Continuous on bath
Air Filters dry fiber	5-10 micron	Porous mat of 5-10 micron glass or synthetic fibers pleated into cells	2	>3	<2	0-180	none	5-25 met	<0.3	initial	>50	increasing yet	Discarded with filters	Continuous renewal
	2-5 micron	Porous mat of 2-5 micron glass or synthetic fibers pleated into cells	2	>0.5	<2	0-180	none	5-25 met	<0.5	initial	<95	increasing yet	Discarded with filters	Continuous renewal
Absolute Filters	Paper	Porous paper of <1 micron glass, ceramic or other fibers pleated into cells	Special	<1	<2	0-180	none	4-6 paper	<1	initial	>99.25	declining yet	Discarded with filters	
Industrial Filters	Shake/reverse to Pulse-jet	Cloth bags or envelopes (low air-to-cloth ratio)	3	>0.3	>200	0-180	none	1-6 fabric	>4	dry	>89	declining yet	Dry dust on bags or envelopes	Continuous or intermittent dislodging
Cleanable		Cloth tubes or bags (high air-to-cloth ratio)	3	>0.3	>200	0-180	none	5-30 fabric	>4	dry	>80	declining yet	dislodged to hopper	Continuous or intermittent dislodging

AIR CLEANING

No.	Category	Subtype	Description										Disposal	Notes
7	Electrostatic Precipitator high voltage	Single stage plate	Ionizing (+) wires between parallel collecting (-) plates	8	<2	>800	0-700	0.2-0.6 kW/1000cfm	180-800	plates	<1	<95 declining vel	Dry dust on plates or pipes dislodged to hopper	Continuous or intermittent dislodging
		Single stage pipe	Ionizing (-) wires inside concentric collecting (+) pipes	8	<2	>800	0-700	0.2-0.6 kW/1000cfm	180-800	pipes	<1	<99 declining vel	Dry dust (or liquid) in hopper	(or on shelves)
8	Dry Inertial Collectors	Settling Chamber	Straight horizontal chamber, some with shelves	3	>50	>10,000	0-700	none	300-600	chamber	<0.1	<50 declining vel	Dry dust (or liquid) in hopper	
		Baffled Chamber	Chamber with one baffle or numerous baffles in parallel	3	>50	>10,000	0-700	none	1000-2000	inlet	<0.5	<50 increasing vel	Dry dust (or liquid) in hopper	
		Skimming Chamber	Sand-shaped chamber with peripheral slots	3	>20	>2,000	0-700	none	2000-4000	inlet	<1	<70 increasing vel	Dry dust (or liquid) in hopper	
		Cyclone	Chamber with provisions for spiral flow	8	>10	>2,000	0-700	none	2000-4000	inlet	<2	<80 increasing vel	Dry dust (or liquid) in hopper	
		Multiple Cyclone	Numerous small cyclones in parallel	8	>5	>2,000	0-700	none	2000-4000	inlet	<4	<80 increasing vel	Dry dust (or liquid) in hopper	
		Impingement	Alternate stages of nozzles and baffles	8	>10	>2,000	0-700	none	2000-6000	nozzles	<4	<80 increasing vel	Dry dust (or liquid) in hopper	
		Dynamic	Power-driven centrifugal fan with skimming slots	8		>2,000	0-700	1-2 hp/1000 cfm	2000-4000	inlet	to 6" developed	<80 increasing vel	Dry dust (or liquid) in hopper	Intermittent dislodging
9	Scrubbers	Cyclone	Cyclone collector with coarse radial sprays	3 & 4	>10	>2,000	40-700	3-5 gpm/1000 cfm	2000-4000	inlet	>2	<80 increasing vel	Slurry (or solution) in water	
		Impingement	Impingement collector with wetted baffles	3 & 4	>5	>2,000	40-700	3-5 gpm/1000 cfm	3000-6000	nozzles	>2	<80 increasing vel	Slurry (or solution) in water	Continuous sludge ejection
		Dynamic	Dynamic collector with coarse sprays	3 & 4	>10	>2,000	40-700	3-5 gpm & 2-4 hp/1000cfm	2000-4000	inlet	to 8" developed	<80 increasing vel	Slurry (or solution) in water	
		Pebble bed	Tower with countercurrently wetted coarse packing	3 & 4	>5	>200	40-700	3-5 gpm /1000 cfm	500-1000	bed	>4	<90 increasing vel	Slurry (or solution) in water	
		Multidynamic	Power-driven turmoil- and reverse-flow fan stages with coarse sprays	3 & 4	<1	>200	40-700	3-5gpm & 20-30 hp/1000cfm	2000-3000	inlet	to 4" developed	<99 increasing vel	Slurry (or solution) in water	
		Submerged nozzle	Nozzle partially submerged in water	3 & 4	>2	>200	40-700	no pumping	2000-4000	nozzles	>2	<90 increasing vel	Slurry (or solution) in water	Continuous sludge ejection
		Jet	Water activated jet pump	3 & 4	>5	>200	40-700	50-100 gpm/1000cfm	2000-3000	inlet	to 8" developed	<90 increasing vel	Slurry (or solution) in water	Continuous sludge ejection
		Fog	Cyclone collector with fine tangential sprays	3 & 4	>2	>200	40-700	3-5 gpm/1000 cfm	3000-4000	inlet	>2	<99 increasing vel	Slurry (or solution) in water	
		Venturi	Venturi with coarse sprays at throat	3 & 4	<2	>200	40-700	3-5 gpm/1000 cfm	12000-24000	throat	>10	<99 increasing vel	Slurry (or solution) in water	
10	Incinerators	Direct	Combustion chamber with supplemental fuel firing	4	any (Combustible)		2,000	none	500-1000	chamber	<1	<95 declining vel	Innocuous flue gas out stack	
		Catalytic	Combustion chamber with catalyst plus supplemental fuel	4	(Molecular)	any	1,000	none	500-1000	chamber	>1	<95 declining vel	Innocuous flue gas out stack	
11	Gas Absorbers	Spray tower	Vertical up air-flow chamber with downward spray	4	(Molecular)	>2	40-100	5-15 gpm/1000 cfm	300-600	tower	<10	<95 declining vel	Solution in water, etc.	
		Packed column	Tower with countercurrently wetted floating rings, Berl or saddles, etc.	4	(Molecular)	>2	40-100	5-15 gpm /1000 cfm	500-1000	bed	<10	<95 declining vel	Solution in water, etc.	
		Fibre cell	One or more stages of concurrently wetted fiber cells	4	(Molecular)	>2	40-100	5-15 gpm /1000 cfm	200-300	cells	<4	<95 declining vel	Solution in water, etc.	
		Crossflow	One or more stages of crossflow wetted packing	4	(Molecular)	>2	40-100	1-5 gpm /1000 cfm	300-600	fill	<2	<90 declining vel	Solution in water, etc.	
12	Gas Adsorbers	Deep bed	Activated charcoal beds in regenerative recovery equipment	4	(Molecular)	>2	0-100	none	20-120	bed	<10	<100 declining vel	Adsorbed on charcoal/recovered	Intermittent extraction

absorption characteristics of the gases measured, because the various gases absorb different spectral wavelengths.

1. Spectroscopic Types

Spectroscopic analysis functions on the basis that gas molecules absorb energy at known wavelengths from beams of light transmitted through the gas. The spectrum is complex, and it is necessary to restrict light to a narrow bandwidth to avoid overlapping and interference. Analyzer cells may shine one or two beams of light. In either case, the pattern of measuring cell/reference cell is maintained. Cells with two beams will have one for reference and the other for the variable sample. Cells with one beam measure at two wavelengths: one for reference, the other for the unknown.

A *nondispersive infrared* (NDIR) analyzer may be used for CO and CO_2. One design focuses energy from an infrared source into two beams. One beam passes through a cell in which the sample gas flows; the other passes through a zero-reference cell. Each beam is passed and reflected intermittently by a rotating chopper, then focused by means of optics at the detector. The detector output is a differential pulse that is proportional to the absorption of the sample cell relative to the reference cell. The effect of the beam chopper in front of the infrared source generates an AC signal that helps minimize drift.

Also, SO_2 and NO_x can be analyzed by instruments that rely on ultraviolet light. The two gases are analyzed in sequence in the sample cell. A split-beam arrangement, with optical filters, phototubes, and amplifiers, measures the difference in light-beam absorption at two wavelengths: 280 nm for SO_2 and 436 nm for NO_x. The light source is generally a mercury vapor lamp.

2. Oxygen Analyzers

Analysis of O_2 and CO are the two most important exhaust gas measurements for combustion efficiency control. In situ analyzers are almost always selected for the measurements. Usually CO measurement depends on infrared absorption. The most common O_2 analyzer design is based on a difference in O_2 partial pressures on the two sides of a zirconia wafer or zirconium oxide cell. Palladium/palladium oxide is another possibility. In one variation, a difference in O_2 partial pressures on the two sides of a cell, which is heated and maintained at a constant temperature, creates ion migration that causes a proportional variation in an electric current conducted by the wafer, while the front side is exposed to the flue gas. Because the zirconia or palladium is affected only by oxygen, the back side sees a predetermined amount of O_2 in the instrument air as a reference gas. The difference in partial pressures generates a millivolt output that is representative of the oxygen level in the sample gas.

See Tables 7.1–7.4 for data on cleaning devices, particles, properties, and combustion gases.

TABLE 7.2 Gas Particles

Methods for Particle Size Analysis

- Sieving
- Electroformed Sieves
- Microscope
- Ultramicroscope*
- Electron Microscope*
- Elutriation
- Centrifuge
- Sedimentation
- Ultracentrifuge
- Turbidimetry*
- X-Ray Diffraction*
- Permeability*++
- Adsorption*
- Scanners
- Light Scattering++
- Nuclei Counter
- Machine Tools (Micrometers, Calipers, etc.)
- Electrical Conductivity
- Visible to Eye
- Impingers

*Furnishes average particle diameter but no size distribution.
++Size distribution may be obtained by special calibration.

Types of Gas Cleaning Equipment

- Settling Chambers
- Centrifugal Separators
- Liquid Scrubbers
- Cloth Collectors
- Packed Beds
- Common Air Filters
- Impingement Separators
- Mechanical Separators
- High Efficiency Air Filters
- Thermal Precipitation (used only for sampling)
- Electrical Precipitators
- Ultrasonics (very limited industrial application)

Terminal Gravitational Settling [for spheres, sp. gr. 2.0]	In Air at 25°C, 1 atm. Reynolds Number												
	Settling Velocity, cm/sec.												
	In Water at 25°C Reynolds Number												
	Settling Velocity, cm/sec.												
Particle Diffusion Coefficient,* cm²/sec.	In Air at 25°C, 1 atm.												
	In Water at 25°C												

*Stokes-Cunningham factor included in values given for air but not included for water

Particle Diameter, microns (μ)

PREPARED BY C. E. LAPPLE

0.0001   0.001 (1mμ)   0.01   0.1   1   10   100   1,000 (1mm.)   10,000 (1cm.)

TABLE 7.3

| Boiler House Notes | | | | | | | | | Page No: | |

Gas Property: Cp

The specific heat at constant pressure, Cp, is the heat required, during a constant pressure process, to raise the temperature of a unit mass of gas one degree.

Example: Determine Cp, in Btu/lb, for gas mixture @ 1500 deg. F. & 14.696 psia.

Gas	Volume %	Cp at temp.	Molecular Weight
N2	80%	0.2819	28
O2	12%	0.2630	32
CO2	8%	0.3013	44

Formula:

$$Cp = \frac{N2(MW \times \%Vol. \times Cp) + O2(MW \times \%Vol. \times Cp) + CO2(MW \times \%Vol. \times Cp)}{N2(MW \times \%Vol.) + O2(MW \times \%Vol.) + CO2(MW \times \%Vol.)}$$

$$= \frac{(28 \times 0.8 \times 0.283) + (32 \times 0.12 \times 0.2637) + (44 \times 0.08 \times 0.3025)}{(28 \times 0.8) + (32 \times 0.12) + (44 \times 0.08)}$$

= 0.28176 Btu/Lb.-deg.F.

Specific Heats of Various Gases
in Btu/lbm-deg.F.

Temperature deg. Rankin	deg. Fahr.	Gas Mol.Wt.	Air 28.965	N2 28.013	O2 31.999	CO2 44.01	CO 28.01	H2 2.016	H2O 18.015	SO2 64.059
519.67	60		0.2394	0.2477	0.2187	0.1996	0.2476	3.4199	0.4476	0.1474
559.67	100		0.2401	0.2482	0.2206	0.2046	0.2484	3.4206	0.4503	0.1505
659.67	200		0.2422	0.2496	0.2251	0.2162	0.2506	3.4239	0.4579	0.1578
759.67	300		0.2444	0.2512	0.2293	0.2270	0.2529	3.4294	0.4667	0.1644
859.67	400		0.2468	0.2531	0.2332	0.2369	0.2553	3.4370	0.4767	0.1704
959.67	500		0.2492	0.2552	0.2370	0.2460	0.2579	3.4466	0.4880	0.1758
1059.67	600		0.2517	0.2574	0.2404	0.2543	0.2605	3.4582	0.5007	0.1806
1159.67	700		0.2543	0.2598	0.2437	0.2619	0.2633	3.4715	0.5147	0.1849
1259.67	800		0.2569	0.2624	0.2468	0.2688	0.2661	3.4866	0.5301	0.1887
1359.67	900		0.2596	0.2651	0.2496	0.2750	0.2689	3.5032	0.5471	0.1921
1459.67	1000		0.2623	0.2678	0.2523	0.2807	0.2717	3.5214	0.5655	0.1950
1609.67	1150		0.2663	0.2720	0.2559	0.2881	0.2760	3.5511	0.5962	0.1986
1759.67	1300		0.2703	0.2763	0.2592	0.2944	0.2802	3.5836	0.6306	0.2015
1909.67	1450		0.2742	0.2805	0.2621	0.2997	0.2842	3.6185	0.6689	0.2037
2059.67	1600		0.2779	0.2846	0.2647	0.3041	0.2881	3.6654	0.7112	0.2054
2209.67	1750		0.2813	0.2885	0.2671	0.3079	0.2916	3.6940	0.7578	0.2066
2359.67	1900		0.2845	0.2921	0.2692	0.3111	0.2949	3.7339	0.8087	0.2074

(Values from formulas, Engineering Thermodynamics)

Temperature-Dependent Molar Specific Heats of Gases @ Zero Pressure - USCS Units

(Note: to get Cp in Btu/lb, divide answer by molecular wt. of substance.)   (deg. R = 459.67 + deg. F.)

Substance	Temp.Range deg. F.	deg. R.	Cp,Btu/(lbmol . deg. R) Cp = a + bT + cT^2 + dT^3			
-----------	--------------------	---------	a	b	c	d
			a	b	c	d
Air	33 - 2780	493-3240	6.713	.02609*10^-2	.0354*10^-5	(-0.08052*10^-9)
Sulfur Dioxide	33 - 2780	493-3240	6.157	.7689*10^-2	(-.281*10^-5)	.3527*10^-9
Hydrogen	33 - 2780	493-3240	6.952	(-.02542*10^-2)	.02952*10^-5	(-.03565*10^-9)
Oxygen	33 - 2780	493-3240	6.085	.2017*10^-2	(-.05275*10^-5)	.05372*10^-9
Nitrogen	33 - 2780	493-3240	6.903	(-.02085*10^-2)	.05957*10^-5	(-.1176*10^-9)
Carbon Monoxide	33 - 2780	493-3240	6.726	.02222*10^-2	.3960*10^-5	(-.091*10^-9)
H2O	33 - 2780	493-3240	7.7	.02552*10^-2	.07781*10^-5	(-.1472*10^-9)
Carbon Dioxide	33 - 2780	493-3240	5.316	.79361*10^-2	(-.2581*10^-5)	.3059*10^-9

Ralph L. Vandagriff   4/1/97

TABLE 7.4

By: Ralph L. Vandagriff								
Date: 11/1/95	HEAT CONTENTS OF COMBUSTION GASES, IN Btu per Lb.							
File: combgasBtu/lb								
re: # 4								
	AIR	CO	CO2	H2	H2O	N2	O2	SO2
Gas Temperature, deg. F.								
60.00	0.00	0.00	0.00	0.00	0.00	0.00	0.00	0.00
100.00	9.70	9.80	8.30	139.10	17.70	9.80	8.70	6.00
150.00	21.85	22.25	18.95	312.95	40.05	22.20	19.80	13.75
200.00	34.00	34.70	29.60	486.80	62.40	34.60	30.90	21.50
250.00	46.20	47.25	40.80	660.85	85.20	47.15	42.20	29.60
300.00	58.40	59.80	52.00	834.90	108.00	59.70	53.50	37.70
350.00	69.05	72.55	63.60	1,009.25	131.20	72.35	65.00	46.05
400.00	79.70	85.30	75.20	1,183.60	154.40	85.00	76.50	54.40
450.00	92.85	98.20	87.30	1,358.30	178.05	97.80	88.20	63.05
500.00	106.00	111.10	99.40	1,533.00	201.70	110.60	99.90	71.70
550.00	117.50	124.10	111.85	1,708.10	225.70	123.50	111.75	80.60
600.00	129.00	137.10	124.30	1,883.20	249.70	136.40	123.60	89.50
650.00	141.50	150.38	137.35	2,059.13	274.38	149.50	135.73	98.75
700.00	154.00	163.65	150.40	2,235.05	299.05	162.60	147.85	108.00
750.00	166.50	176.93	163.45	2,410.98	323.73	175.70	159.98	117.25
800.00	179.00	190.20	176.50	2,586.90	348.40	188.80	172.10	126.50
825.00	185.75	196.96	183.36	2,675.53	361.15	195.49	178.31	131.34
850.00	192.50	203.73	190.23	2,764.15	373.90	202.18	184.53	136.18
868.00	197.36	208.59	195.17	2,827.96	383.08	206.99	189.00	139.66
875.00	199.25	210.49	197.09	2,852.78	386.65	208.86	190.74	141.01
900.00	206.00	217.25	203.95	2,941.40	399.40	215.55	196.95	145.85
910.00	208.70	219.96	206.70	2,976.85	404.50	218.23	199.44	147.79
925.00	212.75	224.01	210.81	3,030.03	412.15	222.24	203.16	150.69
950.00	219.50	230.78	217.68	3,118.65	424.90	228.93	209.38	155.53
975.00	226.25	237.54	224.54	3,207.28	437.65	235.61	215.59	160.36
1,000.00	233.00	244.30	231.40	3,295.90	450.40	242.30	221.80	165.20
1,050.00	246.75	258.13	245.73	3,474.73	476.75	255.93	234.50	175.20
1,100.00	260.50	271.95	260.05	3,653.55	503.10	269.55	247.20	185.20
1,150.00	274.25	285.78	274.38	3,832.38	529.45	283.18	259.90	195.20
1,200.00	288.00	299.60	288.70	4,011.20	555.80	296.80	272.60	205.20
1,250.00	301.50	313.70	303.50	4,191.93	582.98	310.68	285.53	215.45
1,300.00	315.00	327.80	318.30	4,372.65	610.15	324.55	298.45	225.70
1,350.00	328.50	341.90	333.10	4,553.38	637.33	338.43	311.38	235.95
1,400.00	342.00	356.00	347.90	4,734.10	664.50	352.30	324.30	246.20
1,450.00	355.75	370.33	363.08	4,917.03	692.53	366.43	337.43	256.65
1,500.00	369.50	384.65	378.25	5,099.95	720.55	380.55	350.55	267.10
1,550.00	383.25	398.98	393.43	5,282.88	748.58	394.68	363.68	277.55
1,600.00	397.00	413.30	408.60	5,465.80	776.60	408.80	376.80	288.00
1,650.00	410.75	427.88	424.08	5,651.23	805.50	423.18	390.08	298.58
1,700.00	424.50	442.45	439.55	5,836.65	834.40	437.55	403.35	309.15
1,750.00	438.25	457.03	455.03	6,022.08	863.30	451.93	416.63	319.73
1,800.00	452.00	471.60	470.50	6,207.50	892.20	466.30	429.90	330.30
1,850.00	466.50	486.40	486.18	6,395.70	921.93	480.90	443.28	340.90
1,860.00	469.40	489.36	489.31	6,433.34	927.87	483.82	445.95	343.02
1,870.00	472.30	492.32	492.45	6,470.98	933.82	486.74	448.63	345.14
1,880.00	475.20	495.28	495.58	6,508.62	939.76	489.66	451.30	347.26
1,890.00	478.10	498.24	498.72	6,546.26	945.71	492.58	453.98	349.38
1,900.00	481.00	501.20	501.85	6,583.90	951.65	495.50	456.65	351.50
1,935.00	491.15	511.56	512.82	6,715.64	972.46	505.72	466.01	358.92
1,950.00	495.50	516.00	517.53	6,772.10	981.38	510.10	470.03	362.10
2,000.00	510.00	530.80	533.20	6,960.30	1,011.10	524.70	483.40	372.70

8

Flows

Power Plant Piping; Flow Formulas; Liquid Flow Through Valves and Fittings; Insulated Pipe Heat Loss; Pumping Formulas; Formulas for Estimating Steam Properties; Reciprocating Pumps; Compressed Air.

I. POWER PLANT PIPING

A. Reasonable Design Velocities for Flow of Fluids in Pipes

Fluid	Pressure (psig)	Use	Reasonable velocity (ft/min)	Reasonable velocity (ft/sec)
Water	25–40	City water	120–300	2–5
Water	50–150	General service	300–600	5–10
Water	Above 150	Boiler feed	600–1200	10–20
Saturated steam	0–15	Heating	4,000–6,000	66.7–100
Saturated steam	Above 50	Miscellaneous	6,000–10,000	100–167
Superheated steam	Above 200	Large turbine and boiler leads	10,000–20,000	167–333

Source: Ref. 37.

B. Velocities Common in Steam-Generating Systems

Nature of service	Velocity (ft/min)	Velocity (ft/sec)	Velocity (m/sec)
Compressed air lines	1,500–2,000	25–33.3	7.6–10.2
Forced draft air ducts	1,500–3,600	25–60	7.6–18.3
Forced draft air ducts, burner entrance	1,500–2,000	25–33.3	7.6–10.2
Ventilating ducts	1,000–3,000	16.67–50	5.1–15.2
Crude oil lines, (6–30 in.)	60–3,600	1–60	0.3–1.8
Flue gas, air heater	1,000–5,000	16.67–83.3	5.1–25.4
Flue gas, boiler gas passes	3,000–6,000	50–100	15.2–30.5
Flue gas, induced draft flues and breaching	2,000–3,500	33.3–58.3	10.2–17.8

B. Continued

Nature of service	Velocity		
	(ft/min)	(ft/sec)	(m/sec)
Stacks and chimneys	2,000–5,000	33.3–83.3	10.2–25.4
Steam lines: high-pressure	8,000–12,000	133.3–200	40.6–61.0
Steam lines: low-pressure	12,000–15,000	200–250	61.0–76.2
Steam lines: vacuum	20,000–40,000	333.3–666.7	101.6–203.2
Super-heater tubes	2,000–5,000	33.3–83.3	10.2–25.4
Water lines: general	500–750	8.33–12.5	2.5–3.8
Water piping: boiler circulation	70–700	1.16–11.67	0.4–3.6
Water piping: economizer tubes	150–300	2.5–5	0.8–1.5

Source: Ref. 1.

II. FLOW FORMULAS

A. Darcy's Formulas [46]

1. Darcy's Formula for Liquids

$$\Delta P = 0.000000359 \times \frac{fL\rho V^2}{d}$$

$$\Delta P = 0.000216 \times \frac{fL\rho Q^2}{d^5} \quad \text{(gal/min)}$$

$$\Delta P = 0.00000336 \times \frac{fLW^2V^*}{d^5} \quad \text{(lb/hr)}$$

Note: The Darcy formula may be used without restriction for the flow of water, oil, and other liquids in pipe. However, when extreme velocities occurring in pipe cause the downstream pressure to fall to the vapor pressure of the liquid, cavitation occurs, and calculated flow rates are inaccurate.

2. Darcy's Formula for Compressed Air

$$\Delta P = 0.00129 \times \frac{f\rho LV^2}{d}$$

where
f = friction factor, L = length of pipe in feet
d = inside diameter of pipe in inches
V = mean velocity in feet per minute
V* = specific volume of fluid, in cubic feet per pound
ρ = weight density of fluid (lb/ft^3)
Q = rate of flow in gallons/minute
W = rate of flow, in pounds per hour

B. Empirical Formulas for the Flow of Water, Steam, and Gas

1. Hazen and Williams Formula for Water

$$hf = 0.002083 \times L \times \left(\frac{100}{C}\right)^{1.85} \times \frac{gpm^{1.85}}{d^{4.8655}}$$

C = 130 for new steel or cast iron pipe [C = 100 (commonly used for design purposes)]

Water at 60°F and 31.5 SSU viscosity

2. Babcock Formula for Steam Flow

$$\Delta P = 0.0000000363 \times \left(\frac{d + 3.6}{d^6}\right) \times W^2 \times L \times V*$$

3. Weymouth Formula for Natural Gas Lines

$$Qs = 433.45 \times \frac{Ts}{Ps} \times d^{2.667} \times \left(\frac{P_1^2 - P_2^2}{L'ST}\right)^{1/2}$$

Note: Weymouth formula for short pipelines and gathering systems agrees more closely with metered rates than those calculated by any other formula; however, the degree of error increases with pressure.

Q_s = rate of gas flow (ft³/24 hr) measured at standard conditions

d = internal diameter of pipe in inches

P_1 = initial pressure (psia)

L = length of pipe in feet

S = specific gravity flowing gas (air = 1.0)

T = absolute temp. flowing gas (°F + 460)

V* = specific volume (ft³/lb)

P_s = standard pressure (psia)

P_2 = terminal pressure (psia)

L' = length of gas line in miles

hf = head in feet

T_s = standard absolute temp (°F + 460)

W = rate of flow (lb/hr)

Liquid flow through a pipe is said to be laminar (viscous) or turbulent, depending on the liquid velocity, pipe size, and liquid viscosity. For any given liquid and pipe size these factors can be expressed in terms of a dimensionless number called the Reynolds number R [11].

$$R = \frac{VD}{\upsilon}$$

where

V = average velocity (ft/sec)

υ = kinematic viscosity of the fluid (ft²/sec) (pure water at 60°F, υ = 0.00001211 ft²/sec)

D = average internal diameter of pipe (ft)

4. Friction Factor Calculation: Turbulent Flow (R Over 2000) [47]

For turbulent flow (R higher than 2000) the friction factor is affected by both the roughness of the pipe's interior surface and R and can be determined from an equation developed by P. K. Swamee and A. K. Jain in 1976.

$$f = \frac{0.25}{\left[\log \left(\frac{1}{3.7(D/\varepsilon)} + \frac{5.74}{R^{0.9}} \right) \right]^2}$$

where

f = friction factor

D = inside diameter of pipe (ft)

R = Reynolds number

ε = relative roughness of pipe interior

For laminar (viscous) flow (R lower than 2000) the roughness of the pipe's interior surface has no effect, and the friction factor f becomes:

$$f = \frac{64}{R}$$

5. Pipe Roughness: Design Values [11]

Material	Roughness, ε
Glass, plastic	Smooth
Copper, brass, lead (tubing)	0.000005
Cast iron—uncoated	0.0008
Cast iron—asphalt coated	0.0004
Commercial steel or welded steel	0.00015
Wrought Iron	0.00015
Concrete	0.004

6. Properties of Water [11]

Temperature (°F)	Kinematic viscosity (v, ft²/sec)	Temperature (°F)	Kinematic viscosity (v, ft²/sec)
32	0.0000189	120	0.00000594
40	0.0000167	140	0.00000468
50	0.0000140	160	0.00000438
60	0.0000121	180	0.00000384
70	0.0000105	190	0.00000362
80	0.00000915	200	0.00000335
100	0.00000737	212	0.00000317

For more information see Table 8.1 for viscosities of miscellaneous fluids.

III. LIQUID FLOW THROUGH VALVES AND FITTINGS [46; p 2–2]

A. Pressure Drop Chargeable to Valves and Fittings

When a fluid is flowing steadily in a long, straight pipe of uniform diameter, the flow pattern, as indicated by the velocity distribution across the pipe diameter, will assume a certain characteristic form. Any impediment in the pipe that changes the direction of the whole stream, or even part of it, will alter the characteristic flow pattern and create turbulence, causing an energy loss greater than that normally accompanying flow in straight pipe. Because valves and fittings in a pipe line disturb the flow pattern, they produce an additional pressure drop.

The loss of pressure produced by a valve (or fitting) consists of

1. The pressure drop within the valve itself.
2. The pressure drop in the upstream piping in excess of that which would normally occur if there were no valve in the line. This effect is small.
3. The pressure drop in the downstream piping in excess of that which would normally occur if there were no valve in the line. This effect may be comparatively large.

B. Laminar Flow Conditions [46; p 2–11]

One of the problems in flow of fluids, which confronts engineers from time to time and for which there is very meager information, is the resistance of valves and fittings under laminar flow conditions. Flow through straight pipe is ade-

TABLE 8.1

	Viscosities of Miscellaneous Fluids										

Ralph L. Vandagriff   Boiler House Notes: Page No:

re: (11) & (43).

Liquid	Specific Gravity based on water = 1 at 60 deg. F.	Temp. deg. F.	SSU	Centistokes	Temp. deg. F.	Liquid	Specific Gravity based on water = 1 at 60 deg. F.	Temp. deg. F.	SSU	Centistokes	Temp. deg. F.
Acetaldehyde	0.7620	68		0.295	68	Fuel Oil, #1	.82 - .95	60	34 - 40	2.39 - 4.28	70
Acetone	0.7920	68		0.41	68	Fuel Oil, #2	.82 - .95	60	36 - 50	3.0 - 7.4	70
Alcohol, ethyl	0.7890	68	31.7	1.52	68	Fuel Oil, #3	.82 - .95	60	35 - 45	2.69 - 5.84	70
Alcohol, methyl	0.7900	68		0.74	59	Fuel Oil, # 5A	.82 - .95	60	50 - 125	7.4 - 26.4	70
Alcohol, propyl	0.8040	68	35	2.8	68	Fuel Oil, # 5B	.82 - .95	60	125 -	26.4 -	70
Ammonia	0.6620	0		0.3	0	Fuel Oil, # 6	.82 - .95	60	450 - 3000	97.4 - 660	122
Asphalt emulsion, Fed#1	1.0+	60	350-1700	75-367	100	Gasoline - a	0.7400	60		0.88	60
Asphalt emulsion, Fed#2	1.0+	60	90-350	19-75	100	Gasoline - b	0.7200	60		0.64	60
Auto Crankcase Oil						Gasoline - c	0.6800	60		0.46	60
SAE 5W	.88-.94	60	6000 - max	1295 max	0	Glycerine, 100%	1.2600	68	2950	648	68.6
SAE 10W	.88-.94	60	6000-12000	1295-2590	0	Glycerine, 50% water	1.1300	68	43	5.29	68.
SAE 20W	.88-.94	60	12000-48000	2590-10350	0	Glucose	1.35 - 1.44	60	35000-100000	7700-22000	100
SAE 20	.88-.94	60	45-58	5.7-9.6	210	Honey			340	73.6	100
SAE 30	.88-.94	60	58-70	9.6-12.9	210	Kerosene	.78 - .82	60	35	2.71	68
SAE 40	.88-.94	60	70-85	12.9-16.8	210	Linseed Oil	.92 - .94	60	143	30.5	100
SAE 50	.88-.94	60	85-110	16.8-22.7	210	Milk	1.02 - 1.05	60	31.5	1.13	68
Automotive Gear Oils						Molasses, first run	1.40 - 1.46	60	1300 - 23500	281 - 5070	100
SAE 75W	.88-.94	60	40 min.	4.2 min.	210	Molasses, second run	1.43 - 1.48	60	6535 - 61180	1410 - 13200	100
SAE 80W	.88-.94	60	49 min.	7.0 min.	210	Molasses, blackstrap	1.46 - 1.49	60	12190 - 255000	2630 - 55000	100
SAE 85W	.88-.94	60	63 min.	11.0 min.	210	Peanut Oil	0.9200	60	200	42	100
SAE 90	.88-.94	60	74-120	14 - 25	210	Petroleum ether	0.6400	60	1.1	31 (est)	60
SAE 140	.88-.94	60	120-200	25 - 43	210	Propylene glycol	1.0380	68	241	52	70
SAE 150	.88-.94	60	200 min.	43 min.	210	Rosin Oil	0.9800	60	1500	324.7	100
						Rosin (wood)	1.09 average	60	1000-50000	216 - 11000	100
Benzene	0.8990	0	31	1	0	Sodium chloride, 5%	1.0370	39	31.1	1.097	68
Bromine	2.9000	68		0.34	68	Sodium chloride, 25%	1.1960	39	34	2.4	60
Butane-n	0.5840	60		0.52	-50	Sodium hydroxide,20%	1.2200	60	39.4	4	65
Calcium chloride, 5%	1.0400	65		1.156	65	Sodium hydroxide, 30%	1.3300	60	58.1	10	65
Calcium chloride, 25%	1.2300	60	39	4	60	Sodium hydroxide,40%	1.4300	60	110.1		65
Carbolic acid (phenol)	1.0800	65	65	11.83	65	Soya bean oil	.924 - .926	60	165	35.4	100
Carbon disulphide (CS2)	1.2630	68		0.298	68	Sugar solutions					
Corn Oil	0.9240	60	135	28.7	130	Corn syrup, 86.4 Brix	1.4590	60		180000cp	100
Corn Starch, 22 Baume	1.1800	60	150	32.1	70	Corn syrup, 84.4 Brix	1.4450	60		48000cp	100
Corn Starch, 24 Baume	1.2000	60	600	129.8	70	Corn syrup, 82.3 Brix	1.4310	60		17000cp	100
Corn Starch, 25 Baume	1.2100	60	1400	303	70	Corn syrup, 80.3 Brix	1.4180	60		6900cp	100
Cotton Seed Oil	.88-.93	60	176	37.9	100	Corn syrup, 78.4 Brix	1.4050	60		3200cp	100
Creosote	1.04-1.10	60				Sugar solutions					
Crude Oil, 48 deg. API	0.7900	60	39	3.8	60	Sucrose, 60 Brix	1.2900	60	92	18.7	100
Crude Oil, 40 deg. API	0.8250	60	55.7	9.7	60	Sucrose, 64 Brix	1.3100	60	148	31.8	100
Crude Oil, 35.6 deg. API	0.8470	60	88.4	17.8	60	Sucrose, 68 Brix	1.3380	60	275	59.5	100
Crude Oil, 32.6 deg. API	0.8620	60	110	23.2	60	Sucrose, 72 Brix	1.3600	60	640	138.6	100
Diethylene glycol	1.1200	60	149.7	32	70	Sucrose, 74 Brix	1.3760	60	1100	238	100
Diethyl ether	0.7140	68		0.32	68	Sucrose, 76 Brix	1.3900	60	2000	440	100
Diesel Fuel Oil, #2	.82-.95	60	32.6-45.5	2.0 - 6.0	100	Sulphuric acid, 100%	1.8390	68	76	14.6	68
Diesel Fuel Oil, #3	.82-.95	60	45.5 - 65	6.0 - 11.75	100	Sulphuric acid, 95%	1.8390	68	75	14.5	68
Diesel Fuel Oil, #4	.82-.95	60	140 max	29.8 max	100	Sulphuric acid, 60%	1.5000	68	41	4.4	68
Diesel Fuel Oil, #5	.82-.95	60	400 max	86.6 max	100	Sulphuric acid, 20%	1.1400	68			
Dowtherm	1.0560	77				Tar, Pine	1.06+	60	2500	559	100
Ethyl Bromide	1.4500	59		0.27	68	Toluene	0.8660	231		0.68	68
Ethylene Bromide	2.1800	68		0.787	68	Triethylene glycol	1.1250	68		0.68	68
Ethylene Chloride	1.2460	68		0.668	68	Turpentine	.86 - .87	320	400 - 440	86.6 - 95.2	100
Ethylene Glycol	1.1250	60	88.4	17.8	70	Water, distilled	1.0000	60	31	1.0038	68
						Water, fresh	1.0000	60	31.5	1.13	60
						Water, sea	1.0300	60	31.5	1.15	

quately covered by the basic flow equation, $h_L = fL/D \; v^2/2g$, which is identical with Poiseuille's law for laminar flow when the equation for f in this flow range, $f = 64/R_e$, is included in the formula.

For solution of these problems, we have developed, on the basis of data presented in *Principles of Chemical Engineering* by Walker, Lewis, McAdams, and Gilliland, the empirical relation between equivalent length in the laminal flow region (for $R_e < 1000$) to that in the turbulent region, namely:

$$\frac{L}{D_s} = \frac{R_e}{1000} \frac{L}{D_t}$$

Subscript "s" refers to the equivalent length in pipe diameters under laminar flow conditions where the Reynolds number is less than 1000. Subscript "t" refers to the equivalent length in pipe diameters determined from tests in the turbulent flow range. Representative values of equivalent length are given in Table 8.2.

The minimum equivalent length is the length in pipe diameters of the centerline of the actual flow path through the valve or fitting. Although laboratory test data supporting this method is meager, reports of field experience indicate that the results obtained agree closely with observed conditions.

C. Equivalent Resistance of Fittings and Valves [37; p 3–129]

An actual piping installation consists of straight pipe, bends, elbows, tees, valves, and various other obstructions to flow. Thus, it is necessary to take into account the frictional resistance of the fittings involved. The usual approach is to express the loss through a fitting as being the equivalent of the loss through a certain number of linear feet of straight pipe.

D. Friction Losses, Valves and Fittings, and Viscous Liquids [11; p 3–122]

Very little reliable test data on losses through valves and fittings for viscous liquids is available. In the absence of meaningful data, some engineers assume the flow is turbulent and use the equivalent length method (i.e., where friction losses through valves and fittings are expressed in terms of equivalent length of straight pipe). Calculations made on the basis of turbulent flow will give safe

TABLE 8.2

BOILER HOUSE NOTES:
Subject: Losses in Equivalent Feet of Pipe
Ralph L. Vandagriff   March 17, 1997   All pipe to be new, clean commercial steel pipe.
Note: Valves & fittings (except Mitre Elbows) calculated from formula, pg 264, Applied Fluid Mechanics, 4th Ed, 1994, Merrill/Macmillan

		L/D		0.50	0.75	1.00	1.25	1.50	2.00	2.50	3.00	3.50	4	5	6	8	10	12	14	16	18	20	24
Pipe Sch. or Wt:				40	40	40	40	40	40	40	40	40	40	40	40	40	Std.	40	Std.	Std.	Std.	Std.	Std.
Pipe Wall: Inches				0.109	0.113	0.133	0.140	0.145	0.154	0.203	0.216	0.226	0.237	0.258	0.280	0.322	0.365	0.406	0.438	0.375	0.375	0.375	0.375
Pipe I.D.:Feet				0.052	0.069	0.087	0.115	0.134	0.172	0.206	0.256	0.296	0.336	0.421	0.505	0.665	0.835	0.995	1.094	1.271	1.438	1.604	1.937
Friction factor: fr				0.027	0.025	0.023	0.022	0.021	0.019	0.018	0.018	0.017	0.017	0.016	0.015	0.014	0.013	0.013	0.013	0.013	0.012	0.012	0.012
Valve Gate full open		8	K=	0.22	0.20	0.18	0.18	0.17	0.15	0.14	0.14	0.14	0.14	0.13	0.12	0.11	0.11	0.10	0.10	0.10	0.10	0.10	0.10
			Equiv.Ft.Pipe=	0.41	0.55	0.70	0.82	1.07	1.38	1.65	2.05	2.37	2.68	3.36	4.04	5.32	6.68	7.96	8.75	10.17	11.50	12.83	15.50
3/4 open		35	K=	0.95	0.86	0.81	0.77	0.74	0.67	0.63	0.63	0.60	0.60	0.56	0.53	0.49	0.49	0.46	0.46	0.46	0.42	0.42	0.42
			Equiv.Ft.Pipe=	1.81	2.40	3.08	4.03	4.70	6.03	7.20	8.95	10.35	11.74	14.72	17.69	23.28	29.23	34.82	38.29	44.48	50.31	56.14	67.80
1/2 open		160	K=	4.32	4.00	3.68	3.52	3.36	3.04	2.88	2.88	2.72	2.72	2.56	2.40	2.24	2.24	2.08	2.08	2.08	1.92	1.92	1.92
			Equiv.Ft.Pipe=	8.29	10.99	13.98	18.40	21.47	27.57	32.93	40.91	47.31	53.88	67.30	80.86	106.4	133.6	159.2	175.0	203.3	230.0	256.6	309.9
1/4 open		900	K=	24.30	22.50	20.70	19.80	18.90	17.10	16.20	16.20	15.30	15.30	14.40	13.50	12.60	12.60	11.70	11.70	11.70	10.80	10.80	10.80
			Equiv.Ft.Pipe=	46.62	61.83	78.66	103.50	120.78	155.07	185.22	230.13	268.13	301.95	378.54	454.86	598.8	751.5	895.3	984.6	1143.7	1293.8	1443.6	1743.3
Butterfly full open		45	K=	1.22	1.13	1.04	0.99	0.95	0.86	0.81	0.81	0.77	0.77	0.72	0.68	0.63	0.63	0.59	0.59	0.59	0.54	0.54	0.54
			Equiv.Ft.Pipe=	2.33	3.09	3.93	5.18	6.04	7.75	9.26	11.51	13.31	15.10	18.93	22.74	29.9	37.6	44.8	49.2	57.2	64.7	72.2	87.2
Globe full open		340	K=	9.18	8.50	7.82	7.48	7.14	6.46	6.12	6.12	5.78	5.78	5.44	5.10	4.76	4.76	4.42	4.42	4.42	4.08	4.08	4.08
			Equiv.Ft.Pipe=	17.61	23.36	29.72	39.10	45.63	58.58	69.97	86.94	100.54	114.07	143.00	171.84	226.1	283.9	338.2	372.0	432.1	488.8	545.4	658.6
Angle full open		150	K=	4.05	3.75	3.45	3.30	3.15	2.85	2.70	2.70	2.55	2.55	2.40	2.25	2.10	2.10	1.95	1.95	1.95	1.80	1.80	1.80
			Equiv.Ft.Pipe=	7.77	10.31	13.11	17.25	20.13	25.85	30.87	38.36	44.36	50.33	63.09	75.81	99.8	125.3	149.2	164.1	190.6	215.6	240.6	290.6
Swing Check		100	K=	2.70	2.50	2.30	2.20	2.10	1.90	1.80	1.80	1.70	1.70	1.60	1.50	1.40	1.40	1.30	1.30	1.30	1.20	1.20	1.20
			Equiv.Ft.Pipe=	5.18	6.87	8.74	11.50	13.42	17.23	20.58	25.57	29.57	33.55	42.06	50.54	66.5	83.5	99.5	109.4	127.1	143.8	160.4	193.7
Ball Check		150	K=	4.05	3.75	3.45	3.30	3.15	2.85	2.70	2.70	2.55	2.55	2.40	2.25	2.10	2.10	1.95	1.95	1.95	1.80	1.80	1.80
			Equiv.Ft.Pipe=	7.77	10.31	13.11	17.25	20.13	25.85	30.87	38.36	44.36	50.33	63.09	75.81	90.8	125.3	149.2	164.1	190.6	215.6	240.6	290.6
FITTINGS																							
Elbows: 90 deg. Std.	30		K=	0.81	0.75	0.69	0.66	0.63	0.57	0.54	0.54	0.51	0.51	0.48	0.45	0.42	0.42	0.39	0.39	0.39	0.36	0.36	0.36
			Equiv.Ft.Pipe=	1.55	2.06	2.62	3.45	4.03	5.17	6.17	7.67	8.87	10.07	12.62	15.16	19.95	25.05	29.84	32.82	38.12	43.13	46.12	58.11
90 deg. LR	20		K=	0.54	0.50	0.46	0.44	0.42	0.38	0.36	0.36	0.34	0.34	0.32	0.30	0.28	0.28	0.26	0.26	0.26	0.24	0.24	0.24
			Equiv.Ft.Pipe=	1.04	1.37	1.75	2.30	2.68	3.45	4.12	5.11	5.91	6.71	8.41	10.11	13.30	16.70	19.90	21.88	25.42	28.75	32.08	38.74
45 deg. Std.	16		K=	0.43	0.40	0.37	0.35	0.34	0.30	0.29	0.29	0.27	0.27	0.26	0.24	0.22	0.22	0.21	0.21	0.21	0.19	0.19	0.19
			Equiv.Ft.Pipe=	0.83	1.10	1.40	1.84	2.15	2.76	3.29	4.09	4.73	5.37	6.73	8.09	10.64	13.36	15.92	17.50	20.33	23.00	25.66	30.99
Return Bends 180 deg.	50		K=	1.35	1.25	1.15	1.10	1.05	0.95	0.90	0.90	0.85	0.85	0.80	0.75	0.70	0.70	0.65	0.65	0.65	0.60	0.60	0.60
			Equiv.Ft.Pipe=	2.59	3.44	4.37	5.75	6.71	8.62	10.29	12.79	14.79	16.78	21.03	25.27	33.26	41.75	49.74	54.70	63.64	71.88	80.20	96.85
Tees: Branch Flow	60		K=	1.62	1.50	1.38	1.32	1.26	1.14	1.08	1.08	1.02	1.02	0.96	0.90	0.84	0.84	0.78	0.78	0.78	0.72	0.72	0.72
			Equiv.Ft.Pipe=	3.11	4.12	5.24	6.90	8.05	10.34	12.35	15.34	17.74	20.13	25.24	30.32	39.91	50.10	59.69	65.64	76.25	86.25	96.24	116.22
Straight Flow	20		K=	0.54	0.50	0.46	0.44	0.42	0.38	0.36	0.36	0.34	0.34	0.32	0.30	0.28	0.28	0.26	0.26	0.26	0.24	0.24	0.24
			Equiv.Ft.Pipe=	1.04	1.37	1.75	2.30	2.68	3.45	4.12	5.11	5.91	6.71	8.41	10.11	13.30	16.70	19.90	21.88	25.42	28.75	32.08	38.74
Mitre Elbows (re: Mark's Standard Handbook for Mechanical Engineers, 8th Edition, 1978, McGraw-Hill) 90 degree			Equiv.Ft.Pipe=	3.00	4.00	5.00	7.00	8.00	10.00	12.00	15.00	18.00	21.00	25.00	30.00	40.00	50.00	60.00	68.00	78.00	85.00	100.00	115.00
60 degree			Equiv.Ft.Pipe=	1.30	1.60	2.10	3.00	3.40	4.50	5.20	6.40	7.30	8.50	11.00	13.00	17.00	21.00	25.00	29.00	31.00	37.00	41.00	49.00
45 degree			Equiv.Ft.Pipe=	0.70	0.90	1.00	1.50	1.80	2.30	2.80	3.20	4.00	4.50	6.00	7.00	9.00	12.00	13.00	15.00	17.00	19.00	22.00	25.00
30 degree			Equiv.Ft.Pipe=	0.40	0.50	0.70	0.90	1.10	1.30	1.70	2.00	2.40	2.70	3.20	4.00	5.10	7.20	8.00	9.00	10.00	11.00	13.00	16.00

results because friction losses for turbulent flow are higher than for laminar (viscous) flow.

E. Valves and Fittings: Resistance Equal Length of Pipe [47; p 284]

1. Formulas

   1. Resistance coefficient (K)

$$K = f\text{r} \times \frac{L_e}{D}$$

   2. Equivalent length (L_e)

$$L_e = K \times \frac{D}{f\text{r}}$$

   3. D = inside diameter of pipe, in feet:

Resistance in valves and fittings expressed as equivalent length in pipe diameters, L_e/D			Friction factor, $f\text{r}$ turbulence, new pipe	
Type		Le/D	Pipe size (in.)	$f\text{r}$
Gate valve	Fully open	8	1/2	0.027
	3/4 open	35		
	1/2 open	160	3/4	0.025
	1/4 open	900		
Butterfly valve	Fully open	45	1	0.023
Globe valve	Fully open	340	1 1/4	0.022
Angle valve	Fully open	150	1 1/2	0.021
Swing check valve		100	2	0.019
Ball check valve		150	2 1/2, 3	0.018
Elbow	90° std.	30	4	0.017
Elbow	90° long radius	20	5	0.016
Elbow	90° street	50		
Elbow	45° std	16	6	0.015
Elbow	45° street	26		
Close return bend		50	8–10	0.014
Standard tee	Flow through run	20	12–16	0.013
	Flow through branch	60	18–24	0.012

Example. The equivalent length of pipe for a 6-in. fully open globe valve, Sch 40 pipe, is

K = 340 × 0.015 = 5.10

L_e = 5.10 × 0.5054 (ft)/0.015 = 172 ft

For information on flow loss see Tables 8.2 and 8.3

F. Piping Heat Loss to Air

1. Uninsulated

a. Formulas

Radiation Loss

$$\frac{Q}{A} = 0.1713 \times \varepsilon \times \left[\left(\frac{T_s}{100}\right)^4 - \left(\frac{T_a}{100}\right)^4 \right]$$

Example.   3 1/2 in. steam line, length 50 ft, steam at 320°F, emissivity is 0.8

Q = heat loss (Btu/hr)
Ts = absolute temperature of surface in degrees Rankine
Ta = absolute temperature of the air
A = surface (ft^2)
ε = emissivity of the pipe

$$Q/A = 0.1713 \times 0.8 \times [(720/100)^4 - (528/100)^4]$$
$$= 401 \text{ Btu/hr ft.}^{-2}$$

Convection Loss   Calm air.

$$Q/A = \frac{0.27 \, \Delta T^{1.25}}{D^{0.25}}$$

ΔT = T_s − T_a (°F)
D = pipe diameter in feet

Example.

$$Q/A = \frac{0.27 \times 252^{1.25}}{(3.5/12)^{0.25}} = 369 \text{ Btu/hr ft}^{-2}$$

Total heat loss for 50 ft of 3 1/2-in. steam pipe in calm air is

$$(401 + 369) \times 45.81 \text{ ft}^2 = 35,270 \text{ Btu/hr}$$

Convection Heat Loss in Wind: 15 MPH Wind

BOILER HOUSE NOTES:
Subject: Losses in Equivalent Feet of Pipe - Sch 80 & .5 wall
Ralph L. Vandagriff   March 17, 1997   All pipe to be new, clean commercial steel pipe.
Note: Valves & fittings calculated from formula, pg 264, Applied Fluid Mechanics, 4th Ed, 1994, Merrill/Macmillan

Page No.

	Position	Equivalent Length Pipe Dia. L/D		0.50	0.75	1.00	1.25	1.50	2.00	2.50	3.00	3.50	4	5	6	8	10	12	14	16	18	20	24
Pipe Size: inches				0.50	0.75	1.00	1.25	1.50	2.00	2.50	3.00	3.50	4	5	6	8	10	12	14	16	18	20	24
Pipe Wall: inches				0.147	0.154	0.179	0.191	0.200	0.218	0.276	0.300	0.318	0.337	0.375	0.432	0.500	0.500	0.500	0.500	0.500	0.500	0.500	0.500
Pipe I.D.:feet				0.046	0.062	0.080	0.107	0.125	0.162	0.194	0.242	0.280	0.319	0.401	0.480	0.635	0.812	0.979	1.083	1.250	1.417	1.583	1.917
Friction factor: fr				0.027	0.025	0.023	0.022	0.021	0.019	0.018	0.018	0.017	0.017	0.016	0.015	0.014	0.014	0.013	0.013	0.013	0.012	0.012	0.012
Valve																							
Gate	full open	8	K=	0.22	0.20	0.18	0.18	0.17	0.15	0.14	0.14	0.14	0.14	0.13	0.12	0.11	0.11	0.10	0.10	0.10	0.10	0.10	0.10
			Equiv.Ft.Pipe=	0.36	0.49	0.64	0.85	1.00	1.29	1.55	1.93	2.24	2.55	3.21	3.84	5.06	6.50	7.83	8.66	10.00	11.34	12.66	15.34
	3/4 open	35	K=	0.95	0.88	0.81	0.77	0.74	0.67	0.63	0.63	0.60	0.60	0.56	0.53	0.49	0.49	0.46	0.46	0.46	0.42	0.42	0.42
			Equiv.Ft.Pipe=	1.59	2.16	2.79	3.73	4.38	5.66	6.78	8.46	9.81	11.16	14.04	16.80	22.24	28.44	34.27	37.91	43.75	49.60	55.41	67.10
	1/2 open	160	K=	4.32	4.00	3.68	3.52	3.36	3.04	2.86	2.88	2.72	2.72	2.56	2.40	2.24	2.24	2.08	2.08	2.08	1.92	1.92	1.92
			Equiv.Ft.Pipe=	7.28	9.89	12.75	17.04	20.00	25.86	30.98	38.67	44.85	51.01	64.18	76.82	101.7	130.0	156.6	173.3	200.0	226.7	253.3	306.7
	1/4 open	900	K=	24.30	22.50	20.70	19.80	18.90	17.10	16.20	16.20	15.30	15.30	14.40	13.50	12.60	12.60	11.70	11.70	11.70	10.80	10.80	10.80
			Equiv.Ft.Pipe=	40.95	55.62	71.73	95.85	112.50	145.44	174.24	217.53	252.27	286.92	360.99	432.09	571.9	731.3	881.1	974.7	1125.0	1275.3	1424.7	1725.3
Butterfly	full open	45	K=	1.22	1.13	1.04	0.99	0.95	0.86	0.81	0.81	0.77	0.77	0.72	0.68	0.63	0.63	0.59	0.59	0.59	0.54	0.54	0.54
			Equiv.Ft.Pipe=	2.05	2.78	3.59	4.79	5.63	7.27	8.71	10.88	12.61	14.35	18.05	21.60	28.6	39.6	44.1	48.7	56.3	63.8	71.2	86.3
Globe	full open	340	K=	9.18	8.50	7.82	7.48	7.14	6.46	6.12	6.12	5.78	5.78	5.44	5.10	4.76	4.76	4.42	4.42	4.42	4.08	4.08	4.08
			Equiv.Ft.Pipe=	15.47	21.01	27.10	36.21	42.50	54.94	65.82	82.18	95.30	108.39	136.37	163.23	210.0	276.3	332.9	368.2	425.0	481.6	538.2	651.8
Angle	full open	150	K=	4.05	3.75	3.45	3.30	3.15	2.85	2.70	2.70	2.55	2.55	2.40	2.25	2.10	2.10	1.95	1.95	1.95	1.80	1.80	1.80
			Equiv.Ft.Pipe=	6.83	9.27	11.96	15.98	18.75	24.24	29.04	36.26	42.02	47.82	60.17	72.02	95.3	121.9	148.9	182.5	187.5	212.6	237.5	267.6
Swing Check		100	K=	2.70	2.50	2.30	2.20	2.10	1.90	1.80	1.80	1.70	1.70	1.60	1.50	1.40	1.40	1.30	1.30	1.30	1.20	1.20	1.20
			Equiv.Ft.Pipe=	4.55	6.18	7.97	10.65	12.50	16.16	19.36	24.17	28.03	31.86	40.11	48.01	63.5	81.3	97.9	108.3	125.0	141.7	158.3	191.7
Ball Check		150	K=	4.05	3.75	3.45	3.30	3.15	2.85	2.70	2.70	2.55	2.55	2.40	2.25	2.10	2.10	1.95	1.95	1.95	1.80	1.80	1.80
			Equiv.Ft.Pipe=	6.83	9.27	11.96	15.98	18.75	24.24	29.04	36.26	42.02	47.82	60.17	72.02	95.3	121.9	146.9	182.5	187.5	212.6	237.5	267.6
FITTINGS																							
Elbows: 90 deg. Std.		30	K=	0.81	0.75	0.69	0.66	0.63	0.57	0.54	0.54	0.51	0.51	0.48	0.45	0.42	0.42	0.39	0.39	0.39	0.36	0.36	0.36
			Equiv.Ft.Pipe=	1.37	1.85	2.39	3.20	3.75	4.85	5.81	7.25	8.41	9.58	12.03	14.40	19.06	24.38	29.37	32.49	37.50	42.51	47.49	57.51
90 deg. LR		20	K=	0.54	0.50	0.46	0.44	0.42	0.38	0.36	0.36	0.34	0.34	0.32	0.30	0.28	0.28	0.26	0.26	0.26	0.24	0.24	0.24
			Equiv.Ft.Pipe=	0.91	1.24	1.59	2.13	2.50	3.23	3.87	4.83	5.61	6.38	8.02	9.60	12.71	16.25	19.58	21.66	25.00	28.34	31.66	38.34
45 deg. Std.		16	K=	0.43	0.40	0.37	0.35	0.34	0.30	0.29	0.29	0.27	0.27	0.26	0.24	0.22	0.22	0.21	0.21	0.21	0.19	0.19	0.19
			Equiv.Ft.Pipe=	0.73	0.99	1.28	1.70	2.00	2.59	3.10	3.87	4.48	5.10	6.42	7.68	10.17	13.00	15.66	17.33	20.00	22.87	25.33	30.67
Return Bends 180 deg.		50	K=	1.35	1.25	1.15	1.10	1.05	0.95	0.90	0.90	0.85	0.85	0.80	0.75	0.70	0.70	0.65	0.65	0.65	0.60	0.60	0.60
			Equiv.Ft.Pipe=	2.28	3.09	3.99	5.33	6.25	8.08	9.68	12.09	14.02	15.94	20.06	24.01	31.77	40.63	48.95	54.15	62.50	70.85	79.15	95.85
Tees: Branch Flow		60	K=	1.62	1.50	1.38	1.32	1.26	1.14	1.08	1.08	1.02	1.02	0.96	0.90	0.84	0.84	0.78	0.78	0.78	0.72	0.72	0.72
			Equiv.Ft.Pipe=	2.73	3.71	4.78	6.39	7.50	9.70	11.62	14.50	16.82	19.13	24.07	28.81	38.12	48.75	58.74	64.98	75.00	85.02	94.98	115.02
Straight Flow		20	K=	0.54	0.50	0.46	0.44	0.42	0.38	0.36	0.36	0.34	0.34	0.32	0.30	0.28	0.28	0.26	0.26	0.26	0.24	0.24	0.24
			Equiv.Ft.Pipe=	0.91	1.24	1.59	2.13	2.50	3.23	3.87	4.83	5.61	6.38	8.02	9.60	12.71	16.25	19.58	21.66	25.00	28.34	31.66	38.34
Mitre Elbows (re: Mark's Standard Handbook for Mechanical Engineers, 6th Edition, 1978, McGraw-Hill)																							
90 degree			Equiv.Ft.Pipe=	3.00	4.00	5.00	7.00	8.00	10.00	12.00	15.00	18.00	21.00	25.00	30.00	40.00	50.00	60.00	68.00	78.00	85.00	100.00	115.00
60 degree			Equiv.Ft.Pipe=	1.30	1.80	2.10	3.00	3.40	4.50	5.20	6.40	7.30	8.50	11.00	13.00	17.00	21.00	25.00	29.00	31.00	37.00	41.00	49.00
45 degree			Equiv.Ft.Pipe=	0.70	0.90	1.00	1.50	1.80	2.30	2.80	3.20	4.00	4.50	6.00	7.00	9.00	12.00	15.00	17.00	19.00	19.00	22.00	25.00
30 degree			Equiv.Ft.Pipe=	0.40	0.50	0.70	0.90	1.10	1.70	2.00	2.40	2.70	3.20	4.00	5.10	7.20	9.00	11.00	13.00	15.00	16.00	18.00	18.00

1. Mass velocity of air:

$G = pv \qquad p = \text{density} = .075 \text{ lb/ft}^3$
$V = \text{velocity} = 15 \text{ mile/hr} = 79{,}200 \text{ ft/hr}$
$G = 0.075 \times 79{,}200 = 5{,}940 \text{ lb/(h)(ft}^2)$

2. Heat transfer coefficient formula:

$h = 0.11 \times c \times G^{0.6}/D^{0.4}$
$[c = \text{specific heat in Btu/lb/}^\circ\text{F} \quad \text{Air} = 0.24]$
$h = 0.11 \times 0.24 \times (5940)^{0.6}/(3.5/12)^{0.4}$
$h = 7.94 \text{ Btu/(h)(ft}^2)(^\circ\text{F})$

3. Convection heat loss:

$Q/A = h \times (T_s - T_a)$
$\qquad = 7.94 \times (320 - 68) = 2000.9 \text{ Btu/hr ft}^{-2}$

Total heat loss for 50 ft of 3.5-in. pipe $Q = (401 + 2000.9) \times 45.81 = 110{,}030$
Btu/hr.

Also, see Table 8.4 on estimated piping heat loss, bare pipe.

IV. INSULATED PIPE HEAT LOSS [50]

A. Formula for Pipe Heat Loss (One Layer of Insulation– Calm Wind)

$$Q = \frac{T_s - T_a}{\left[\left(\dfrac{1}{f_1}\right) + \left(\dfrac{r_3 \log_e \dfrac{r_2}{r_1}}{k_1}\right) + \left(\dfrac{r_3 \log_e \dfrac{r_3}{r_2}}{k_2}\right) + \left(\dfrac{1}{f_2}\right)\right]}$$

$$\left(\frac{1}{f_1}\right) \qquad \left(\frac{r_3 \log_e \dfrac{r_2}{r_1}}{k_1}\right) \qquad \left(\frac{r_3 \log_e \dfrac{r_3}{r_2}}{k_2}\right) \qquad \left(\frac{1}{f_2}\right)$$

1. Resistance through fluid to pipe.	2. Resistance through pipe wall.	3. Resistance through insulation.	4. Resistance through outer air film.

Ralph L. Vandagriff

Date: 3/18/98

Formula: pg 2-12, Handbook of Chemical Engineering Calculations, McGraw-Hill, 1984

Heat Loss, Btu, Non-Insulated Saturated Steam Piping : Calm Air & with Wind

Variables:

Air Velocity, mph	15
Ambient Temp, F	68
Air density, lb/cu ft	0.075
Air mass velocity	5940

	50	75	100	125	150	175	200	225	250	300	330	400	500	600	700
Steam Pressure, psig	50	75	100	125	150	175	200	225	250	300	330	400	500	600	700
Steam Temp, deg F	297.7	320.1	337.9	352.8	365.9	377.5	387.7	397.3	406.1	421.7	435.7	448.1	470	488.8	505.4
Ambient Temp, deg F	68	68	68	68	68	68	68	68	68	68	68	68	68	68	68
Temp Diff, deg F	229.7	252.1	269.9	284.8	297.9	309.5	319.7	329.3	338.1	353.7	367.7	380.1	402	420.8	437.4
emissivity	0.8	0.8	0.8	0.8	0.8	0.8	0.8	0.8	0.8	0.8	0.8	0.8	0.8	0.8	0.8
Length of pipe, feet	1	1	1	1	1	1	1	1	1	1	1	1	1	1	1

ESTIMATED BTU LOSS per HOUR PER LINEAR FOOT OF BARE PIPE

Pipe Size	O.D.		Outside Surf sq ft	50	75	100	125	150	175	200	225	250	300	330	400	500	600	700
1/2" Pipe-no wind	0.840	lb	0.220	179	204	225	243	260	275	288	301	314	336	356	375	410	441	469
		WIND	14.055	786	867	933	988	1,038	1,081	1,120	1,157	1,191	1,252	1,307	1,336	1,445	1,521	1,590
3/4" Pipe-no wind	1.050	lb	0.275	217	247	273	295	315	333	350	366	381	408	433	456	498	536	571
		WIND	12.855	907	1,001	1,077	1,142	1,199	1,250	1,295	1,338	1,377	1,448	1,513	1,570	1,673	1,763	1,844
1" Pipe - no wind	1.315	lb	0.344	263	300	331	358	383	405	425	445	463	496	527	555	607	654	696
		WIND	11.746	1,048	1,158	1,246	1,321	1,388	1,447	1,500	1,550	1,596	1,679	1,754	1,822	1,942	2,048	2,142
1.5" pipe-no wind	1.900	lb	0.497	362	413	456	494	528	559	587	614	639	686	729	768	840	905	965
		WIND	10.140	1,330	1,471	1,585	1,681	1,767	1,843	1,912	1,976	2,036	2,143	2,240	2,328	2,485	2,622	2,745
2" pipe-no wind	2.375	lb	0.622	440	502	554	600	642	680	714	747	778	835	887	936	1,024	1,104	1,177
		WIND	9.274	1,539	1,703	1,836	1,948	2,048	2,138	2,217	2,293	2,363	2,488	2,603	2,705	2,889	3,051	3,196
2.5" pipe - no wind	2.875	lb	0.753	520	594	656	710	759	804	845	884	921	989	1,051	1,109	1,214	1,309	1,397
		WIND	8.392	1,745	1,932	2,083	2,212	2,326	2,429	2,520	2,607	2,687	2,831	2,962	3,079	3,291	3,477	3,644
3" pipe - no wind	3.500	lb	0.916	617	706	780	844	903	957	1,006	1,053	1,097	1,178	1,253	1,321	1,448	1,562	1,667
		WIND	7.842	1,988	2,202	2,375	2,523	2,655	2,772	2,877	2,977	3,069	3,235	3,386	3,522	3,767	3,982	4,176
4" pipe - no wind	4.500	lb	1.178	770	881	974	1,055	1,129	1,196	1,258	1,317	1,372	1,474	1,568	1,655	1,814	1,958	2,090
		WIND	7.182	2,350	2,606	2,813	2,989	3,146	3,286	3,413	3,533	3,644	3,843	4,025	4,189	4,484	4,743	4,978
5" pipe - no wind	5.563	lb	1.456	929	1,063	1,175	1,274	1,363	1,445	1,520	1,591	1,659	1,782	1,897	2,002	2,196	2,371	2,532
		WIND	6.598	2,710	3,007	3,247	3,453	3,636	3,801	3,948	4,087	4,217	4,450	4,663	4,855	5,201	5,506	5,782
6" pipe - no wind	6.625	lb	1.734	1,084	1,241	1,373	1,488	1,594	1,690	1,777	1,861	1,940	2,085	2,220	2,343	2,571	2,777	2,967
		WIND	6.153	3,050	3,386	3,659	3,892	4,100	4,287	4,454	4,613	4,761	5,026	5,269	5,488	5,883	6,232	6,548
6" pipe - no wind	6.625	lb	2.258	1,372	1,571	1,738	1,885	2,019	2,141	2,252	2,360	2,461	2,645	2,817	2,975	3,266	3,529	3,772
		WIND	5.537	3,651	4,057	4,386	4,671	4,924	5,151	5,354	5,548	5,728	6,051	6,348	6,616	7,100	7,528	7,916
10" pipe - no wind	10.750	lb	2.814	1,670	1,913	2,118	2,297	2,461	2,612	2,748	2,879	3,003	3,229	3,440	3,634	3,991	4,314	4,613
		WIND	5.070	4,249	4,726	5,114	5,447	5,745	6,014	6,253	6,482	6,694	7,078	7,429	7,746	8,321	8,830	9,291
12" pipe - no wind	12.750	lb	3.338	1,946	2,230	2,470	2,680	2,871	3,047	3,207	3,361	3,505	3,771	4,018	4,245	4,664	5,043	5,393
		WIND	4.735	4,783	5,323	5,765	6,143	6,481	6,787	7,060	7,320	7,562	8,000	8,401	8,763	9,420	10,003	10,531
14" pipe - no wind	14.000	lb	3.865	2,117	2,426	2,687	2,916	3,125	3,317	3,491	3,659	3,816	4,106	4,376	4,623	5,081	5,495	5,877
		WIND	4.561	5,105	5,684	6,158	6,563	6,927	7,255	7,548	7,826	8,088	8,558	8,990	9,380	10,088	10,715	11,285
16" pipe - no wind	16.000	lb	4.189	2,387	2,737	3,032	3,291	3,527	3,744	3,941	4,131	4,310	4,637	4,943	5,224	5,742	6,212	6,645
		WIND	4.324	5,606	6,246	6,769	7,218	7,621	7,984	8,309	8,619	8,908	9,430	9,909	10,342	11,129	11,828	12,463
18" pipe - no wind	18.000	lb	4.712	2,655	3,045	3,373	3,662	3,926	4,168	4,387	4,599	4,799	5,164	5,506	5,819	6,398	6,923	7,407
		WIND	4.125	6,564	7,320	7,940	8,472	8,950	9,382	9,768	10,137	10,481	11,103	11,675	12,193	13,134	13,971	14,733
20" pipe - no wind	20	lb	5.236	3,183	3,652	4,048	4,395	4,713	5,005	5,269	5,525	5,765	6,207	6,619	6,997	7,697	8,330	8,916
		WIND	3.955	6,092	6,790	7,362	7,853	8,294	8,692	9,048	9,388	9,704	10,277	10,803	11,279	12,144	12,912	13,610
22" pipe - no wind	22	lb	5.760	2,920	3,350	3,712	4,030	4,321	4,588	4,830	5,064	5,284	5,687	6,064	6,410	7,050	7,629	8,164
		WIND	3.607	7,025	7,837	8,504	9,076	9,591	10,056	10,472	10,871	11,241	11,912	12,529	13,088	14,105	15,010	15,833
24" pipe - no wind	24	lb	6.283	3,183	3,652	4,048	4,395	4,713	5,103	5,706	5,983	6,244	6,723	7,170	7,580	8,340	9,028	9,664
		WIND	3.677	7,475	8,344	9,056	9,668	10,219	10,717	11,163	11,589	11,987	12,706	13,367	13,967	15,059	16,030	16,915

Q = Btu/ft^2 h (outer surface)
r_1 = Inside radius of pipe (in.)
r_2 = Outside radius of pipe (in.)
r_2 = Inside radius of pipe insulation (in.)
r_3 = Outside radius of pipe insulation (in.)
T_s = Steam temperature (°F)
T_a = Ambient air temperature (°F)
f_1 = Inside steam to steel conductance
f_2 = Outer air film conductance:
K_1 = Conductivity of steel at temperature
K_2 = Conductivity of insulation at mean temperature
Q_L = Btu/linear Ft/hr heat loss

Example: 4″-Sch. 40 pipe, 100 psig saturated steam, 68°F ambient air, 3″ cal. silicate insulation

r_1 = 2.013
r_2 = 2.25
r_3 = 5.25
T_s = 337.9
T_a = 68
f_1 = 190
f_2 = 1.9
K_1 = 350
K_2 = .48

$$Q = \frac{337.9 - 68}{\left[\left(\dfrac{1}{190}\right) + \left(\dfrac{5.25 \log_e \dfrac{2.25}{2.013}}{350}\right) + \left(\dfrac{5.25 \log_e \dfrac{5.25}{2.25}}{.48}\right) + \left(\dfrac{1}{1.9}\right)\right]} = 27.54$$

$$Q_L = Q \times \left(\frac{2\, r_3\, \Pi}{12}\right) = 75.7$$

Note: In heat-transfer calculations covering losses from insulated pipes, Eqs (1 and 2) can be ignored.

    See Tables 8.5–8.7 for related information on heat loss and thermal conductivity.

TABLE 8.5

Ralph L. Vandagriff
Date: 6/1/98
re: # 50, page 31
Steam Conditions

ESTIMATED Piping Heat Loss

Heat Loss, Btu, Insulated Steam Pipe – Calm Air

VARIABLES

			h1	ho	k1	k2	f2	
Pressure, psig			100	337.9	68	269.9	350	0.48	1.9
Temperature, deg. F								
Ambient Temp, deg.F								
Pipe Conductivity, k1								
Insulation Conductivity, k2								
Outside air film conductance, f2								

Insulation: Calcium Silicate
Insulation Thickness, inches
Length of pipe, feet

ESTIMATED BTU LOSS per HOUR per LINEAR FOOT OF INSULATED PIPE

(Columns C1–C12 correspond to increasing insulation thickness; each column also lists outer radius r3.)

Pipe Size	O.D.	r2	C1	C2	C3	C4	C5	C6	C7	C8	C9	C10	C11	C12
.5" pipe	0.840	0.420	84.07	46.59	41.08	36.55	33.49	31.25	29.52	28.14	27.00	26.05	25.23	24.52
.75" pipe	1.050	0.525	74.09	55.06	46.00	40.60	36.97	34.33	32.31	30.70	29.38	28.27	27.33	26.51
1" pipe	1.315	0.658	86.54	62.96	51.97	45.47	41.13	38.00	35.61	33.71	32.17	30.88	29.78	28.83
1.5" pipe	1.900	0.950	113.61	79.93	64.57	55.66	49.77	45.56	42.37	39.86	37.83	36.14	34.71	33.48
2" pipe	2.375	1.188	135.37	93.38	74.46	63.60	56.45	51.37	47.54	44.54	42.11	40.10	38.41	36.95
2.5" pipe	2.875	1.438	158.16	107.38	84.72	71.76	63.29	57.29	52.78	49.27	46.43	44.09	42.12	40.43
3" pipe	3.500	1.750	186.36	126.73	97.35	81.78	71.63	64.50	59.15	54.99	51.65	48.89	46.58	44.61
4" pipe	4.500	2.250	231.86	152.28	117.32	97.54	84.73	75.76	69.06	63.67	59.71	56.30	53.44	51.01
5" pipe	5.563	2.782	279.96	181.40	138.35	114.09	98.44	87.49	79.35	73.06	68.03	63.92	60.48	57.56
6" pipe	6.625	3.313	327.94	210.40	159.23	130.47	111.96	99.03	89.47	82.08	76.19	71.37	67.35	63.95
8" pipe	8.625	4.313	418.22	264.87	198.37	161.11	137.22	120.57	108.27	98.79	91.26	85.11	80.00	75.68
10" pipe	10.750	5.375	514.09	322.63	239.80	193.46	163.84	143.21	128.00	116.31	107.02	99.44	93.19	87.88
12" pipe	12.750	6.375	604.29	376.92	278.70	223.85	188.78	164.40	146.44	132.65	121.72	112.82	105.44	99.21
14" pipe	14.000	7.000	660.65	412.84	302.99	242.60	204.33	177.60	157.93	142.82	130.85	121.12	113.05	106.24
16" pipe	16.000	8.000	750.82	465.08	341.43	273.07	229.17	198.88	176.25	159.04	145.41	134.34	125.16	117.43
18" pipe	18.000	9.000	840.96	519.30	380.64	303.32	253.97	219.71	194.32	175.21	159.92	147.50	137.22	128.35
20" pipe	20	10.000	931.13	573.51	419.42	333.54	278.74	240.72	212.77	191.34	174.39	160.63	149.23	139.63
22" pipe	22	11.000	1,021.28	627.72	458.20	363.74	303.49	261.70	230.98	207.44	188.82	173.72	161.21	150.68
24" pipe	24	12.000	1,111.43	681.91	496.96	393.93	328.23	282.66	249.17	223.52	203.24	186.78	173.16	161.70

TABLE 8.6

BOILER HOUSE NOTES:													Page No.	
Subject: Thermal Conductivity of Pipe Insulation														
Ralph L. Vandagriff Date: 6/2/98													re:	#50
Generic Type	Form	Density	Pipe Temperature											
		Lbs./cu.ft.	Deg.F.	Deg.F	Deg.F	Deg.F	Deg.F	Deg.F	Deg.F	Deg.F	Deg.F.	Deg.F.	Deg.F.	Deg.F.
		(maximum)	100	200	300	400	500	600	700	800	900	1000	1100	1200
			Thermal Conductivity of Insulation - Btu,inch/ft.,hour,deg.F.											
Calcium-Silicate	Pipe Covering	14	0.35	0.43	0.50	0.55	0.60	0.66	0.72					
		15				0.66	0.70	0.76	0.82	0.94	1.00	1.08	1.16	1.25
		24					0.70	0.72	0.75	0.77	0.80	0.83	0.87	0.91
Diatomaceous-silica	Pipe Covering	17 to 24					0.72	0.74	0.76	0.78	0.80	0.83	0.86	0.90
		26					0.80	0.82	0.84	0.87	0.90	0.93	0.96	0.99
Glass Fiber w/ binder	Pipe Covering	3 to 5	0.28	0.32	0.39	0.48								
		5 to 10	0.28	0.32	0.42	0.50								
Mineral Fiber	Pipe Covering	9 to 11	0.27	0.33	0.42									
		11 to 14	0.33	0.39	0.46	0.53								
		16 to 20	0.36	0.42	0.49	0.56	0.64							
Perlite-expanded	Pipe Covering	13 to 15	0.40	0.45	0.50	0.55	0.60	0.65	0.71	0.77	0.83	0.90		
Phenolic foam-expanded	Pipe Covering	2 to 3	0.26											
Polystyrene-expanded	Pipe Covering	1.0 to 1.5	0.28											
		1.75 to 2.0	0.24											
		2.0 to 2.2	0.24											
Polyurethane-expanded	Pipe Covering	up to 1.7	0.27											
		1.7 to 2.5	0.26											
		2.5 to 5.0	0.24											
Silica - expanded	Pipe Covering	13	0.40	0.42	0.45	0.50	0.55	0.60	0.65	0.70	0.76	0.82		
Elastomeric-Flexible	Pipe Covering	8.5 to 9.5	0.32											
Urethane Foam	Pipe Covering	up to 1.5	0.31											
		over 1.5	0.29											
Silica fibers	blanket	6.0 to 9.5						0.47	0.50	0.56	0.67	0.77	0.90	1.02
Mineral fibers w/ binders	blanket	7 to 9	0.30	0.34	0.38	0.44	0.52							
Mineral Fibers	blanket	8 to 12			0.49	0.57	0.68	0.82						
Glass fiber,fine,no binder	blanket	11 to 12	0.30	0.34	0.40	0.47	0.55	0.62	0.76	0.87	0.98	1.10		
Glass fiber,fine,w/ binder	blanket	0.5 to 2.0	0.24	0.30	0.38									
Glass fiber w/ binder	blanket	1 to 2	0.38	0.54										
Glass fiber w/ binder	blanket	2.5 to 3.5	0.35	0.48										
Alumina fiber	blanket	3 to 4				0.40	0.49	0.60	0.74	0.90	1.05	1.25	1.42	1.65
		6 to 8				0.33	0.38	0.44	0.53	0.62	0.72	0.83	0.94	1.06
		12				0.30	0.36	0.42	0.49	0.57	0.65	0.72	0.82	0.92

TABLE 3.1

BOILER HOUSE NOTES:
Subject: Thermal Expansion Data - METALS
Ralph L. Vandagriff
re # 16, page5-196 & others
Date: 1/14/2000
Page No:

Linear Thermal Expansion, Inches per 100 feet of run (in going from 70 deg. to indicated temperature)

Temperature Range: 70 deg. F. to 1400 deg. F.

Material	70 F.	100 F.	150 F.	200 F.	250F.	300 F.	350F.	400 F.	450F.	500 F.	550F.	600 F.	650F.	700 F.	750F.	800 F.	850F.	900 F.	950F.	1000F.	1100 F.	1200 F.	1300 F.	1400 F.
Carbon Steel: Carbon-moly steel & Low chrome steels (through 3% Cr)	0	0.23	0.61	0.99	1.40	1.82	2.26	2.70	3.16	3.62	4.11	4.60	5.11	5.63	6.16	6.70	7.25	7.81	8.35	8.89	10.04	11.10	12.22	13.34
Intermediate alloy steels 5 Cr Mo to 9 Cr Mo	0	0.22	0.56	0.94	1.33	1.71	2.10	2.50	2.93	3.35	3.80	4.24	4.69	5.14	5.62	6.10	6.59	7.07	7.56	8.06	9.05	10.00	11.06	12.05
Austenitic stainless steels 18 Cr; 8 Ni	0	0.34	0.90	1.46	2.03	2.61	3.20	3.80	4.41	5.01	5.62	6.24	6.87	7.50	8.15	8.80	9.46	10.12	10.80	11.48	12.84	14.20	15.56	16.92
Straight Chromium Stainless Steels: 12 Cr, 17 Cr and 27 Cr	0	0.20	0.53	0.86	1.21	1.56	1.93	2.30	2.69	3.08	3.49	3.90	4.31	4.73	5.16	5.60	6.05	6.49	6.94	7.40	8.31	9.20	10.11	11.01
25 Chrome - 20 Nickel	0	0.28	0.74	1.21	1.70	2.18	2.69	3.20	3.72	4.24	4.79	5.33	5.88	6.44	7.02	7.60	8.19	8.78	9.37	9.95	11.12	12.31	13.46	14.65
Monel 67 Ni-30 Cu	0	0.26	0.75	1.22	1.71	2.21	2.66	3.25	3.79	4.33	4.90	5.46	6.05	6.64	7.25	7.85	8.48	9.12	9.77	10.42	11.77	13.15	14.58	16.02
Monel 66 Ni - 29 CuAl	0			1.17		2.12		3.13		4.17		5.28		6.43		7.62		8.86		10.16	11.50	13.00	14.32	15.78
Aluminum	0	0.46	1.23	2.00	2.63	3.66	4.52	5.39	6.28	7.17	8.10	9.03												
Gray cast iron	0	0.21	0.55	0.90	1.27	1.64	2.03	2.42	2.83	3.24	3.67	4.11	4.57	5.03	5.50	5.98	6.47	6.97	7.50	8.02				
Bronze	0	0.38	0.96	1.56	2.17	2.79	3.42	4.05	4.69	5.33	5.98	6.64	7.29	7.95	8.62	9.30	9.99	10.68	11.37	12.05	13.47	14.92		
Brass	0	0.35	0.94	1.52	2.14	2.76	3.41	4.05	4.72	5.40	6.10	6.80	7.53	8.26	9.02	9.78	10.57	11.35	12.16	12.98	14.65	16.39		
Wrought iron	0	0.28	0.70	1.14	1.60	2.08	2.53	3.01	3.50	3.99	4.50	5.01	5.53	6.06	6.59	7.12	7.69	8.26	8.81	9.36				
Copper - nickel (70/30)	0	0.31	0.82	1.33	1.86	2.40	2.96	3.52																

V. PUMPING FORMULAS

A. Liquid Velocity Formulas

$$V = \frac{0.4085 \times gpm}{d^2} = \frac{0.2859 \times bph}{d^2} = \frac{0.0028368 \times gpm}{D^2}$$

$$= \frac{0.001985 \times bph}{D^2}$$

where

V = velocity of flow (ft/sec)
D = diameter of pipe (ft)
bph = barrels per hour (42 gal—oil)
d = diameter of pipe (in.)
gpm = U.S. gal/min.

B. Pump Horsepower Formulas

1. Centrifugal Pump Terminology

$$bhp = \frac{gpm \times tdh \times sp\ gr}{3960 \times efficiency} \qquad bhp = \frac{bph \times tdh \times sp\ gr}{5657 \times efficiency}$$

2. Positive Displacement Pump Terminology

$$bhp = \frac{gpm \times psi}{1714 \times efficiency} \qquad bhp = \frac{bph \times psi}{2449 \times efficiency}$$

where

gpm = U.S. gal/min delivered
tdh = total dynamic head (ft)
psi = lb/in.^{2} differential
bph = barrels (42 gall) per hour delivered
sp gr = specific gravity of liquid pumped
efficiency = pump efficiency, expressed as a decimal

C. Head and Pressure Formulas

$$Head\ (ft) = \frac{tdh\ (in\ psi) \times 2.31}{specific\ gravity} \qquad Head\ in\ psi = \frac{tdh\ (ft) \times sp\ gr}{2.31}$$

tdh = total dynamic head = the total discharge head minus the total suction head or plus the total suction lift.

Note: Suction head exists when the liquid supply level is above the pump centerline or impeller eye.

Note: Suction lift exists when the liquid supply level or suction source is below the pump centerline or impeller eye.

D. Electric Motor Formulas

$$\text{Electrical hp input to motor} = \frac{\text{pump bhp}}{\text{motor efficiency}}$$

$$\text{kW input to motor} = \frac{\text{pump bhp} \times 0.7457}{\text{motor efficiency}}$$

E. Specific Gravity in Pump Application

In any consideration of centrifugal pump application, the specific gravity of the liquid being pumped must be taken into account.

A centrifugal pump impeller converts mechanical energy into capacity and head in feet. In converting mechanical energy into capacity and head, the specific gravity has a direct effect on the amount of horsepower required. Whether we are considering a pump as producing a given head in feet of liquid or in pressure in pounds per square inch, the specific gravity will still affect the results directly.

Liquids heavier than water will require more power, and liquids lighter than water will require less horsepower, see the three examples of water, gas, and brine in Fig. 8.1

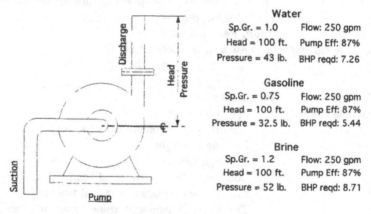

Water
Sp.Gr. = 1.0	Flow: 250 gpm
Head = 100 ft.	Pump Eff: 87%
Pressure = 43 lb.	BHP reqd: 7.26

Gasoline
Sp.Gr. = 0.75	Flow: 250 gpm
Head = 100 ft.	Pump Eff: 87%
Pressure = 32.5 lb.	BHP reqd: 5.44

Brine
Sp.Gr. = 1.2	Flow: 250 gpm
Head = 100 ft.	Pump Eff: 87%
Pressure = 52 lb.	BHP reqd: 8.71

FIGURE 8.1 Schematic of the effect of specific gravity on the amount of horsepower required to convert a pumps mechanical energy to its capacity and head.

1. Centrifugal Pump Formulas

$$\text{HP} = \frac{\text{gpm} \times \text{ft head} \times \text{sp gr}}{3960 \times \text{efficiency}}$$

$$\text{Feet of head} = \frac{\text{lb/in.}^2 \times 2.31}{\text{sp gr}}$$

$$\text{lb/in.}^2 = \frac{\text{ft head} \times \text{sp gr}}{2.31}$$

F. Centrifugal Pump Troubleshooting

1. Trouble: **Liquid not delivered**
    Possible causes: Pump not primed
                            Air or vapor pocket in suction line
                            Pump not up to rated speed
                            Wrong rotation
                            Impeller or passages clogged

2. Trouble: **Failure to deliver rated capacity and pressure**
    Possible causes: Available NPSH not sufficient
                            Pump not up to rated speed
                            Wrong rotation
                            Impeller or passages partially clogged
                            Wear rings worn or impeller damaged
                            Air or gases in liquid
                            Viscosity or specific gravity not as specified
                            Air or vapor pocket in suction line
                            Air leak in stuffing box
                            Total head greater than head for which pump designed
                            Injection of low-vapor–pressure oil in lantern ring
                                of hot pump

3. Trouble: **Pump loses prime**
    Possible causes: Air leak in suction line
                            Air leak in stuffing box
                            Air or gases in liquid

4. Trouble: **Pump overloads driver**
    Possible causes: Speed too high
                            Specific gravity or viscosity too high
                            Packing too tight
                            Misalignment
                            Total head lower than rated head
                            Low voltage or other electrical trouble
                            Trouble with engine, turbine, gear, or other allied
                                equipment

5. Trouble: **Pump vibration**
    Possible causes: Available NPSH not sufficient
                            Air or gases in liquid
                            Misalignment
                            Worn bearings
                            Damaged rotating element
                            Foundation not rigid
                            Pump operating below minimum recommended capacity
                            Impeller clogged

6. Trouble: **Stuffing box overheats**
    Possible causes: Packing too tight
                              Packing not lubricated
                              Incorrect type packing
                              Gland cocked
7. Trouble: **Bearings overheat or wear rapidly**
    Possible causes: Incorrect oil level
                              Misalignment or piping strains
                              Insufficient cooling water
                              Bearings too tight or preloaded
                              Oil rings not functioning
                              Suction pressure appreciably different from specified
                              Improper lubrication
                              Vibration
                              Dirt or water in bearings

VI. FORMULAS FOR ESTIMATING STEAM PROPERTIES

A. Saturated Steam, Dry

Y = property
X = steam pressure (psia)
Property formula: $Y = Ax + B/x + Cx^{1/2} + D \ln x + Ex^2 + Fx^3 + G$

1. Properties

1. Temperature (°F)
2. Liquid specific volume (ft^3/lb)
3. Vapor specific volume (ft^3/lb—1–200 psia)
4. Vapor specific volume (ft^3/lb—200–1500 psia)
5. Liquid enthalpy (Btu/lb)
6. Vaporization enthalpy (Btu/lb)
7. Vapor enthalpy (Btu/lb)
8. Liquid entropy (Btu/lb °R)
9. Vaporization entropy (Btu/lb°R)
10. Vapor entropy (Btu/lb °R)
11. Liquid internal energy (Btu/lb)
12. Vapor internal energy (Btu/lb)

See Tables 8.8 and 8.9 for formulas for estimating properties and spreadsheet examples of piping runs.

2. Constants

	A	B	C	D	E	F	G
1.	-0.17724	3.83986	11.48345	31.1311	8.762969×10^{-5}	-2.78794×10^{-8}	86.594
2.	-5.280126×10^{-7}	2.99461×10^{-5}	1.521874×10^{-4}	6.62512×10^{-5}	8.408856×10^{-10}	1.86401×10^{-14}	0.01596
3.	-0.48799	304.717614	9.8299035	-16.455274	9.474745×10^{-4}	-1.363366×10^{-8}	19.53953
4.	2.662×10^{-3}	457.5802	-0.176959	0.826862	-4.601876×10^{-7}	6.35×10^{-11}	-2.3928
5.	-0.15115567	3.671404	11.622558	30.832667	8.74117	-2.62306×10^{-8}	54.55
6.	0.008676153	-1.3049844	-8.2137368	-16.37649	-4.3043×10^{-5}	9.763×10^{-9}	$1,045.81$
7.	-0.14129	2.258225	3.4014802	14.438078	4.222624×10^{-5}	-1.569916×10^{-8}	$1,100.5$
8.	-1.67772×10^{-4}	4.272688×10^{-3}	0.01048048	0.05801509	9.101291×10^{-8}	-2.7592×10^{-11}	0.11801
9.	3.454439×10^{-3}	-2.75287×10^{-3}	-7.33044×10^{-3}	-0.14263733	-3.49366×10^{-8}	7.433711×10^{-12}	1.85565
10.	-1.476933×10^{-4}	1.2617946×10^{-3}	3.44201×10^{-3}	-0.08494128	6.89138×10^{-8}	-2.4941×10^{-11}	1.97364
11.	-0.154939	3.662121	11.632628	30.82137	8.76248×10^{5}	-2.646533×10^{-8}	54.56
12.	-0.0993951	1.93961	2.428354	10.9818864	2.737201×10^{-5}	-1.057475×10^{-8}	$1,040.03$

Table 8.8

Spreadsheet Example:	This spreadsheet is set up to calculate six (6) different saturated steam piping runs for pressure drop.				Boiler House Notes:	Page:
	The only variables that need to be entered are the length of pipe run, steam flow in lb./hr. and steam pressure in psig.					
	The answers are: Pressure drop per linear foot in psi, Total Pressure drop for pipe run(in psi) and Steam Velocity in Feet per Second.					

SATURATED STEAM PIPING
(Dry and Saturated)

	Run # 1	Run # 2	Run # 3	Run # 4	Run # 5	Run # 6
ENTER VARIABLES:						
Total Linear Feet of Pipe Run:	45.40	102.35	89.50	214.87	166.00	33.75
Steam Flow: Lb. per Hour:	1,500	2,200	13,000	17,000	22,000	30,000
Steam Pressure: PSIG:	15	15	15	15	15	300
CONVERSIONS						
Input: Steam Pressure. PSIA:	29.696	29.696	29.696	29.696	314.696	314.696
(.5 power: Steam Pressure)	5.4494036	5.4494036	5.4494036	5.4494036	17.739673	17.739673
(Log: Steam Pressure)	3.3910124	3.3910124	3.3910124	3.3910124	5.7516071	5.7516071
(Steam Pressure Squared)	881.85242	881.85242	881.85242	881.85242	99033.572	99033.572
(Steam Pressure Cubed)	26187.489	26187.489	26187.489	26187.489	31165469	31165469
SOLUTIONS						
Temperature: Degree F.:	249.68043	249.68043	249.68043	249.68043	421.40539	421.40539
Liquid specific volume. Cu.Ft./lb.:	0.0170002	0.0170002	0.0170002	0.0170002	0.018959	0.018959
Vapor specific volumn. Cu.Ft./lb.:						
1 - 200 psia	13.876318	13.876318	13.876318	13.876318	0	0
200 - 1,500 psia	0	0	0	0	1.4719551	1.4719551
Liquid enthalpy. BTU/Lb.	218.15128	218.15128	218.15128	218.15128	398.35055	398.35055
Vaporization enthalpy. BTU/Lb.	945.69315	945.69315	945.69315	945.69315	804.67783	804.67783
Vapor enthalpy. BTU/Lb.	1163.9129	1163.9129	1163.9129	1163.9129	1203.1196	1203.1196
Liquid entropy. BTU/Lb.(deg.R)	0.3670835	0.3670835	0.3670835	0.3670835	0.5929701	0.5929701
Vaporization enthropy. BTU/Lb.(deg.R)	1.3328959	1.3328959	1.3328959	1.3328959	0.9128427	0.9128427
Vapor enthropy. BTU/Lb.(deg.R)	1.6964367	1.6964367	1.6964367	1.6964367	1.5020843	1.5020843
Liquid Internal energy. BTU/Lb.	218.06522	218.06522	218.06522	218.06522	397.29583	397.29583
Vapor Internal energy. BTU/lb.	1087.8403	1087.8403	1087.8403	1087.8403	1117.3796	1117.3796

STEAM PIPING:

Linear Feet of Piping Run:

Steam Flow: Pounds per Hour:	1500 ΔP/Lin. Ft	45.40 Total ΔP	Vel. FPS	2200 ΔP/Lin. Ft	102.35 Total ΔP	Vel. FPS	13000 ΔP/Lin. Ft	89.50 Total ΔP	Vel. FPS	17000 ΔP/Lin. Ft	214.67 Total ΔP	Vel. FPS	22000 ΔP/Lin. Ft	168.00 Total ΔP	Vel. FPS	30000 ΔP/Lin. Ft	33.75 Total ΔP	Vel. FPS
IPS: 1"-Sch 40	3.9489	179.28	983.35	8.4944	869.40	1412.91	298.6030	26,545.97	8949.02	507.2087	106,882.50	10917.95	90.1062	14,857.62	1468.77	167.5528	5,654.91	2043.78
1"-Sch 80	6.7139	304.81	1157.47	14.4423	1,478.16	1667.63	504.2854	45,133.54	10031.43	862.3576	185,122.35	13118.02	153.1088	25,431.00	1800.79	294.8738	9,814.49	2455.62
1.5"-Sch 40	0.3386	15.37	408.96	0.7283	74.54	599.81	25.4303	2,278.01	3544.34	43.4873	9,335.42	4634.90	7.7256	1,282.44	638.26	14.3657	484.84	667.83
1.5"-Sch 80	0.5067	23.01	471.14	1.0001	111.57	691.01	38.0620	3,406.55	4083.23	65.0883	13,972.50	5339.61	11.5630	1,919.46	733.00	21.5014	725.67	999.55
2"-Sch 40	0.0822	3.73	248.12	0.1799	18.11	363.90	6.1770	552.84	2150.33	10.5931	2,267.58	2811.97	1.8765	311.51	386.02	3.4694	117.77	526.39
2"-Sch 80	0.1180	5.36	261.95	0.2537	25.97	413.53	8.8900	792.97	2443.61	15.1511	3,252.49	3195.48	2.6916	446.81	438.66	5.0051	168.92	598.18
3"-Sch 40	0.0090	0.41	112.82	0.0195	1.99	165.16	0.6797	60.84	978.06	1.1624	249.53	1276.38	0.2065	34.28	175.22	0.3840	12.96	238.93
3"-Sch 80	0.0124	0.56	126.05	0.0266	2.72	184.67	0.9290	83.14	1092.42	1.5866	341.02	1428.55	0.2822	46.85	196.11	0.5248	17.71	267.42
4"-Sch 40	0.0020	0.09	65.40	0.0044	0.45	95.92	0.1522	13.63	569.81	0.2603	55.89	741.22	0.0462	7.68	101.75	0.0990	2.90	138.75
4"-Sch 80	0.0027	0.12	72.42	0.0058	0.59	106.21	0.2013	18.01	627.62	0.3442	73.88	820.73	0.0611	10.15	112.67	0.1137	3.84	153.64
6"-Sch 40	0.0002	0.01	28.82	0.0005	0.05	42.27	0.0185	1.48	249.76	0.0282	6.08	326.61	0.0050	0.83	44.84	0.0093	0.31	61.14
6"-Sch 80	0.0003	0.01	31.84	0.0006	0.06	46.85	0.0218	1.95	276.82	0.0372	7.99	361.99	0.0066	1.10	49.66	0.0123	0.42	67.76
8"-Sch 40	0.0001	0.00	18.64	0.0001	0.01	24.41	0.0038	0.34	144.24	0.0065	1.40	188.62	0.0012	0.19	25.89	0.0022	0.07	35.31
8"-Sch 80	0.0001	0.00	18.23	0.0001	0.01	26.74	0.0049	0.43	158.02	0.0083	1.78	206.64	0.0015	0.24	28.37	0.0027	0.09	38.66
10"-Sch 40	0.0000	0.00	10.56	0.0000	0.00	15.49	0.0011	0.10	91.51	0.0020	0.42	119.66	0.0003	0.06	16.43	0.0006	0.02	22.40
10"-Sch 80	0.0000	0.00	11.59	0.0000	0.00	17.00	0.0015	0.13	100.44	0.0026	0.54	131.34	0.0004	0.07	18.03	0.0008	0.03	24.59
12"-.375 wall	0.0000	0.00	8.04	0.0000	0.00	8.86	0.0003	0.02	52.33	0.0005	0.10	68.43	0.0001	0.01	9.39	0.0001	0.01	12.81
12"-.500 wall	0.0000	0.00	6.27	0.0000	0.00	9.20	0.0003	0.03	54.39	0.0005	0.11	71.09	0.0001	0.01	9.76	0.0002	0.01	13.31
14"-.375 wall	0.0000	0.00	6.04	0.0000	0.00	8.66	0.0003	0.02	52.33	0.0005	0.10	68.43	0.0001	0.01	9.39	0.0001	0.01	12.81
14"-.438 wall	0.0000	0.00	6.15	0.0000	0.00	9.03	0.0003	0.02	53.34	0.0005	0.10	69.75	0.0001	0.01	9.58	0.0002	0.01	13.06
16"-.375 wall	0.0000	0.00	4.56	0.0000	0.00	6.69	0.0001	0.01	39.50	0.0002	0.05	51.66	0.0000	0.01	7.09	0.0001	0.00	9.67
16"-.375 wall	0.0000	0.00	4.71	0.0000	0.00	6.91	0.0001	0.01	40.83	0.0002	0.05	53.40	0.0000	0.01	7.33	0.0001	0.00	10.00
16"-.375 wall	0.0000	0.00	3.56	0.0000	0.00	5.23	0.0001	0.01	30.88	0.0001	0.02	40.38	0.0000	0.00	5.54	0.0000	0.00	7.56
16"-.500 wall	0.0000	0.00	3.67	0.0000	0.00	5.38	0.0001	0.01	31.79	0.0001	0.03	41.57	0.0000	0.00	5.71	0.0000	0.00	7.78

TABLE 8.9

#	A	B	E	F
1	Spreadsheet Example	Spreadsheet software used Excel® by Microsoft®	Lb of steam per second	=B35/3600
2	FORMULAS	Formulas shown for Columns B, C and D are common to each run. Column designations change	Cu Ft of steam per second	=IF(B21=0,B22*F1,B21*F1)
3			Pipe Inside Diameter:Inches	Area:Sq.Ft.
4				
5	Program: Steam Properties(Dry & Sat.)		1.049	=((E4/2)^2*3.1416)/144
6		Run # 1	0.957	=((E5/2)^2*3.1416)/144
7	ENTER VARIABLES:		1.61	=((E6/2)^2*3.1416)/144
8	Total Linear Feet of Pipe Run:	200	1.5	=((E7/2)^2*3.1416)/144
9	Steam Flow: Lb. per Hour:	80000	2.067	=((E8/2)^2*3.1416)/144
10	Steam Pressure: PSIG:	225	1.939	=((E9/2)^2*3.1416)/144
11		CONVERSIONS	3.068	=((E10/2)^2*3.1416)/144
12	Input: Steam Pressure PSIA:	=B10+14.696	2.9	=((E11/2)^2*3.1416)/144
13	(.5 power: Steam Pressure)	=(B12)^0.5	4.026	=((E12/2)^2*3.1416)/144
14	(Log. Steam Pressure)	=LN(B12)	3.826	=((E13/2)^2*3.1416)/144
15	(Steam Pressure Squared)	=(B12)^2	6.065	=((E14/2)^2*3.1416)/144
16	(Steam Pressure Cubed)	=(B12)^3	5.761	=((E15/2)^2*3.1416)/144
17		SOLUTIONS	7.981	=((E16/2)^2*3.1416)/144
18	Temperature Degree F:	=(-0.1772*4*B12)+(3.89966/B12)+(11.48345*B13)+(31.1311*B14)+(0.00098762989*B15)+(-0.0000002378784*B16)+86.594	7.625	=((E17/2)^2*3.1416)/144
19	Liquid specific volume, Cu Ft./lb	=(-0.000000026601*B12)+(0.0000299461/B12)+(0.000152191*B13)+(0.0000663*B14)+(0.0000000000002*B16)+0.01596	10.02	=((E18/2)^2*3.1416)/144
20	Vapor specific volume, Cu Ft./lb	=IF(B12<200,(-0.48798*B12)+(304.71764/B12)+(9.8299035*B13)+(-16.45527*4*B14)+(0.0004747457*B15)+(-0.0000013633666*B16)+19.53953,0)	9.564	=((E19/2)^2*3.1416)/144
21	1 - 200 psia	=IF(B12<200,(0.0026652*B12)+(11.622656*B13)+(30.832667*B13)+(-16.37649*B13)+(0.0000874117*B15)+(-0.00000002822061*B16)-54.55	13.25	=((E20/2)^2*3.1416)/144
22	200 - 1,500 psia	=IF(B12>=200,(0.0026652*B12)+(11.622656*B13)+(30.832667*B13)+(0.0000874117*B15)+(-0.00000002822061*B16)-54.55	13	=((E21/2)^2*3.1416)/144
23	Liquid enthalpy, BTU/lb.	=(-0.16115567*B12)+(3.671404*B12)+(11.622656*B13)+(30.832667*B13)+(-16.37649*B14)+(-0.0000430437*B15)+(0.0000009763*B16)+1045.81	13.25	=((E22/2)^2*3.1416)/144
24	Vaporization enthalpy, BTU/lb.	=(0.006878153*B12)+(-1.30498644*B12)+(-6.2137369*B13)+(14.43678*B14)+(0.00004222824*B15)+(-0.0000001569916*B16)+1100.5	13.124	=((E23/2)^2*3.1416)/144
25	Vapor enthalpy, BTU/lb.	=(-0.14126*B12)+(2.258225*B12)+(3.401602*B13)+(14.43678*B13)+(0.0580150*B14)+(-0.000000910291*B15)+(-0.000000009763*B16)+1045.81	15.25	=((E24/2)^2*3.1416)/144
26	Liquid entropy, BTU/Lb.(deg.R)	=(-0.00016777*B12)+(0.0042726898/B12)+(0.01048046*B13)+(-0.14263*B13)+(0.00773304*B14)+(0.000000910291*B15)+(-0.000000000027592*B16)+0.118	15	=((E25/2)^2*3.1416)/144
27	Vaporization entropy, BTU/Lb.(deg.R)	=(0.00000345439*B12)+(-0.002752087/B12)+(0.00126179465/B12)+(0.03344201*B13)+(0.08494129*B14)+(0.000000048366*B15)+(0.0000000000246366*B16)+1.6556	17.25	=((E26/2)^2*3.1416)/144
28	Vapor entropy, BTU/Lb.(deg.R)	=(-0.00014766933*B12)+(0.0012617/B12)+(-0.82137*B13)+(30.82137*B14)+(0.0000067624*B15)+(-0.0000000264533*B16)+1.97	17	=((E27/2)^2*3.1416)/144
29	Liquid internal energy, BTU/lb.	=(-0.1549439*B12)+(3.662121/B12)+(11.632628*B13)+(30.82137*B14)+(0.000067624*B15)+(-0.0000000264533*B16)+54.56		
31	Vapor internal energy, BTU/lb.	=(-0.093395*B12)+(1.93961/B12)+(2.428354*B13)+(10.981864*B14)+(0.0002737201*B15)+(-0.0000001057475*B16)+1040.03		

		Δ P / Lin Ft	Total Δ P	Vel FPS
32	STEAM PIPING :			
33	Linear Feet of Piping Run:			
34		=B9	=B8	
35	Steam Flow Pounds per Hour			
36		=IF($B$21=0,(($B$25*(1+(3.6E4))/1000000000000)*$B$21*1*($B$36)/2)/(E4)^5,(($B$25*(1+(3.6E4))/1000000000000)*$B$21*1*($B$35)/2)/(E4)^5)	Total Δ P	Vel FPS
37	IPS: 1" Sch 40	=IF($B$21=0,(($B$25*(1+(3.6E4))/1000000000000)*$B$22*1*($B$35)/2)/(E4)^5,(($B$25*(1+(3.6E4))/1000000000000)*$B$21*1*($B$35)/2)/(E4)^5)	=$C$33*B37	=$F$2/F4
38	1" Sch 80	=IF($B$21=0,(($B$25*(1+(3.6E5))/1000000000000)*$B$22*1*($B$35)/2)/(E5)^5,(($B$25*(1+(3.6E5))/1000000000000)*$B$21*1*($B$35)/2)/(E5)^5)	=$C$33*B38	=$F$2/F5
39	1.5" Sch 40	=IF($B$21=0,(($B$25*(1+(3.6E6))/1000000000000)*$B$22*1*($B$35)/2)/(E6)^5,(($B$25*(1+(3.6E6))/1000000000000)*$B$21*1*($B$35)/2)/(E6)^5)	=$C$33*B39	=$F$2/F6
40	1.5" Sch 80	=IF($B$21=0,(($B$25*(1+(3.6E7))/1000000000000)*$B$22*1*($B$35)/2)/(E7)^5,(($B$25*(1+(3.6E7))/1000000000000)*$B$21*1*($B$35)/2)/(E7)^5)	=$C$33*B40	=$F$2/F7
41	2" Sch 40	=IF($B$21=0,(($B$25*(1+(3.6E8))/1000000000000)*$B$22*1*($B$35)/2)/(E8)^5,(($B$25*(1+(3.6E8))/1000000000000)*$B$21*1*($B$35)/2)/(E8)^5)	=$C$33*B41	=$F$2/F8
42	2" Sch 80	=IF($B$21=0,(($B$25*(1+(3.6E9))/1000000000000)*$B$22*1*($B$35)/2)/(E9)^5,(($B$25*(1+(3.6E9))/1000000000000)*$B$21*1*($B$35)/2)/(E9)^5)	=$C$33*B42	=$F$2/F9
43	3" Sch 40	=IF($B$21=0,(($B$25*(1+(3.6E10))/1000000000000)*$B$22*1*($B$35)/2)/(E10)^5,(($B$25*(1+(3.6E10))/1000000000000)*$B$21*1*($B$35)/2)/(E10)^5)	=$C$33*B43	=$F$2/F10
44	3" Sch 80	=IF($B$21=0,(($B$25*(1+(3.6E11))/1000000000000)*$B$22*1*($B$35)/2)/(E11)^5,(($B$25*(1+(3.6E11))/1000000000000)*$B$21*1*($B$35)/2)/(E11)^5)	=$C$33*B44	=$F$2/F11
45	4" Sch 40	=IF($B$21=0,(($B$25*(1+(3.6E12))/1000000000000)*$B$22*1*($B$35)/2)/(E12)^5,(($B$25*(1+(3.6E12))/1000000000000)*$B$21*1*($B$35)/2)/(E12)^5)	=$C$33*B45	=$F$2/F12
46	4" Sch 80	=IF($B$21=0,(($B$25*(1+(3.6E13))/1000000000000)*$B$22*1*($B$35)/2)/(E13)^5,(($B$25*(1+(3.6E13))/1000000000000)*$B$21*1*($B$35)/2)/(E13)^5)	=$C$33*B46	=$F$2/F13
47	6" Sch 40	=IF($B$21=0,(($B$25*(1+(3.6E14))/1000000000000)*$B$22*1*($B$35)/2)/(E14)^5,(($B$25*(1+(3.6E14))/1000000000000)*$B$21*1*($B$35)/2)/(E14)^5)	=$C$33*B47	=$F$2/F14
48	6" Sch 80	=IF($B$21=0,(($B$25*(1+(3.6E15))/1000000000000)*$B$22*1*($B$35)/2)/(E15)^5,(($B$25*(1+(3.6E15))/1000000000000)*$B$21*1*($B$35)/2)/(E15)^5)	=$C$33*B48	=$F$2/F15
49	8" Sch 40	=IF($B$21=0,(($B$25*(1+(3.6E16))/1000000000000)*$B$22*1*($B$35)/2)/(E16)^5,(($B$25*(1+(3.6E16))/1000000000000)*$B$21*1*($B$35)/2)/(E16)^5)	=$C$33*B49	=$F$2/F16
50	8" Sch 80	=IF($B$21=0,(($B$25*(1+(3.6E17))/1000000000000)*$B$22*1*($B$35)/2)/(E17)^5,(($B$25*(1+(3.6E17))/1000000000000)*$B$21*1*($B$35)/2)/(E17)^5)	=$C$33*B50	=$F$2/F17
51	10" Sch 40	=IF($B$21=0,(($B$25*(1+(3.6E18))/1000000000000)*$B$22*1*($B$35)/2)/(E18)^5,(($B$25*(1+(3.6E18))/1000000000000)*$B$21*1*($B$35)/2)/(E18)^5)	=$C$33*B51	=$F$2/F18
52	10" Sch 80	=IF($B$21=0,(($B$25*(1+(3.6E19))/1000000000000)*$B$22*1*($B$35)/2)/(E19)^5,(($B$25*(1+(3.6E19))/1000000000000)*$B$21*1*($B$35)/2)/(E19)^5)	=$C$33*B52	=$F$2/F19
53	12".375 wall	=IF($B$21=0,(($B$25*(1+(3.6E20))/1000000000000)*$B$22*1*($B$35)/2)/(E20)^5,(($B$25*(1+(3.6E20))/1000000000000)*$B$21*1*($B$35)/2)/(E20)^5)	=$C$33*B53	=$F$2/F20
54	12".500 wall	=IF($B$21=0,(($B$25*(1+(3.6E21))/1000000000000)*$B$22*1*($B$35)/2)/(E21)^5,(($B$25*(1+(3.6E21))/1000000000000)*$B$21*1*($B$35)/2)/(E21)^5)	=$C$33*B54	=$F$2/F21
55	14".375 wall	=IF($B$21=0,(($B$25*(1+(3.6E22))/1000000000000)*$B$22*1*($B$35)/2)/(E22)^5,(($B$25*(1+(3.6E22))/1000000000000)*$B$21*1*($B$35)/2)/(E22)^5)	=$C$33*B55	=$F$2/F22
56	14".438 wall	=IF($B$21=0,(($B$25*(1+(3.6E23))/1000000000000)*$B$22*1*($B$35)/2)/(E23)^5,(($B$25*(1+(3.6E23))/1000000000000)*$B$21*1*($B$35)/2)/(E23)^5)	=$C$33*B56	=$F$2/F23
57	16".375 wall	=IF($B$21=0,(($B$25*(1+(3.6E24))/1000000000000)*$B$22*1*($B$35)/2)/(E24)^5,(($B$25*(1+(3.6E24))/1000000000000)*$B$21*1*($B$35)/2)/(E24)^5)	=$C$33*B57	=$F$2/F24
58	16".500 wall	=IF($B$21=0,(($B$25*(1+(3.6E25))/1000000000000)*$B$22*1*($B$35)/2)/(E25)^5,(($B$25*(1+(3.6E25))/1000000000000)*$B$21*1*($B$35)/2)/(E25)^5)	=$C$33*B58	=$F$2/F25
59	18".375 wall	=IF($B$21=0,(($B$25*(1+(3.6E26))/1000000000000)*$B$22*1*($B$35)/2)/(E26)^5,(($B$25*(1+(3.6E26))/1000000000000)*$B$21*1*($B$35)/2)/(E26)^5)	=$C$33*B59	=$F$2/F26
60	18".500 wall	=IF($B$21=0,(($B$25*(1+(3.6E27))/1000000000000)*$B$22*1*($B$35)/2)/(E27)^5,(($B$25*(1+(3.6E27))/1000000000000)*$B$21*1*($B$35)/2)/(E27)^5)	=$C$33*B60	=$F$2/F27

B. Superheated Steam

P = pressure, atmospheres (psia/14.696)

T = temperature, Kelvin (°C + 273.15) [(°F + 459.67)/1.8]

1. Formulas

H = enthalpy (Btu/lb)

$$= 775.596 + (0.63296 \times T) + (0.000162467 \times T^2)$$
$$+ (47.3635 \times \log T) + 0.043557 \{C_7 P$$
$$+ 0.5\, C_4[C_{11} + C_3(C_{10} + C_9 C_4)]\}$$

v = Specific volume (ft³/lb)

$$= \{[(C_8 C_4 C_3 + C_{10})(C_4/P) + 1]\, C_3 + 4.55504\, (T/P)\}\, 0.016018$$

S = entropy (Btu/lb °F)

$$= 1/T\{[[(C_8 C_3 - 2\, C_9)C_3 C_4/2 - C_{11}]C_4/2 + (C_3 - C_7)P\}$$
$$\times (-0.0241983) - 0.355579 - 11.4276/T + 0.00018052T$$
$$- 0.253801 \log P + 0.809691 \log T$$

2. Constants

$C_1 = 80{,}870/T^2$

$C_2 = (-2641.62/T) \times 10 C_1$

$C_3 = 1.89 + C_2$

$C_4 = C_3(P^2/T^2)$

$C_5 = 2 + (372{,}420/T^2)$

$C_6 = C_5 \times C_2$

$C_7 = 1.89 + C_6$

$C_8 = 0.21878T - 126{,}970/T$

$C_9 = 2C_8 C_7 - (C_3/T)(126{,}970)$

$C_{10} = 82.546 - 162{,}460/T$

$C_{11} = 2C_{10} C_7 - (C_3/T)(162{,}460)$

See Tables 8.10 and 8.11 spreadsheet examples and formulas for steam properties and piping runs. Table 8.12 presents formulas for estimating superheated steam properties.

TABLE 8.10

Spreadsheet Example: This spreadsheet is set up to calculate six (6) superheated steam piping runs for pressure drop and velocity
Variables to be entered are; length of pipe run in feet, steam flow in lb/hr., steam pressure in psig and steam temperature in deg. F
Answers calculated: Steam velocity in feet per second and total pipe run pressure drop in psi

Superheated Steam Piping

Ralph L. Vandagriff	Date:1/31/96				Boiler House Notes:	Page:
			VARIABLES			
Length of Pipe Run; Feet	100	50	100	50	100	50
Steam Flow; lb/hour	200,000	200,000	200,000	25000	50,000	100,000
Steam Pressure: psig	850.00	600.00	600.00	400.00	900.00	250.00
Steam Temperature: deg. F.	825.00	750.00	825.00	650.00	825.00	500.00
			Conversions			
Pressure: psia	864.6960	614.6960	614.6960	414.6960	914.6960	264.6960
Pressure: atmospheres	58.8389	41.8274	41.8274	28.2183	62.2412	18.0114
Temperature: deg. K.	713.7056	672.0389	713.7056	616.4833	713.7056	533.1500
			CONSTANTS			
c_1 =	0.158763	0.17906002	0.158763	0.21278684	0.158763	0.28450416
c_2 =	-5.33475235	-5.93657502	-5.33475235	-6.99416451	-5.33475235	-9.53948847
c_3 =	-3.44475235	-4.04657502	-3.44475235	-5.10416451	-3.44475235	-7.64948847
c_4 =	-0.02341254	-0.0156755	-0.01183157	-0.01069408	-0.02619842	-0.00873032
c_5 =	2.7311304	2.82460162	2.7311304	2.97991931	2.7311304	3.31018966
c_6 =	-14.5699043	-16.7684595	-14.5699043	-20.8420459	-14.5699043	-31.5775161
c_7 =	-12.6799043	-14.8784595	-12.6799043	-18.9520459	-12.6799043	-29.6875161
c_8 =	-21.7579949	-41.9038507	-21.7579949	-71.0843045	-21.7579949	-121.508057
c_9 =	1164.60863	2011.4591	1164.60863	3745.63221	1164.60863	9036.27519
c_{10} =	-145.082885	-159.195963	-145.082885	-180.980995	-145.082885	-222.171247
c_{11} =	4463.39934	5715.40834	4463.39934	8205.00536	4463.39934	15522.356

	Solutions					
Specific Volume: cu.ft./lb.	0.82662243	1.10333645	1.18751203	1.50652394	0.7780531	2.02353448
H, enthalpy: Btu/lb.	1410.17726	1378.93963	1420.8601	1334.22384	1407.98498	1261.71307
Lb. per Hour of Steam	200000	200000	200000	25000	50000	100000
Pipe Size	Velocity Ft./Second / Pressure Drop psi	Velocity Ft./Second / Pressure Drop psi	Velocity Ft./Second / Pressure Drop psi	Velocity Ft./Second / Pressure Drop psi	Velocity Ft./Second / Pressure Drop psi	Velocity Ft./Second / Pressure Drop psi
1" - Sch 40	7512.01 / 418,774.68	10028.49 / 270,530.96	10791.63 / 601,604.74	1711.34 / 5,982.65	1767.66 / 24,635.56	9194.53 / 128,142.70
1.5"-Sch 80	3189.01 / 35,906.11	4257.31 / 23,966.56	4581.27 / 51,580.68	726.50 / 511.23	750.41 / 2,112.22	3903.27 / 10,966.78
2"-Sch 80	2198.63 / 12,509.45	2935.16 / 8,350.03	3158.51 / 17,970.87	500.88 / 178.11	517.36 / 735.90	2691.07 / 3,827.82
2.5"-Sch 80	1531.82 / 4,623.98	2044.98 / 3,019.74	2200.59 / 6,499.07	348.97 / 64.41	360.45 / 266.14	1874.92 / 1,384.31
3"-Sch 80	982.90 / 1,311.59	1312.17 / 875.46	1412.02 / 1,884.21	223.92 / 18.67	231.29 / 77.16	1203.05 / 401.34
3"-Sch 160	1198.72 / 2,278.84	1600.28 / 1,521.12	1722.06 / 3,273.75	273.08 / 32.45	282.07 / 134.09	1467.21 / 697.31
4"-Sch 80	564.70 / 284.16	753.87 / 189.67	811.24 / 408.22	128.65 / 4.05	132.88 / 18.72	691.18 / 88.95
4"-Sch 160	699.35 / 511.55	933.63 / 341.46	1004.68 / 734.88	159.32 / 7.28	164.57 / 30.09	855.99 / 156.53
6"-Sch 80	249.06 / 30.73	332.50 / 20.51	357.80 / 44.15	56.74 / 0.44	58.61 / 1.81	304.85 / 9.40
6"-Sch 160	307.00 / 54.04	409.84 / 36.07	441.03 / 77.63	69.94 / 0.77	72.24 / 3.18	375.76 / 16.54
8"-.5" wall	142.18 / 6.88	189.80 / 4.58	204.25 / 9.85	32.39 / 0.10	33.46 / 0.40	174.02 / 2.10
8"-.875" wall	174.89 / 11.91	233.48 / 7.95	251.24 / 17.10	39.84 / 0.17	41.15 / 0.70	214.06 / 3.64
10"-.5" wall	86.96 / 1.87	116.09 / 1.25	124.92 / 2.68	19.81 / 0.03	20.46 / 0.11	106.43 / 0.57
10"-.718" wall	95.29 / 2.37	127.21 / 1.58	136.89 / 3.41	21.71 / 0.03	22.42 / 0.14	116.63 / 0.73
12"-.5" wall	59.87 / 0.70	79.93 / 0.47	86.01 / 1.01	13.64 / 0.01	14.09 / 0.04	73.28 / 0.21
12"-.843" wall	67.53 / 0.98	90.15 / 0.64	97.01 / 1.38	15.38 / 0.01	15.89 / 0.06	82.65 / 0.29
14"-.5 " wall	48.91 / 0.41	65.30 / 0.28	70.27 / 0.59	11.14 / 0.01	11.51 / 0.02	59.87 / 0.13
14"-.75" wall	52.90 / 0.51	70.63 / 0.34	76.00 / 0.73	12.05 / 0.01	12.45 / 0.03	64.75 / 0.16
16"-.5 wa...	36.74 / 0.20	49.05 / 0.13	52.78 / 0.28	8.37 / 0.00	8.65 / 0.01	44.97 / 0.08
16"-.75" wall	39.32 / 0.23	52.49 / 0.16	56.48 / 0.34	8.96 / 0.00	9.25 / 0.01	48.12 / 0.07
18"-.5" wall	28.60 / 0.10	38.18 / 0.07	41.09 / 0.15	6.52 / 0.00	6.73 / 0.01	35.01 / 0.03
18"-.75" wall	30.36 / 0.12	40.53 / 0.08	43.62 / 0.17	6.92 / 0.00	7.14 / 0.01	37.16 / 0.04
20"-.5" wall	22.90 / 0.06	30.57 / 0.04	32.90 / 0.08	5.22 / 0.00	5.39 / 0.00	28.03 / 0.02
20"-.75" wall	24.15 / 0.07	32.24 / 0.04	34.70 / 0.10	5.50 / 0.00	5.68 / 0.00	29.56 / 0.02
24"-.5" wall	15.63 / 0.02	20.86 / 0.01	22.45 / 0.03	3.56 / 0.00	3.68 / 0.00	19.13 / 0.01
24"-.75" wa...	16.33 / 0.02	21.80 / 0.02	23.46 / 0.03	3.72 / 0.00	3.84 / 0.00	19.99 / 0.01

TABLE 8.11

	A	B	C
1	Spreadsheet Example:	Spreadsheet software used: Excel® by Microsoft®	
2	FORMULAS.	Formulas shown for Columns B and C are common to each run. Column designation changes.	
3			
4			S
5	Ralph L. Vandagriff	Date:1/31/96	
6			URR
7	Length of Pipe Run: Feet	100	
8	Steam Flow; lb/hour	200000	
9	Steam Pressure: psig	850	
10	Steam Temperature: deg. F.	825	
11			Con
12	Pressure: psia	=B9+14.696	
13	Pressure: atmospheres	=B12/14.696	
14	Temperature: deg. K.	=(5/9)*(B10+459.67)	
15			CON
16	c1 =	=80870/(B14^2)	
17	c2 =	=(-2641.62/B14)*(10^B16)	
18	c3 =	=1.89+B17	
19	c4 =	=B18*(B19^2/B14^2)	
20	c5 =	=2+(372420/B14^2)	
21	c6 =	=B17*B20	
22	c7 =	=1.89-B21	
23	c8 =	=(0.21878*B14)-(126970/B14)	
24	c9 =	=(2*(B22*B23))-((B18/B14)*126970)	
25	c10 =	=82.546-(162460/B14)	
26	c11 =	=(2*(B22*825))-((B18/B14)*162460)	
27			Bol
28	Specific Volume: cu.ft./lb.	=(((((B23*B19*B18+B25)*(B19/B13)+1)*B18)+(4.55504*(B14/B13))))*0.016018	
29			
30	H, enthalpy: Btu/lb.	=(775.596+(0.63296*B14)+(0.000162467*(B14^2))+(47.3635*(LOG(B14))))+	
31		(0.5*B19)*(B26+B18*(825+(B24*B19)))))	
32			
33	Lb. per Hour of Steam	=B8	
34		Velocity	Pressure Drop
35	Pipe Size	Ft./Second	psi
36	1" - Sch 40	=0.05*B33*(B28/($P$36^2))	=0.0000000363*((P36+3.6)/P36^6)*($B$33^2)*$B$7*$B28
37	1.5"-Sch 80	=0.05*$B$33*($B$28/(P37^2))	=0.0000000363*((P37+3.6)/P37^6)*($B$33^2)*$B$7*$B28
38	2"-Sch 80	=0.05*$B$33*($B$28/(P38^2))	=0.0000000363*((P38+3.6)/P38^6)*($B$33^2)*$B$7*$B28
39	2.5"-Sch 80	=0.05*$B$33*($B$28/(P39^2))	=0.0000000363*((P39+3.6)/P39^6)*($B$33^2)*$B$7*$B28
40	3"-Sch 80	=0.05*$B$33*($B$28/(P40^2))	=0.0000000363*((P40+3.6)/P40^6)*($B$33^2)*$B$7*$B28
41	3"-Sch 160	=0.05*$B$33*($B$28/(P41^2))	=0.0000000363*((P41+3.6)/P41^6)*($B$33^2)*$B$7*$B28
42	4"-Sch 80	=0.05*$B$33*($B$28/(P42^2))	=0.0000000363*((P42+3.6)/P42^6)*($B$33^2)*$B$7*$B28
43	4"-Sch 160	=0.05*$B$33*($B$28/(P43^2))	=0.0000000363*((P43+3.6)/P43^6)*($B$33^2)*$B$7*$B28
44	6"-Sch 80	=0.05*$B$33*($B$28/(P44^2))	=0.0000000363*((P44+3.6)/P44^6)*($B$33^2)*$B$7*$B28
45	6"-Sch 160	=0.05*$B$33*($B$28/(P45^2))	=0.0000000363*((P45+3.6)/P45^6)*($B$33^2)*$B$7*$B28
46	8"- .5" wall	=0.05*$B$33*($B$28/(P46^2))	=0.0000000363*((P46+3.6)/P46^6)*($B$33^2)*$B$7*$B28
47	8"- .875" wall	=0.05*$B$33*($B$28/(P47^2))	=0.0000000363*((P47+3.6)/P47^6)*($B$33^2)*$B$7*$B28
48	10"- .5" wall	=0.05*$B$33*($B$28/(P48^2))	=0.0000000363*((P48+3.6)/P48^6)*($B$33^2)*$B$7*$B28
49	10"- .718" wall	=0.05*$B$33*($B$28/(P49^2))	=0.0000000363*((P49+3.6)/P49^6)*($B$33^2)*$B$7*$B28
50	12"-.5" wall	=0.05*$B$33*($B$28/(P50^2))	=0.0000000363*((P50+3.6)/P50^6)*($B$33^2)*$B$7*$B28
51	12"- .843" wall	=0.05*$B$33*($B$28/(P51^2))	=0.0000000363*((P51+3.6)/P51^6)*($B$33^2)*$B$7*$B28
52	14" - .5 " wall	=0.05*$B$33*($B$28/(P52^2))	=0.0000000363*((P52+3.6)/P52^6)*($B$33^2)*$B$7*$B28
53	14" - .75" wall	=0.05*$B$33*($B$28/(P53^2))	=0.0000000363*((P53+3.6)/P53^6)*($B$33^2)*$B$7*$B28
54	16" - .5" wall	=0.05*$B$33*($B$28/(P54^2))	=0.0000000363*((P54+3.6)/P54^6)*($B$33^2)*$B$7*$B28
55	16" - .75" wall	=0.05*$B$33*($B$28/(P55^2))	=0.0000000363*((P55+3.6)/P55^6)*($B$33^2)*$B$7*$B28
56	18" - .5" wall	=0.05*$B$33*($B$28/(P56^2))	=0.0000000363*((P56+3.6)/P56^6)*($B$33^2)*$B$7*$B28
57	18" - .75" wall	=0.05*$B$33*($B$28/(P57^2))	=0.0000000363*((P57+3.6)/P57^6)*($B$33^2)*$B$7*$B28
58	20" - .5" wall	=0.05*$B$33*($B$28/(P58^2))	=0.0000000363*((P58+3.6)/P58^6)*($B$33^2)*$B$7*$B28
59	20" - .75" wall	=0.05*$B$33*($B$28/(P59^2))	=0.0000000363*((P59+3.6)/P59^6)*($B$33^2)*$B$7*$B28
60	24" - .5" wall	=0.05*$B$33*($B$28/(P60^2))	=0.0000000363*((P60+3.6)/P60^6)*($B$33^2)*$B$7*$B28
61	24" - .75" wall	=0.05*$B$33*($B$28/(P61^2))	=0.0000000363*((P61+3.6)/P61^6)*($B$33^2)*$B$7*$B28

TABLE **8.11** Continued

	O	P	Q
30			
31	Superheated Steam		
32	FORMULAS, continued:		
33			
34		Inside Dia.	Inside area
35	Pipe Size	inches	square inches
36	1" - Sch 40	1.049	0.864
37	1.5"-Sch 80	1.61	2.036
38	2"-Sch 80	1.939	2.953
39	2.5"-Sch 80	2.323	4.24
40	3"-Sch 80	2.9	6.61
41	3"-Sch 160	2.626	5.42
42	4"-Sch 80	3.826	11.5
43	4"-Sch 160	3.438	9.28
44	6"-Sch 80	5.761	26.07
45	6"-Sch 160	5.189	21.15
46	8"- .5" wall	7.625	45.7
47	8"- .875" wall	6.875	37.1
48	10"- .5" wall	9.75	74.7
49	10"- .718" wall	9.314	68.1
50	12"-.5" wall	11.75	108.4
51	12"- .843" wall	11.064	96.1
52	14" - .5 " wall	13	132.7
53	14" - .75" wall	12.5	122.7
54	16" - .5" wa..	15	176.7
55	16" - .75" wall	14.5	165.1
56	18" - .5" wall	17	227
57	18" - .75" wall	16.5	213.8
58	20" - .5" wall	19	283.5
59	20" - .75" wall	18.5	268.8
60	24" - .5" wall	23	413
61	24" - .75" wa..	22.5	398
62			

TABLE 8.12

Boiler House Notes:						Page No:
Ralph L. Vandagriff	Date: 7/6/98	Steam Desuperheater - Direct Contact				
	Example:	A direct contact desuperheater is used to remove the superheat from 150,000 lb./hour of 600 psig - 730 deg. F. steam.				
		Deaerated water at 230 deg. F. is available for desuperheating the steam.				
		How many pounds per hour of water is required for desuperheating the steam?				
Superheated Steam Conditions:		Lb/hour:	150,000	Pressure,psig:	600	
		Temperature,deg.F.:	730	Saturation temp.,deg.F.	488.81	
		Enthalpy,btu/lb	1366.748	Sensible Heat	474.8	btu/lb.
				Enthalpy of Vaporization	728.63	btu/lb.
Desuperheating water temp.,deg.F.			230			
Sensible heat, btu/lb.			198.1			
Sensible Heat absorbed by water:	equals sensible heat of superheated steam		474.8	btu/lb.		
	minus sensible heat of water		198.1	btu/lb.		
			276.7	btu/lb.		
Vaporization btu @ 600 psig	equals enthalpy of vaporization of steam		728.63	btu/lb.		
Btu/lb required to change	equals superheated steam enthalpy		1366.75	btu/lb		
superheated steam to saturated	minus saturated steam enthalpy		1203.5	btu/lb		
			163.25	btu/lb		
The weight of water evaporated by 1 (one) lb. of steam while it is being desuperheated equals:						
Heat absorbed by saturated water divided by heat required to evaporate 1 lb. of water entering the desuperheater at 230 deg. F.						
		163.25 / (276.7 + 728.63) =	0.1624	lb. of water/lb. superheated steam		
Desuperheating water required:	Steam flow per hour times water equals:		24,357	lb/hour water		
Total Lb./hour steam leaving desuperheater:			174,357			
	Steam Conditions:	600 psig @ 488.81 deg. F. (saturated)				
Note: Desuperheating water comes from either the boiler or the deaerator.						
re: # 49						

VII. RECIPROCATING PUMPS [43; p. 181]

A. Suction Piping

The suction piping should be as direct and as short as possible and at least one or two sizes larger than the pump suction. Length and size are determined by the maximum allowable suction lift, which should never exceed 22 ft (friction included). Where changing from one pipe size to another, standard ASME suction reducers should be used. Hot liquids must flow to pump suction by gravity. Piping should be laid out so that air pockets are eliminated. Piping should be pressure tested.

1. Foot Valve

When working on a suction lift, a foot valve placed in the suction line will keep pump primed as long as the foot valve is leak-tight. The net area of the foot valve should be at least equal to the area of suction pipe and, preferably, larger.

2. Strainer

To protect the pump from being clogged with foreign matter, a strainer should be installed with a net area of at least three or four times the area of the suction pipe.

3. Surge Chambers

When the suction or discharge lines, or both, are of considerable length, under a static head, if the pump speed in revolutions per minute is high, or if the liquid handled is hot, an air chamber or desurging device of suitable size on the suction or discharge lines, or both, may be necessary to ensure smooth, quiet, operation of the unit.

In general, suction surge chambers are more frequently required than are discharge surge chambers.

B. Pump Questions and Answers [Worthington Pump and Machinery Corp., 1949, p. 269]

Question. Because a direct-acting pump is a positive-displacement machine, capable of priming itself at considerable lift, does it have good "sucking" qualities?

Answer. No machine or device can "suck" fluid from a lower level. Piston motion can do no more than lower the pressure within the cylinder to the point where atmospheric pressure on an open suction supply can force the liquid up the suction pipe into the pump cylinder.

Question. What is the maximum practical suction lift of a direct-acting pump?

Answer. With cold water at sea level, an average pump can operate at a

TABLE 8.13

Boiler House Notes:						Page No:
Pneumatic Conveying of Materials						
Ralph L. Vandagriff	5/2/00					

Pneumatic Conveying

Material	Conveying Velocities Feet per Minute	Average Bulk Density lbm/cu.ft.	Coefficient of Friction Sliding on Steel	Source
Cinders		42	0.7	1
Castor beans	5000			1
Cement, portland	6000	95	0.8	2
Cement	7000			1
Coal, powdered	4000			1
Coal, pulverized, bituminous	4000	30	0.8	2
Coffee Beans	3500	42	0.5	2
Cork, ground	3000			1
Corn	5600	45	0.4	1
Cotton	4500			1
Cotton	4000			2
Flour	3500	37	0.8	2
Flyash, powdered		45	0.9	1
Grain	5000			2
Ground Feed	5000			2
Hemp	4500			2
Hog Waste	4500			2
Iron Oxide	6500	25	0.8	1
Jute	4500			2
Knots, blocks	5000			1
Limestone, pulverized	5000	85	0.9	1
Oats	4500	26	0.4	1
Paper	5000			1 & 2
Pulp Chips	4500			2
Rags	4500			1 & 2
Resin & Wood Flour, powdered		19	0.8	1
Rubber, scrap, ground	4500	23	0.7	1
Salt, granulated	5500	81	0.6	1
Salt	6000			2
Sand	7000	105	0.7	1
Sand	7500			2
Sawdust, dry	3000	20	0.7	1
Sawdust	4000			2
Shavings	3500			1
Sugar	6000			2
Sulphur, pulverized		50	1.0	1
Wheat	5800	48	0.4	1
Wood Blocks	5000			2
Wood Chips	4500	23	0.4	2
Wood Flour	4000	20	1.0	2
Wool	5000			1
Vegetable pulp, dry	4500			1 & 2
		(Source #1)		

Dust Collecting and Fume Removal

Material	Velocity Ft./minute
Bakelite Moulding Powder	3500
Bakelite Moulding Dust	2500
Buffing lint, dry	3000
Buffing lint, wet	4000
Carbon black	3500
Cotton	3000
Cotton lint	2000
Foundry dust	4500
Grain dust	3000
Grinding dust	5000
Hog waste	4500
Jute lint	3000
Jute dust	3500
Lead dust	5000
Melting pot and furnace	2000
Metallizing booth	3500
Metal turnings	5000
Oven hood	2000
Paint Spray	2000
Paper	3500
Rubber dust	3500
Rubber buffings	4500
Sand blast dust	4000
Sander dust	2000
Sawdust, dry	3000
Sawdust, wet	4000
Shavings, dry	3000
Shavings, wet	4000
Shoe dust	4000
Soldering fumes	2000
Tail pipe exhaust	3000
Wood blocks	4500
Wood flour	2000
Wool	4000
(Source # 2)	

Sources:
1: re: #44
2: re: #54

lift of about 22 ft. The difference between the maximum theoretical lift, 34 ft, and the maximum practical lift represents losses through the valves and entrance losses, velocity head, and pipe friction in the suction line.

Question. Are all power pumps capable of operation with a suction lift?

Answer. No. Suction requirements vary widely with the design and the service for which the pumps are intended. Piston pumps for low and moderate pressures are usually capable of operating with a suction lift of 10–20 ft of water, depending on the speed and relative valve area. Horizontal duplex sidepot pumps are usually offered with a considerable range of liner and piston sizes for each size of pump cylinder. Thus, with a small liner and piston, the valve area is relatively large, and a higher suction lift is permissible. With the largest liner and piston for a given cylinder, the relative valve area is less; hence, the pump must be operated with less suction lift.

Question. Are all power pumps self-priming with a suction lift?

Answer. No. This depends on the clearance volume and on the required valve loading. Moderate-speed piston pumps for operation at moderate pressures are usually self-priming with some suction lift. If the clearance volume is filled with liquid, the pump should be self-priming at any reasonable operating lift. Where priming is no problem, plunger pumps for higher pressures are usually designed to operate with flooded suction.

Question. What types of power pumps require the greatest net positive suction head?

Answer. Speed and pressure influence the suction head required. High-pressure pumps necessarily use heavy valves, and valve and passage areas cannot be made overly large without greatly increasing pump weight and cost. Thus, high-pressure hydraulic pumps usually require considerable suction head, which is provided by bringing all returns to an elevated suction tank. High-speed pumps also require more suction head, unless made with unusually large suction valve area. High-speed hydraulic or boiler feed pumps may require as much as 15–20 psi net positive suction head.

Question. Should every power pump have a discharge relief valve?

Answer. Yes. This cannot be overemphasized.

See Table 8.13 on the pneumatic conveying of materials.

VIII. COMPRESSED AIR

A. Definitions and Formulas [37,45]

1. Free Air

This is defined as air at atmospheric conditions at any specific location. Because the altitude, barometer, and temperature may vary at different localities and at different times, it follows that this term does not mean air under identical or standard conditions.

2. Standard Air

This is defined as air at a temperature of 68°F, a pressure of 14.70 psia, and a relative humidity of 36% (0.075 density). This is in agreement with definitions adopted by ASME, but in the gas industries the temperature of "standard air" is usually given as 60°F.

Ratings for equipment using compressed air and for compressors delivering the air are given in terms of *free air*. This gives the quantity of air delivered per unit time, assuming that the air is at standard atmospheric conditions of 14.7 psia and 60°F.

To determine the flow rate at other conditions, the following equation can be used:

$$Q_a = Q_s \times \frac{14.7 \text{ psia}}{P_{atm} + P_a} \times \frac{(t_a + 460)°R}{520°R}$$

Q_a = volume flow rate at actual conditions
Q_s = volume flow rate at standard conditions
P_{atm} = actual absolute atmospheric pressure
P_a = actual gage pressure
T_a = actual absolute temperature

Example. An air compressor has a rating of 500 cfm free air. Compute the flow rate in a pipe line in which the pressure is 100 psig and the temperature is 80°F.

Assuming that the local atmospheric pressure is 14.7.

$$Q_a = 500 \text{ cfm} \times \frac{14.7 \text{ psia}}{14.7 + 100} \times \frac{(80 + 460)}{520} = 66.54 \text{ cfm}$$

3. Pressure Drop, ΔP (Harris Formula)

$$\Delta P = \frac{L \times Q_a^2}{2390 \times p_c \times d^{5.31}}$$

Example.   L = 1000, Q_a = 3000, p_c = 119, d = 6.065, ΔP = 2.21 psi

4. Pipe Diameter, d, Required for a Specific Flow (Harris Formula)

$$d = 0.255 \times \left(\frac{L \times Q_a^2}{(p_1^2 - p_2^2)} \right)^{0.188}$$

Example. L = 10,000, Q_a = 3000, p_1 = 120, p_2 = 100, d = 6.04 in.

L = length of pipe run (ft)
Q_a = cfm of free air
d = inside diameter of pipe (in.)
p_c = average pressure (psia) in pipe $(p_1 + p_2)/2$
p_1 = pipe inlet pressure (psia)
p_2 = pipe outlet pressure (psia)

See Table 8.14 concerning compressed airflow through an orifice or air leak.

B. Dried and Oil-Free Air

1. Why Moisture in Compressed Air

Air entering the first stage of any air compressor carries with it a certain amount of native moisture. This is unavoidable, although the quantity carried will vary widely with the ambient temperature and relative humidity. For the purpose of this discussion, relative humidity is assumed to be the same as degree of saturation. A maximum error of less than 2% is involved.

In any air–vapor mixture, each component has its own partial pressure, and the air and the vapor are each indifferent to the existence of the other. It follows that the conditions of either component may be studied without reference to the other. In a certain volume of mixture, each component fills the full volume at its own partial pressure. The water vapor may saturate this space (be at its saturation pressure and temperature) or it may be superheated (above saturation temperature for its partial pressure).

As this vapor is compressed, its volume is reduced while, at the same time, the temperature automatically increases and the vapor may become superheated. More pounds of vapor are now contained in 1 ft^3 than when originally entering the compressor.

Under the laws of vapors, the maximum quantity of a particular vapor a given space can contain is solely dependent on the vapor temperature. As the compressed water vapor is cooled it will eventually reach the temperature at which the space becomes saturated, now containing the maximum it can hold. Any further cooling will force part of the vapor to condense into the liquid form. It is clearly evident that the lower the temperature and the greater the pressure of compressed air, the greater will be the amount of vapor condensed.

Example. Given that 1000 ft^3 of saturated free air drawn into a compressor at atmospheric pressure and at a temperature of 70°F contains 1.12 lb of moisture. After this air has been compressed to 100 psig pressure and then cooled to its original temperature of 70°F its moisture content will be reduced to 0.15 lb. If its temperature is reduced an additional 15°F, that is to 55°F, the remaining moisture

TABLE 8.14

Boiler House Notes:										Page No:

Ralph L. Vandagriff

Compressed Air - Flow through Orifice or Air Leak

In cubic feet of free air per minute at standard atmospheric pressure of 14.7 psia and 70 deg. F.

Diameter of Orifice in Inches

Gage Pressure before orifice in psig	0.015625 (1/64)	0.03125 (1/32)	0.0625 (1/16)	0.125 (1/8)	0.25 (1/4)	0.375 (3/8)	0.5 (1/2)	0.625 (5/8)	0.75 (3/4)	0.875 (7/8)	1.00
				Discharge in Cubic Feet of free air per minute							
15	0.105	0.420	1.680	6.72	26.89	60.50	107.5	168.0	242.0	329.4	430.2
16	0.109	0.434	1.737	6.95	27.79	62.53	111.2	173.7	250.1	340.5	444.7
17	0.112	0.448	1.794	7.17	28.70	64.57	114.8	179.4	258.3	351.6	459.2
18	0.116	0.463	1.850	7.40	29.60	66.61	118.4	185.0	266.4	362.6	473.7
19	0.119	0.477	1.907	7.63	30.51	68.64	122.0	190.7	274.6	373.7	488.1
20	0.123	0.491	1.963	7.85	31.41	70.68	125.7	196.3	282.7	384.8	502.6
22	0.130	0.519	2.077	8.31	33.22	74.76	132.9	207.7	299.0	407.0	531.6
24	0.137	0.547	2.190	8.76	35.04	78.83	140.1	219.0	315.3	429.2	560.6
26	0.144	0.576	2.303	9.21	36.85	82.90	147.4	230.3	331.6	451.4	589.5
28	0.151	0.604	2.416	9.66	38.66	86.98	154.6	241.6	347.9	473.5	618.5
30	0.158	0.632	2.529	10.12	40.47	91.05	161.9	252.9	364.2	495.7	647.5
32	0.165	0.661	2.642	10.57	42.28	95.12	169.1	264.2	380.5	517.9	676.4
34	0.172	0.689	2.756	11.02	44.09	99.20	176.4	275.6	396.8	540.1	705.4
36	0.179	0.717	2.869	11.47	45.90	103.27	183.6	286.9	413.1	562.3	734.4
38	0.186	0.745	2.982	11.93	47.71	107.35	190.8	298.2	429.4	584.4	763.3
40	0.193	0.774	3.095	12.38	49.52	111.42	198.1	309.5	445.7	606.6	792.3
42	0.201	0.802	3.208	12.83	51.33	115.49	205.3	320.8	462.0	628.8	821.3
44	0.208	0.830	3.321	13.29	53.14	119.57	212.6	332.1	478.3	651.0	850.3
46	0.215	0.859	3.434	13.74	54.95	123.64	219.8	343.4	494.6	673.2	879.2
48	0.222	0.887	3.548	14.19	56.76	127.72	227.0	354.8	510.9	695.3	908.2
50	0.229	0.915	3.661	14.64	58.57	131.79	234.3	366.1	527.2	717.5	937.2
55	0.246	0.986	3.944	15.77	63.10	141.97	252.4	394.4	567.9	773.0	1,010
60	0.264	1.057	4.227	16.91	67.63	152.16	270.5	422.7	606.6	826.4	1,082
65	0.282	1.127	4.510	18.04	72.15	162.34	288.6	451.0	649.4	883.9	1,154
70	0.300	1.198	4.792	19.17	76.68	172.53	306.7	479.2	690.1	939.3	1,227
75	0.317	1.269	5.075	20.30	81.21	182.71	324.8	507.5	730.8	994.8	1,299
80	0.335	1.340	5.358	21.43	85.73	192.90	342.9	535.8	771.6	1,050	1,372
85	0.353	1.410	5.641	22.56	90.26	203.08	361.0	564.1	812.3	1,106	1,444
90	0.370	1.481	5.924	23.70	94.78	213.27	379.1	592.4	853.1	1,161	1,517
100	0.406	1.622	6.490	25.96	103.84	233.64	415.4	649.0	934.5	1,272	1,661
110	0.441	1.764	7.056	28.22	112.89	254.00	451.6	705.6	1,016	1,383	1,806
120	0.476	1.905	7.621	30.49	121.94	274.37	487.8	762.1	1,097	1,494	1,951
130	0.512	2.047	8.187	32.75	131.00	294.74	524.0	818.7	1,179	1,605	2,096
140	0.547	2.188	8.753	35.01	140.05	315.11	560.2	875.3	1,260	1,716	2,241
150	0.582	2.330	9.319	37.28	149.10	335.48	596.4	931.9	1,342	1,827	2,386
160	0.618	2.471	9.885	39.54	158.16	355.85	632.6	988.5	1,423	1,937	2,530
170	0.653	2.613	10.451	41.80	167.21	376.22	668.8	1,045	1,505	2,048	2,675
180	0.689	2.754	11.016	44.07	176.26	396.59	705.0	1,102	1,586	2,159	2,820
190	0.724	2.896	11.582	46.33	185.31	416.96	741.3	1,158	1,668	2,270	2,965
200	0.759	3.037	12.148	48.59	194.37	437.33	777.5	1,215	1,749	2,381	3,110
225	0.848	3.391	13.563	54.25	217.00	488.25	868.0	1,356	1,953	2,658	3,472
250	0.936	3.744	14.977	59.91	239.63	539.17	958.5	1,498	2,157	2,936	3,834
275	1.024	4.098	16.392	65.57	262.27	590.10	1,049	1,639	2,360	3,213	4,196
300	1.113	4.452	17.806	71.22	284.90	641.02	1,140	1,781	2,564	3,490	4,558
400	1.467	5.866	23.464	93.86	375.43	844.71	1,502	2,346	3,379	4,599	6,007
500	1.820	7.281	29.122	116.49	465.96	1,048	1,864	2,912	4,194	5,708	7,455
600	2.174	8.695	34.780	139.12	556.49	1,252	2,226	3,478	5,006	6,817	8,904
750	2.704	10.817	43.268	173.07	692.28	1,558	2,769	4,327	6,231	8,480	11,077
1000	3.588	14.353	57.413	229.65	918.61	2,067	3,674	5,741	8,267	11,253	14,898

re: # 38

content, will be 0.09 lb. This is only 0.06 lb less than at 70°F, showing that in cooling air to eliminate moisture a point is reached below which little additional moisture is removed. As a general rule a differential of 15°F between the temperature of the cooling water entering and the temperature of the air leaving the aftercooler should be maintained, with about 1–1.5 gal of water required per 100 ft³ of free air handled.

2. Problems Caused by Water in Compressed Air

Few plant operators need to be told of the problems caused by water in compressed air. They are most apparent to those who operate pneumatic tools, rock drills, automatic pneumatic-powered machinery, paint and other sprays, sand-blasting equipment, and pneumatic controls. However, almost all applications, particularly of 100-psig power, could benefit from the elimination of water carry-over. The principal problems might be summarized as:

1. Washing away of required lubrication
2. Increase in wear and maintenance
3. Sluggish and inconsistent operation of automatic valves and cylinders
4. Malfunctioning and high maintenance of control instruments
5. Spoilage of product by spotting in paint and other types of spraying
6. Rusting of parts that have been sandblasted
7. Freezing of exposed lines during cold weather
8. Further condensation and possible freezing of moisture in the exhaust of those more efficient tools that expand the air considerably

In connection with the last item, in some rock drills there is a 70°F drop in temperature from inlet to exhaust. Most portable pneumatic tools have a considerably lower temperature drop, but the foregoing problem sometimes exists.

The increased use of control systems and automatic machinery has made these problems more serious and has spurred activity toward their reduction. The amount of moisture entering the compressor is widely variable, depending on ambient temperature and relative humidity. The problems are usually the worst when both temperature and humidity are high. Pipeline freezing problems are prevalent only in the winter months.

A fact to remember is that water vapor *as vapor* does no harm in a pneumatic system. It is only when the vapor condenses and remains in the system as liquid that problems exist. The goal, therefore, is to condense and *remove* as much of the vapor as is economically desirable, considering the applications involved.

3. The Conventional System

The air compressor plant should always include a water-cooled aftercooler followed by a receiver. There are few exceptions to this rule, all due to local conditions or a special use of the air.

Aftercoolers alone, or aftercoolers following intercoolers, will under normal summer conditions condense at 100 psig up to 70% or more of the vapor entering the system. This is a substantial portion, some often being collected in the receiver. Therefore, both cooler and receiver must be kept drained. Inevitably, more water will condense in the distribution lines if the air cools further. This must also be removed if the problems outlined earlier are to be reduced. To remove this water, one may

1. Take all feeders off the top of mains and branches
2. Slope mains and branches toward a dead end
3. Drain all low points and dead ends through a water leg using automatic traps to ensure drainage
4. Incorporate strainers and lubricators in the piping to all tools

The temperature of compressed air leaving an aftercooler and receiver will largely depend on the temperature and quantity of the water used in the cooler. Unfortunately, when atmospheric temperature and humidity are highest and condensation in the cooler is most needed, the water temperature is usually also high. Results are not always all that could be desired.

This is the system used most generally throughout industry and, if applied and operated with understanding and care, will give reasonable results. However, because the air–vapor mixture leaving the receiver is at, or very near, the saturation point, and the mixture usually cools further in the system, condensation in lines must be expected and its elimination provided for.

C. The Dried Air System

A dried air system involves processing the compressed air *beyond* the aftercooler and receiver to further reduce moisture content. This requires special equipment, a higher first cost and a higher operating cost. These costs must be balanced against the gains obtained. They may show up as less wear and maintenance of tools and air-operated devices, greater reliability of devices and controls, and greater production through fewer outages for repairs. Frequently, reduction or elimination of product spoilage or a better product quality may result. Many automobile plants are drying air with the high-priority objective of improving car finish by better paint spraying.

The degree of drying desired will vary with the pneumatic equipment and application involved. The air is to eliminate further condensation in the line and tool. Prevailing atmospheric conditions also have an influence.

In many 100-psig installations, a dew point, at line pressure, of from 50° to 35°F is felt to be adequate. Occasional equipment may find lower dew points of value even down to minus 50°F. In such cases this may be obtained, but at higher cost.

Terminology involves drier outlet dew point at the *line* pressure. This is the saturation temperature of the remaining moisture. If the compressed air temperature is never reduced below this dew point at any point beyond the drying equipment, there will be no further condensation.

Another value sometimes involved when the air pressure is reduced before it is used is the dew point at that lower pressure condition. A major example is the use of 100-psig (or higher) air reduced to 15 psig for use in pneumatic instruments and controls. This dew point will be lower because the volume involved increases as the pressure is lowered.

The dew point at atmospheric pressure is often used as a reference point for measurement of drying effect. This is of little interest when handling *compressed* air.

Example. 1000 ft^3 of compressed air at 100 psig at 50°F, or 1000 ft^3 of compressed air at 15 psig at 50°F will hold the same amount of vapor at the dew point. However, 1000 ft^3 at 100 psig and 50°F reduced to 15 psig will become 3860 ft^3 at 50°F, so is capable of holding 3.86 times as much vapor, and the dew point will not be reached until the mixture temperature is lowered materially.

D. General Drying Methods

There are three general methods of drying air, chemical drying, adsorbing, and refrigerating. In all cases, aftercooling and adequate condensate removal must be done ahead of this equipment. The initial and operating costs and the results obtained vary considerably.

These methods are primarily for water vapor removal. Removal of lubrication oil is secondary, although all systems will reduce its carryover. It must be understood that complete elimination of lubricating oil, particularly in the vapor form, is very difficult and that, when absolutely oil-free air is required, some form of nonlubricated compressor is the best guaranteed method.

1. Chemical Drying

Chemical driers use materials that combine with or absorb moisture from air when brought into close contact. There are two general types. One, using deliquescent material in the form of pellets or beads, is reputed to obtain a dew point, with 70°F air to the drier, of between 35°F and 50°F, depending on material. The material turns into a liquid as the water vapor is absorbed. This liquid must be drained off and the pellets or beads replaced periodically. Entering air above 90°F is not generally recommended.

The second type utilizes an ethylene glycol liquid to absorb the moisture. Standard dew point reduction claimed is 40°F, but greater reductions are said to be possible with special equipment. The glycol is regenerated (dried) in a still using fuel gas or steam as a heating agent. The released moisture is vented to atmosphere. The regenerated glycol is recirculated by a pump. Usually driven by compressed air. A water-cooled glycol cooler is also required.

2. Adsorbing

Adsorption is the property of certain extremely porous materials to hold vapors in the pores until the desiccant is either heated or exposed to a drier gas. The material is a solid at all times and operates alternately through drying and reactivation cycles with no change in composition. Adsorbing materials in principal use are activated alumina and silica gel. Molecular sieves are also used. Atmospheric dew points of minus 100°F are readily attained.

Reactivation or regeneration is usually obtained by diverting a portion of the already dried air through a reducing valve or orifice, reducing its pressure to atmospheric, and passing it through the wet desiccant bed. This air, with the moisture it has picked up, is vented to atmosphere. The air diverted may vary from 7 to 17% of the mainstream flow, depending upon the final dew point desired from the apparatus. Heating the activating air before its passing through the bed, or heating the bed itself, is often done. This requires less diverted air because each cubic foot will carry much more moisture out of the system. Other modifications are also available to reduce the diverted air quantity or even eliminate it.

3. Refrigeration

Refrigeration for drying compressed air is growing rapidly. It has been applied widely to small installations, sections of larger plants, and even to entire manufacturing plant systems. Refrigeration has been applied to the airstream both before and after compression. In the before-compression system, the air must be cooled to a lower temperature for a given final line pressure dew point. This takes more refrigeration power for the same end result. Partially off-setting this is a saving in air compressor power per 1000 cfm of atmospheric air compressed, owing to the reduction in volume at compressor inlet caused by the cooling and the removal of moisture. There is also a reduction in discharge temperature on single-stage compressors that may at times have some value. An atmospheric (inlet) dew point of 35°F is claimed. Referred to line pressure (at which condition condensation actually takes place), the foregoing dew point becomes

Line Pressure (psig)	Line Dew Point (°F)
25	61
35	67
50	75
70	84
100	94

These systems are usually custom designed.

When air is refrigerated *following compression*, two systems have been used. Flow of air through directly refrigerated coils is used predominately in the smaller and moderate-sized systems. These are generally standardized for cooling to 35°F, which is the dew point obtained at line pressure. The larger systems chill the water that is circulated through coils to cool the air. A dew point at line pressure of about 50°F is obtainable by this method. When the incoming air is partially cooled by the outgoing airstream, the system is called regenerative. This reduces the size and first cost of the refrigeration compressor and exchanger. It also reduces power cost and reheats the air returning to the line. Reheating of the air after it is dried has several advantages: (1) the air volume is increased and less free air is required to do a job; (2) chance of line condensation is still further reduced; and (3) sweating of the cold pipe leaving the drier is eliminated. Regenerative driers seldom need further reheating. All refrigeration-type driers will remove some oil from the air.

4. Combination Systems

The use of a combination drier should be investigated when a very low dew point is necessary. Placing a refrigeration system ahead of an adsorbent drier will let the more economical refrigeration system remove most of the vapor, and reduce the load on the desiccant.

a. Example of the Effect of Drying Air. As a reasonably typical example of the effect of drying air regardless of the method employed, consider the following:

cfm free air	1000
Hours operated per day	10
Total inlet cubic feet	600,000/10 hr
Atmosphere	75°F, 76% RH, 14.696 psia
Weight of vapor	601 lb/10 hr
Equivalent gallons	72.1/10 hr (if all condensed)

5. Separators

Separators are available from many sources, in many designs, and usually consist of a knockout chamber and condensate removal trap. Some designs include a removable filter of some type. These can remove only contaminants condensed to this point and are meant to be placed at the actual point the compressed air is used.

E. Compressed Air System Fires

The danger of fire is inherent in almost any air compressor system. Although there are few such occurrences for the number of air compressors in operation, there are enough to cause concern. The reasons should be appreciated.

What is known as the "fire triangle" exists in any fire or potential fire. The triangle consists of oxygen, fuel, and an ignition source. In the air compressor system, oxygen is always present. Petroleum oils are used as lubricant. These have fuel value; they and their vapors will burn if ignited. Two sides of the triangle are always present. The third side, an ignition source, is most likely to be brought into action when *too much* or an *improper oil* is used, or when *maintenance* is neglected.

Maintenance is most important because dirty water-cooled intercoolers, dirty fins on air-cooled units, broken or leaky discharge valves, broken piston rings, and the like always tend to increase normal discharge air temperature, sometimes rapidly. These excessive temperatures cause more rapid oil deterioration and formation of deposits, both of which are further accelerated if too much oil or an improper oil is being used.

Based on experience, fires and explosions are seldom if ever caused by reaching the autogenous ignition temperature of the oil. This averages between 600° and 750°F, there appears little opportunity for the existence of such a temperature.

Petroleum oils do decompose and form carbonaceous deposits. They collect on valves, heads, and discharge ports, and in piping. Experiments have shown that, in time, they may absorb some oxygen from the air and, under favorable conditions, will themselves start to decompose, generating heat. This heat might reach a point where the mass glows and becomes a trigger for more violent burning. This action is speeded by high temperatures. It is believed this reaction applies to a majority of reported incidents.

Fires have been known to occur and to burn out, causing no damage. Others cause a high pressure rise owing to the increase in temperature and attempted expansion of the compressed air. Because expansion is impossible, the pressure increases until relieved. This is an ordinary pressure rise followed by mechanical failure. Maximum pressure may be six to ten times initial line pressure.

A special and destructive type of explosion may occur, although its frequency seems to be very low in air compressor systems. This is the rather unpredictable *detonation*, caused by development and propagation of a very high-speed pressure wave. As the originally ignited fuel burns, it becomes hot, expands, and sends pressure waves ahead that push into the unburned gas. These waves compress the unburned gas ahead of the flame and materially heat this gas. As the flame follows through this hot, but unburned, gas, a normal explosion takes place. This builds a pressure front just ahead of the flame and sends ahead additional, faster moving, pressure waves that catch up to the slower waves and build up a *shock wave*. This travels at many times the speed of sound. It may reach extreme pressures of 60–100 times the initial line pressure and ruptures vessels, pipe, and fittings with great violence. Although simplified, this illustrates the mechanism of detonation. The shock wave may move against the actual flow of the air.

There is an obvious approach to the prevention of fires and explosions.

1. Keep the compressor in good repair.
2. Replace broken and leaking valves and parts immediately.
3. Check and record discharge temperatures frequently.
4. Keep the compressor clean internally and externally.
5. See that the coolant is actually flowing and in proper quantity.
6. Drain separators and receiver frequently.
7. Use the proper lubricant.
8. Use only enough lubricant.

An aftercooler should be used with every air compressor. If a fire starts between the compressor and aftercooler, it will go no farther than the cooler, as a rule,

TABLE 8.15 Theoretical Adiabatic Discharge Temperature for Air Compression

Single stage		Two stage[a]		Three stage[a]	
Discharge pressure (psig)	Discharge temperature (°F)	Discharge pressure (psig)	Discharge temperature (°F)	Discharge pressure (psig)	Discharge temperature (°F)
10	154	60	207	400	266
20	216	80	230	600	294
30	266	100	249	800	314
40	309	125	269	1000	331
50	347	150	286	1200	344
60	380	175	301	1400	355
70	410	200	315	1600	366
80	438	225	326	1800	375
90	464	250	338	2000	383
100	488	275	348	2200	391
110	511	300	357	2400	398
120	532	350	375	2600	404
130	553	400	390	2800	410
140	572	450	404	3000	416
150	590	500	416		

[a] Based on 70°F to all stages.

Note: The relation between actual and theoretical discharge temperatures for reciprocation compressors will depend on such variables as cylinder size, compression ratio, rpm, effectiveness of cylinder cooling, use of a dry liner, etc. For water-cooled cylinders of moderate size only, an approximate assumption can be made that at a compression ratio of 3.5, the theoretical and actual temperature will be about the same. Lower ratios will actually exceed theoretical, whereas higher ratios will actually be below theoretical.

Source: Refs. 38 and 39.

and usually there is no explosion. A safety valve upstream from the aftercooler is recommended.

The chapter concludes with a series of tables on the properties of tubing, pipes, fittings, and flanges (Tables 8.16–8.22), and two figures illustrating heat-exchange nomenclature.

TABLE 8.16

O.D. of Tubing	B.W.G. Gauge	Thickness Inches	Internal Area Sq. Inch	Sq. Ft. External Surface Per Foot Length	Sq. Ft. Internal Surface Per Foot Length	Weight Per Ft. Length Steel Lbs.*	I.D. Tubing Inches	Moment of Inertia Inches⁴	Section Modulus Inches³	Radius of Gyration Inches	Constant C**	O.D./I.D.	Metal Area (Transverse Metal Area) Sq. Inch
1/4	22	.028	.0295	.0655	.0508	.066	.194	.00012	.00098	.0792	46	1.289	.0195
1/4	24	.022	.0333	.0655	.0539	.054	.206	.00011	.00083	.0810	52	1.214	.0159
1/4	26	.018	.0360	.0655	.0560	.045	.214	.00009	.00071	.0824	56	1.168	.0131
3/8	18	.049	.0603	.0982	.0725	.171	.277	.00068	.0036	.1164	94	1.354	.0502
3/8	20	.035	.0731	.0982	.0798	.127	.305	.00055	.0029	.1213	114	1.233	.0374
3/8	22	.028	.0799	.0982	.0835	.104	.319	.00046	.0025	.1227	125	1.176	.0305
3/8	24	.022	.0860	.0982	.0867	.083	.331	.00038	.0020	.1248	134	1.133	.0244
1/2	16	.065	.1075	.1309	.0969	.302	.370	.0022	.0086	.1556	168	1.351	.0888
1/2	18	.049	.1269	.1309	.1052	.236	.402	.0018	.0072	.1606	198	1.244	.0694
1/2	20	.035	.1452	.1309	.1126	.174	.430	.0014	.0056	.1648	227	1.163	.0511
1/2	22	.028	.1548	.1309	.1162	.141	.444	.0012	.0046	.1671	241	1.126	.0415
5/8	12	.109	.1301	.1636	.1065	.602	.407	.0061	.0197	.1864	203	1.536	.177
5/8	13	.095	.1486	.1636	.1139	.537	.435	.0057	.0183	.1903	232	1.437	.158
5/8	14	.083	.1655	.1636	.1202	.479	.459	.0053	.0170	.1936	258	1.362	.141
5/8	15	.072	.1817	.1636	.1259	.425	.481	.0049	.0156	.1971	283	1.299	.125
5/8	16	.065	.1924	.1636	.1296	.388	.495	.0045	.0145	.1993	300	1.263	.114
5/8	17	.058	.2035	.1636	.1333	.350	.509	.0042	.0134	.2016	317	1.228	.103
5/8	18	.049	.2181	.1636	.1380	.303	.527	.0037	.0118	.2043	340	1.186	.089
5/8	19	.042	.2298	.1636	.1416	.262	.541	.0033	.0105	.2068	358	1.155	.077
5/8	20	.035	.2419	.1636	.1453	.221	.555	.0028	.0091	.2089	377	1.126	.065
3/4	10	.134	.1825	.1963	.1262	.884	.482	.0129	.0344	.2229	285	1.556	.260
3/4	11	.120	.2043	.1963	.1335	.809	.510	.0122	.0325	.2257	319	1.471	.238
3/4	12	.109	.2223	.1963	.1393	.748	.532	.0116	.0309	.2299	347	1.410	.220
3/4	13	.095	.2463	.1963	.1488	.656	.560	.0107	.0285	.2340	384	1.339	.196
3/4	14	.083	.2679	.1963	.1529	.592	.584	.0098	.0262	.2376	418	1.284	.174
3/4	15	.072	.2884	.1963	.1587	.520	.606	.0089	.0238	.2410	450	1.238	.153
3/4	16	.065	.3019	.1963	.1623	.476	.620	.0083	.0221	.2433	471	1.210	.140
3/4	17	.058	.3157	.1963	.1660	.428	.634	.0076	.0203	.2455	492	1.185	.126
3/4	18	.049	.3339	.1963	.1707	.367	.652	.0067	.0178	.2684	521	1.150	.108
3/4	20	.035	.3632	.1963	.1780	.269	.680	.0050	.0124	.2532	567	1.103	.079
7/8	10	.134	.2892	.2291	.1589	1.061	.607	.0221	.0505	.2662	451	1.441	.312
7/8	11	.120	.3166	.2291	.1662	.969	.635	.0208	.0475	.2703	494	1.372	.285
7/8	12	.109	.3390	.2291	.1720	.891	.657	.0196	.0449	.2736	529	1.332	.262
7/8	13	.095	.3685	.2291	.1793	.792	.685	.0180	.0411	.2778	575	1.277	.233
7/8	14	.083	.3948	.2291	.1856	.709	.709	.0164	.0374	.2815	616	1.234	.207
7/8	16	.065	.4359	.2291	.1950	.561	.745	.0137	.0312	.2873	680	1.174	.165
7/8	18	.049	.4742	.2291	.2034	.432	.777	.0109	.0249	.2925	740	1.126	.127
7/8	20	.035	.5090	.2291	.2107	.313	.805	.0082	.0187	.2972	794	1.087	.092
1	8	.165	.3526	.2618	.1754	1.462	.670	.0392	.0784	.3009	550	1.493	.430
1	10	.134	.4208	.2618	.1916	1.237	.732	.0350	.0700	.3098	656	1.366	.364
1	11	.120	.4536	.2618	.1990	1.129	.760	.0327	.0654	.3140	708	1.316	.332
1	12	.109	.4803	.2618	.2047	1.037	.782	.0307	.0615	.3174	749	1.279	.305
1	13	.095	.5153	.2618	.2121	.918	.810	.0280	.0559	.3217	804	1.235	.270
1	14	.083	.5463	.2618	.2183	.813	.834	.0253	.0507	.3255	852	1.199	.239
1	15	.072	.5755	.2618	.2241	.714	.856	.0227	.0455	.3291	898	1.167	.210
1	16	.065	.5945	.2618	.2278	.649	.870	.0210	.0419	.3314	927	1.149	.191
1	18	.049	.6390	.2618	.2361	.496	.902	.0166	.0332	.3366	997	1.109	.146
1	20	.035	.6793	.2618	.2435	.360	.930	.0124	.0247	.3414	1060	1.075	.106
1-1/4	7	.180	.6221	.3272	.2330	2.057	.890	.0890	.1425	.3836	970	1.404	.605
1-1/4	8	.165	.6648	.3272	.2409	1.921	.920	.0847	.1355	.3880	1037	1.359	.565
1-1/4	10	.134	.7574	.3272	.2571	1.598	.982	.0741	.1186	.3974	1182	1.273	.470
1-1/4	11	.120	.8012	.3272	.2644	1.448	1.010	.0688	.1100	.4018	1250	1.238	.426
1-1/4	12	.109	.8365	.3272	.2702	1.329	1.032	.0642	.1027	.4052	1305	1.211	.391
1-1/4	13	.095	.8825	.3272	.2775	1.173	1.060	.0579	.0928	.4097	1377	1.179	.345
1-1/4	14	.083	.9229	.3272	.2838	1.033	1.084	.0521	.0833	.4136	1440	1.153	.304
1-1/4	16	.065	.9852	.3272	.2932	.823	1.120	.0426	.0682	.4196	1537	1.116	.242
1-1/4	18	.049	1.042	.3272	.3016	.629	1.152	.0334	.0534	.4250	1626	1.085	.185
1-1/4	20	.035	1.094	.3272	.3089	.456	1.180	.0247	.0395	.4297	1707	1.059	.134
1-1/2	10	.134	1.192	.3927	.3225	1.955	1.232	.1354	.1806	.4653	1860	1.218	.575
1-1/2	12	.109	1.291	.3927	.3356	1.618	1.282	.1159	.1546	.4933	2014	1.170	.476
1-1/2	14	.083	1.398	.3927	.3492	1.258	1.334	.0931	.1241	.5018	2181	1.124	.370
1-1/2	16	.065	1.474	.3927	.3587	.996	1.370	.0756	.1008	.5079	2299	1.095	.293
2	11	.120	2.433	.5236	.4608	2.410	1.760	.3144	.3144	.6660	3795	1.135	.709
2	13	.095	2.573	.5236	.4739	1.934	1.810	.2586	.2586	.6744	4014	1.105	.569
2-1/2	9	.148	3.815	.6540	.5770	3.719	2.204	.7592	.6074	.8332	5951	1.134	1.094

*Weights are based on low carbon steel with a density of 0.2833#/inch³. For other metals multiply by the following factors:

Aluminum	0.35
A.I.S.I. 400 Series Stainless Steels	0.99
A.I.S.I. 300 Series Stainless Steels	1.02
Aluminum Bronze	1.04
Aluminum Brass	1.06
Nickel-Chrome-Iron	1.07
Admiralty	1.09
Nickel and Nickel-Copper	1.13
Copper and Cupro-Nickels	1.14

**Liquid Velocity $= \dfrac{\text{Lbs. Per Tube Per Hour}}{C \times \text{SP. GR. of Liquid}}$ in feet per sec. (Sp. Gr. of Water at 60°F. = 1.0)

Ralph L. Vandagriff   Date: 9/18/98

Dimensions and Properties of Tubing

Outside Dia. Inches	Wall Thickness BWG	Wall Thickness Inches	Inside Dia. Inches	Weight per Ft. of Length	Sq.Ft. Outside Surface per Ft.L.	Sq.Ft. Inside Surface per Ft.L.	Inside Cross-Sectional Area Sq.Ft.
2.00	4	0.24	1.52	4.4776	0.5236	0.39794	0.01260
2.00	5	0.22	1.56	4.1570	0.5236	0.40641	0.01327
2.00	6	0.2	1.60	3.7983	0.5236	0.41888	0.01398
2.00	7	0.18	1.64	3.4556	0.5236	0.42935	0.01467
2.00	8	0.165	1.67	3.2122	0.5236	0.43721	0.01521
2.00	9	0.15	1.70	2.9202	0.5236	0.44506	0.01578
2.00	10	0.135	1.73	2.8709	0.5236	0.45291	0.01632
2.00	11	0.12	1.76	2.3848	0.5236	0.46077	0.01689
2.00	12	0.105	1.79	2.0928	0.5236	0.46862	0.01748
2.00	13	0.095	1.81	1.8981	0.5236	0.47386	0.01787
2.25	4	0.24	1.77	5.1055	0.58905	0.46339	0.01709
2.25	5	0.22	1.81	4.7161	0.58905	0.47386	0.01787
2.25	6	0.2	1.85	4.3288	0.58905	0.48433	0.01857
2.25	7	0.18	1.89	3.9374	0.58905	0.4948	0.01948
2.25	8	0.165	1.92	3.6454	0.58905	0.50266	0.02011
2.25	9	0.15	1.95	3.3534	0.58905	0.51051	0.02074
2.25	10	0.135	1.98	3.0127	0.58905	0.51836	0.02138
2.25	11	0.12	2.01	2.7207	0.58905	0.52622	0.02204
2.25	12	0.105	2.04	2.3800	0.58905	0.53407	0.02270
2.25	13	0.095	2.06	2.1853	0.58905	0.53931	0.02315
2.50	4	0.24	2.02	5.7723	0.6545	0.52884	0.02226
2.50	5	0.22	2.06	5.3342	0.6545	0.53931	0.02315
2.50	6	0.2	2.10	4.8962	0.6545	0.54978	0.02405
2.50	7	0.18	2.14	4.4095	0.6545	0.56025	0.02498
2.50	8	0.165	2.17	4.0698	0.6545	0.56811	0.02568
2.50	9	0.15	2.20	3.7281	0.6545	0.57596	0.02640
2.50	10	0.135	2.23	3.3874	0.6545	0.58381	0.02712
2.50	11	0.12	2.26	3.0467	0.6545	0.59167	0.02786
2.50	12	0.105	2.29	2.6574	0.6545	0.59952	0.02860
2.50	13	0.095	2.31	2.4140	0.6545	0.60476	0.02910
3.00	4	0.24	2.52	7.0326	0.7854	0.65974	0.03464
3.00	5	0.22	2.56	6.4974	0.7854	0.67021	0.03574
3.00	6	0.2	2.60	5.9621	0.7854	0.68068	0.03687
3.00	7	0.18	2.64	5.3760	0.7854	0.69115	0.03801
3.00	8	0.165	2.67	4.9400	0.7854	0.89901	0.03888
3.00	9	0.15	2.70	4.5506	0.7854	0.70686	0.03976
3.00	10	0.135	2.73	4.1126	0.7854	0.71471	0.04065
3.00	11	0.12	2.76	3.6746	0.7854	0.72267	0.04155
3.00	12	0.105	2.79	3.2366	0.7854	0.73042	0.04246
3.25	4	0.24	2.77	7.6704	0.85085	0.72519	0.04165
3.25	5	0.22	2.81	7.0864	0.85085	0.73566	0.04307
3.25	6	0.2	2.85	6.4536	0.85085	0.74613	0.04430
3.25	7	0.18	2.89	5.8996	0.85085	0.7566	0.04555
3.25	8	0.165	2.92	5.3929	0.85085	0.76446	0.04650
3.25	9	0.15	2.95	4.9449	0.85085	0.77231	0.04748
3.25	10	0.135	2.98	4.4562	0.85085	0.78016	0.04844
3.25	11	0.12	3.01	3.9715	0.85085	0.78802	0.04942
3.25	12	0.105	3.04	3.4848	0.85085	0.79587	0.05041

Outside Dia. Inches	Wall Thickness BWG	Wall Thickness Inches	Inside Dia. Inches	Weight per Ft. of Length	Sq.Ft. Outside Surface per Ft.L.	Sq.Ft. Inside Surface per Ft.L.	Inside Cross-Sectional Area Sq.Ft.
3.50	4	0.24	3.02	8.3031	0.9163	0.790038	0.04974
3.50	5	0.22	3.06	7.6704	0.9163	0.801106	0.05107
3.50	6	0.2	3.10	6.9960	0.9163	0.81158	0.05241
3.50	7	0.18	3.14	6.3563	0.9163	0.822052	0.05378
3.50	8	0.165	3.17	5.8809	0.9163	0.829900	0.05481
3.50	9	0.15	3.20	5.3342	0.9163	0.83776	0.05585
3.50	10	0.135	3.23	4.7969	0.9163	0.845614	0.05690
3.50	11	0.12	3.26	4.3122	0.9163	0.853466	0.05796
3.50	12	0.105	3.29	3.7786	0.9163	0.861322	0.05904
4.00	4	0.24	3.52	9.560	1.0472	0.921506	0.06758
4.00	5	0.22	3.56	8.6093	1.0472	0.932008	0.06912
4.00	6	0.2	3.60	8.0792	1.0472	0.94246	0.07089
4.00	7	0.18	3.64	7.3005	1.0472	0.953052	0.07227
4.00	8	0.165	3.67	6.7165	1.0472	0.960906	0.07346
4.00	9	0.15	3.70	6.1324	1.0472	0.96866	0.07467
4.00	10	0.135	3.73	5.5464	1.0472	0.976514	0.07586
4.00	11	0.12	3.76	4.9943	1.0472	0.984366	0.07711
4.50	4	0.24	4.02	10.8339	1.1781	1.052436	0.08814
4.50	5	0.22	4.06	10.0066	1.1781	1.062008	0.08990
4.50	6	0.2	4.10	9.1305	1.1781	1.07338	0.09168
4.50	7	0.18	4.14	8.2544	1.1781	1.083852	0.09348
4.50	8	0.165	4.17	7.5731	1.1781	1.091706	0.09484
4.50	9	0.15	4.20	6.9403	1.1781	1.09956	0.09621
4.50	10	0.135	4.23	6.2590	1.1781	1.107414	0.09759
4.50	11	0.12	4.26	5.5778	1.1781	1.115268	0.09898
5.00	4	0.24	4.52	12.1432	1.309	1.183336	0.11143
5.00	5	0.22	4.56	11.1696	1.309	1.193908	0.11341
5.00	6	0.2	4.60	10.1964	1.309	1.20428	0.11541
5.00	7	0.18	4.64	9.2230	1.309	1.214762	0.11743
5.00	8	0.165	4.67	8.4929	1.309	1.222606	0.11895
5.00	9	0.15	4.70	7.7142	1.309	1.23046	0.12048
5.00	10	0.135	4.73	6.9841	1.309	1.238314	0.12203
5.375	4	0.24	4.90	13.0676	1.407175	1.281611	0.13069
5.375	5	0.22	4.94	12.0456	1.407175	1.291983	0.13283
5.375	6	0.2	4.98	10.9751	1.407175	1.302455	0.13499
5.375	7	0.18	5.02	9.9043	1.407175	1.312927	0.13717
5.375	8	0.165	5.05	9.1256	1.407175	1.320761	0.13882
5.375	9	0.15	5.08	8.0962	1.407175	1.328635	0.14048
5.50	4	0.24	5.02	13.4135	1.4399	1.314236	0.13745
5.50	5	0.22	5.06	12.3427	1.4399	1.324708	0.13965
5.50	6	0.2	5.10	11.2233	1.4399	1.33518	0.14186
5.50	7	0.18	5.14	10.1526	1.4399	1.349652	0.14410
5.50	8	0.165	5.17	9.3252	1.4399	1.353506	0.14578
5.50	9	0.15	5.20	8.4978	1.4399	1.36136	0.14748
6.00	4	0.24	5.52	14.6497	1.5708	1.445136	0.16619
6.00	5	0.22	5.56	13.4816	1.5708	1.455906	0.16861
6.00	6	0.2	5.60	12.3135	1.5708	1.46606	0.17104
6.00	7	0.18	5.64	11.0966	1.5708	1.476552	0.17349
6.00	8	0.165	5.67	10.2207	1.5708	1.484406	0.17535

TABLE 8.18

				Table of PROPERTIES of PIPE						

Ralph L. Vandagriff   Boiler House Notes:

Date: 9/23/96   Note: Weight of water is 62.37 lb./cu.ft. @ 60 deg. F.
Note: Flow in GPM is calculated using 10 FPS velocity.   Page No:

Nominal Pipe Size	Schedule No. Carbon & Alloy Steels	Stainless Steels	Weight	Outside Dia. Inches	Inside Dia. Inches	Wall Thickness Inches	Weight per Ft., lb./ft.	Wt. of Water Per Ft. lb./ft.	Sq.Ft. Outside Surface per Ft.	Sq.Ft. Inside Surface per Ft.	Transverse Area Sq.Inches	Flow GPM
		10S		0.405	0.307	0.049	0.186	0.0320	0.106	0.0804	0.0740	2.313
1/8 inch	40	40S	Std.	0.405	0.269	0.068	0.244	0.0246	0.106	0.0705	0.0568	1.775
	80	80S	X-Strong	0.405	0.215	0.095	0.314	0.0157	0.106	0.0563	0.0364	1.138
		10S		0.540	0.410	0.065	0.330	0.0570	0.141	0.1073	0.1320	4.125
1/4 inch	40	40S	Std.	0.540	0.364	0.088	0.424	0.0451	0.141	0.0955	0.1041	3.253
	80	80S	X-Strong	0.540	0.302	0.119	0.535	0.0310	0.141	0.0794	0.0716	2.238
		10S		0.675	0.545	0.065	0.423	0.1010	0.177	0.1427	0.2333	7.290
3/8 inch	40	40S	Std.	0.675	0.493	0.091	0.567	0.0827	0.177	0.1295	0.1910	5.969
	80	80S	X-Strong	0.675	0.423	0.126	0.738	0.0609	0.177	0.1106	0.1405	4.391
		10S		0.840	0.670	0.083	0.671	0.1550	0.220	0.1784	0.3568	11.150
	40	40S	Std.	0.840	0.622	0.109	0.850	0.1316	0.220	0.1637	0.3040	9.500
1/2 inch	80	80S	X-Strong	0.840	0.546	0.147	1.087	0.1013	0.220	0.1433	0.2340	7.313
	160			0.840	0.466	0.187	1.310	0.0740	0.220	0.1220	0.1706	5.331
			XX-Strong	0.840	0.252	0.294	1.714	0.0216	0.220	0.0680	0.0499	1.559
		10S		1.050	0.884	0.083	0.857	0.2660	0.275	0.2314	0.6138	19.180
	40	40S	Std.	1.050	0.824	0.113	1.130	0.2301	0.275	0.2168	0.5330	16.656
3/4 inch	80	80S	X-Strong	1.050	0.742	0.154	1.473	0.1875	0.275	0.1948	0.4330	13.530
	160			1.050	0.675	0.188	1.727	0.1514	0.275	0.1759	0.3570	11.156
				1.050	0.614	0.218	1.940	0.1280	0.275	0.1607	0.2961	9.253
			XX-Strong	1.050	0.434	0.308	2.440	0.0633	0.275	0.1137	0.1479	4.622
		10S		1.315	1.097	0.109	1.404	0.4090	0.344	0.2872	0.9448	29.53
	40	40S	Std.	1.315	1.049	0.133	1.678	0.3740	0.344	0.2740	0.8640	27.00
1 inch	80	80S	X-Strong	1.315	0.957	0.179	2.171	0.3112	0.344	0.2520	0.7190	22.47
	160			1.315	0.877	0.219	2.561	0.2614	0.344	0.2290	0.6040	18.87
				1.315	0.815	0.250	2.850	0.2261	0.344	0.2134	0.5217	16.30
			XX-Strong	1.315	0.599	0.358	3.659	0.1221	0.344	0.1570	0.2818	8.81
		10S		1.660	1.442	0.109	1.806	0.7080	0.434	0.3775	1.6330	51.03
	40	40S	Std.	1.660	1.380	0.140	2.272	0.6471	0.434	0.3620	1.4950	46.72
1 1/4 inch	80	80S	X-Strong	1.660	1.278	0.191	2.996	0.5553	0.434	0.3356	1.2830	40.09
	160			1.660	1.160	0.250	3.764	0.4575	0.434	0.3029	1.0570	33.03
			XX-Strong	1.660	0.896	0.382	5.214	0.2732	0.434	0.2331	0.6305	19.70
		10S		1.900	1.682	0.109	2.085	0.9630	0.497	0.4403	2.2210	69.41
	40	40S	Std.	1.900	1.610	0.145	2.717	0.8820	0.497	0.4213	2.0360	63.62
1 1/2 inch	80	80S	X-Strong	1.900	1.500	0.200	3.631	0.7648	0.497	0.3927	1.7670	55.22
	160			1.900	1.337	0.281	4.862	0.6082	0.497	0.3519	1.4050	43.91
			XX-Strong	1.900	1.100	0.400	6.408	0.4117	0.497	0.2903	0.9500	29.69
		10S		2.375	2.157	0.109	2.638	1.5630	0.622	0.5647	3.6540	114.19
	40	40S	Std.	2.375	2.067	0.154	3.652	1.4520	0.622	0.5401	3.3550	104.84
				2.375	2.041	0.167	3.938	1.4200	0.622	0.5380	3.2800	102.50
				2.375	2.000	0.188	4.380	1.3630	0.622	0.5237	3.1420	98.19
2 inch	80	80S	X-Strong	2.375	1.939	0.218	5.022	1.2790	0.622	0.5074	2.9530	92.28
				2.375	1.875	0.250	5.673	1.1960	0.622	0.4920	2.7610	86.28
				2.375	1.750	0.312	6.883	1.0410	0.622	0.4581	2.4050	75.16
	160			2.375	1.689	0.343	7.450	0.7670	0.622	0.4422	2.2400	70.00
			XX-Strong	2.375	1.503	0.436	9.029	0.7890	0.622	0.3929	1.7740	55.44
		10S		2.875	2.635	0.120	3.530	2.3600	0.753	0.6900	5.4530	170.41
	40	40S	Std.	2.875	2.469	0.203	5.790	2.0720	0.753	0.6462	4.7880	149.63
				2.875	2.441	0.217	6.160	2.0260	0.753	0.6381	4.6800	146.25
2 1/2 inch	80	80S	X-Strong	2.875	2.323	0.276	7.660	1.8340	0.753	0.6095	4.2380	132.44
	160			2.875	2.125	0.375	10.010	1.5350	0.753	0.5564	3.5470	110.84
			XX-Strong	2.875	1.771	0.552	13.690	1.0670	0.753	0.4627	2.4640	77.00

TABLE 8.18 Continued

Table of PROPERTIES of PIPE, continued:											

Ralph L. Vandagriff
Date: 9/23/96

Boiler House Notes:
Page No:

Nominal Pipe Size	Schedule No. Carbon & Alloy Steel	Schedule No. Stainless Steels	Weight	Outside Dia. Inches	Inside Dia. Inches	Wall Thickness Inches	Weight per Ft., lb./ft.	Wt. of Water Per Ft. lb./ft.	Sq.Ft. Outside Surface per Ft.	Sq.Ft. Inside Surface per Ft.	Transverse Area Sq.Inches	Flow GPM Velocity of 10FPS
		10S		3.500	3.260	0.120	4.33	3.62	0.916	0.853	8.346	260.81
				3.500	3.250	0.125	4.52	3.60	0.916	0.851	8.300	259.38
				3.500	3.204	0.148	5.30	3.52	0.916	0.840	8.100	253.13
				3.500	3.124	0.188	6.65	3.34	0.916	0.819	7.700	240.63
	40	40S	Std.	3.500	3.068	0.216	7.57	3.20	0.916	0.802	7.393	231.03
				3.500	3.016	0.241	8.39	3.10	0.916	0.790	7.155	223.59
3 inch				3.500	2.992	0.254	8.80	3.06	0.916	0.785	7.050	220.31
				3.500	2.922	0.289	9.91	2.91	0.916	0.765	6.700	209.38
	80	80S	X-Strong	3.500	2.900	0.300	10.25	2.86	0.916	0.761	6.605	206.41
				3.500	2.875	0.312	10.64	2.81	0.916	0.753	6.492	202.88
	160			3.500	2.667	0.406	13.42	2.46	0.916	0.704	5.673	177.28
				3.500	2.624	0.438	14.32	2.34	0.916	0.687	5.407	168.97
			XX-Strong	3.500	2.300	0.600	18.58	1.80	0.916	0.601	4.155	129.84
		10S		4.000	3.760	0.120	4.97	4.81	1.047	0.984	11.10	346.88
				4.000	3.744	0.128	5.38	4.78	1.047	0.981	11.01	344.06
				4.000	3.732	0.134	5.58	4.66	1.047	0.978	10.95	342.19
				4.000	3.704	0.148	6.26	4.66	1.047	0.971	10.75	335.94
3 1/2 inch				4.000	3.624	0.188	7.71	4.48	1.047	0.950	10.32	322.50
	40	40S	Std.	4.000	3.548	0.226	9.11	4.28	1.047	0.929	9.89	309.06
				4.000	3.438	0.281	11.17	4.02	1.047	0.900	9.28	290.00
	80	80S	X-Strong	4.000	3.364	0.318	12.51	3.85	1.047	0.880	8.89	277.81
				4.000	3.312	0.344	13.42	3.73	1.047	0.867	8.62	269.38
				4.000	3.062	0.469	17.66	3.19	1.047	0.802	7.37	230.31
			XX-Strong	4.000	2.728	0.636	22.85	2.53	1.047	0.716	5.84	182.50
		10S		4.500	4.260	0.120	5.61	6.18	1.178	1.115	14.25	445.31
				4.500	4.244	0.128	5.99	6.14	1.178	1.111	14.15	442.19
				4.500	4.232	0.134	6.26	6.11	1.178	1.110	14.10	440.63
				4.500	4.216	0.142	6.61	6.06	1.178	1.105	13.98	436.88
				4.500	4.170	0.165	7.64	5.92	1.178	1.093	13.67	427.19
				4.500	4.124	0.188	8.66	5.80	1.178	1.082	13.39	418.44
				4.500	4.090	0.205	9.39	5.71	1.178	1.071	13.15	410.94
	40	40S	Std.	4.500	4.026	0.237	10.79	5.51	1.178	1.055	12.73	397.81
				4.500	4.000	0.250	11.35	5.45	1.178	1.049	12.57	392.81
4 inch				4.500	3.958	0.271	12.24	5.35	1.178	1.036	12.31	384.69
				4.500	3.938	0.281	12.67	5.27	1.178	1.031	12.17	380.31
				4.500	3.900	0.300	13.42	5.19	1.178	1.023	11.96	373.75
				4.500	3.876	0.312	14.00	5.12	1.178	1.013	11.80	368.75
	80	80S	X-Strong	4.500	3.826	0.337	14.98	4.96	1.178	1.002	11.50	359.38
				4.500	3.750	0.375	16.52	4.78	1.178	0.982	11.04	345.00
	120			4.500	3.624	0.438	19.00	4.47	1.178	0.949	10.32	322.50
				4.500	3.500	0.500	21.36	4.16	1.178	0.916	9.62	300.63
	160			4.500	3.438	0.531	22.60	4.02	1.178	0.900	9.28	290.00
			XX-Strong	4.500	3.152	0.674	27.54	3.38	1.178	0.826	7.80	243.75
		10S		5.563	5.295	0.134	7.77	9.54	1.456	1.386	22.02	688.13
	40	40S	Std.	5.563	5.047	0.258	14.62	8.66	1.456	1.321	20.01	625.31
				5.563	4.859	0.352	19.50	8.06	1.456	1.272	18.60	581.25
5 inch	80	80S	X-Strong	5.563	4.813	0.375	20.78	7.87	1.456	1.260	18.19	568.44
				5.563	4.688	0.437	23.95	7.47	1.456	1.227	17.26	539.40
	120			5.563	4.563	0.500	27.10	7.08	1.456	1.195	16.35	510.94
	160			5.563	4.313	0.625	32.96	6.32	1.456	1.129	14.61	456.56
			XX-Strong	5.563	4.063	0.750	38.55	5.62	1.456	1.064	12.97	405.31

TABLE 8.18 Continued

				Table of PROPERTIES of PIPE, continued:							
Ralph L. Vandagriff									Boiler House Notes:		
Date: 9/24/96									Page No:		
Nominal Pipe Size	Schedule No. Carbon & Alloy Steels	Stainless Steels	Weight	Outside Dia. Inches	Inside Dia. Inches	Wall Thickness Inches	Weight per Ft. lb./ft.	Wt. of Water Per Ft. lb./ft.	Sq.Ft. Outside Surface per Ft.	Sq.Ft. Inside Surface per Ft.	Transverse Area Sq.Inches	Flow GPM Velocity of 10FPS
---	---	---	---	---	---	---	---	---	---	---	---	---
		10S		6.625	6.357	0.134	9.29	13.73	1.734	1.664	31.74	991.85
				6.625	6.287	0.160	11.56	13.43	1.734	1.646	31.04	970.13
				6.625	6.265	0.180	12.50	13.34	1.734	1.640	30.83	963.35
				6.625	6.249	0.188	12.93	13.27	1.734	1.636	30.67	958.43
				6.625	6.187	0.219	15.02	13.01	1.734	1.620	30.06	939.51
				6.625	6.125	0.250	17.02	12.75	1.734	1.604	29.46	920.77
6 inch				6.625	6.071	0.277	18.96	12.53	1.734	1.589	28.95	904.61
	40	40S	Std.	6.625	6.065	0.280	18.97	12.50	1.734	1.588	28.89	902.82
				6.625	5.875	0.375	25.10	11.73	1.734	1.538	27.11	847.14
	80	80S	X-Strong	6.625	5.761	0.432	28.57	11.28	1.734	1.508	26.07	814.59
				6.625	5.625	0.500	32.79	10.75	1.734	1.473	24.85	776.58
	120			6.625	5.501	0.562	36.40	10.28	1.734	1.440	23.77	742.72
	160			6.625	5.189	0.718	45.30	9.15	1.734	1.358	21.15	660.96
			XX-Strong	6.625	4.897	0.864	53.16	8.15	1.734	1.282	18.83	588.57
		10S		8.625	8.329	0.148	13.40	23.58	2.258	2.181	54.48	1702.65
				8.625	8.309	0.158	14.26	23.46	2.258	2.175	54.22	1694.40
				8.625	8.295	0.165	14.91	23.38	2.258	2.172	54.04	1688.78
				8.625	8.249	0.188	16.90	23.13	2.258	2.160	53.44	1670.10
				8.625	8.219	0.203	18.30	22.96	2.258	2.152	53.06	1657.98
				8.625	8.187	0.219	19.64	22.76	2.258	2.143	52.64	1645.09
				8.625	8.153	0.236	21.43	22.59	2.258	2.134	52.21	1631.46
	20			8.625	8.125	0.250	22.40	22.44	2.258	2.127	51.85	1620.27
	30			8.625	8.071	0.277	24.70	22.14	2.258	2.113	51.16	1598.81
	40	40S	Std.	8.625	7.981	0.322	28.55	21.85	2.258	2.089	50.03	1563.35
8 inch				8.625	7.937	0.344	30.40	21.41	2.258	2.078	49.48	1546.16
				8.625	7.921	0.352	31.00	21.32	2.258	2.074	49.28	1539.93
				8.625	7.875	0.375	33.10	21.08	2.258	2.062	48.71	1522.10
	60			8.625	7.813	0.406	35.70	20.75	2.258	2.045	47.94	1498.22
				8.625	7.687	0.468	40.83	20.08	2.258	2.012	46.41	1450.29
	80	80S	X-Strong	8.625	7.625	0.500	43.39	19.76	2.258	1.996	45.66	1426.98
	100			8.625	7.439	0.593	50.90	18.81	2.258	1.948	43.46	1358.22
				8.625	7.375	0.625	53.40	18.48	2.258	1.931	42.72	1334.95
	120			8.625	7.189	0.718	60.70	17.56	2.258	1.882	40.59	1268.46
	140			8.625	7.001	0.812	67.80	16.66	2.258	1.833	38.50	1202.99
			XX-Strong	8.625	6.875	0.875	72.42	16.06	2.258	1.800	37.12	1160.08
	160			8.625	6.813	0.906	74.70	15.77	2.258	1.784	36.46	1139.25
		10S		10.750	10.420	0.165	18.65	36.90	2.814	2.728	85.28	2664.87
				10.750	10.374	0.188	21.12	36.57	2.814	2.716	84.52	2641.40
				10.750	10.344	0.203	22.86	36.36	2.814	2.708	84.04	2626.14
				10.750	10.312	0.219	24.60	36.14	2.814	2.700	83.52	2609.92
	20			10.750	10.250	0.250	28.03	35.71	2.814	2.683	82.52	2578.63
				10.750	10.192	0.279	31.20	35.30	2.814	2.668	81.58	2549.53
	30			10.750	10.136	0.307	34.24	34.92	2.814	2.654	80.69	2521.59
				10.750	10.054	0.348	38.06	34.35	2.814	2.632	79.30	2480.95
	40	40S	Std.	10.750	10.020	0.365	40.48	34.12	2.814	2.623	78.85	2464.20
10 inch				10.750	9.960	0.395	43.88	33.71	2.814	2.608	77.91	2434.76
	60	80S	X-Strong	10.750	9.750	0.500	54.74	32.31	2.814	2.553	74.66	2333.19
				10.750	9.688	0.531	57.98	31.90	2.814	2.536	73.72	2303.61
	80			10.750	9.564	0.593	64.40	31.09	2.814	2.504	71.84	2245.02
	100			10.750	9.314	0.718	77.00	29.48	2.814	2.438	68.13	2129.18
				10.750	9.250	0.750	80.10	29.08	2.814	2.422	67.20	2100.02
	120			10.750	9.064	0.843	89.20	27.92	2.814	2.373	64.53	2016.42
	140			10.750	8.750	1.000	104.20	26.02	2.814	2.291	60.13	1879.13
				10.750	8.624	1.063	109.90	25.28	2.814	2.258	58.41	1825.40
	160			10.750	8.500	1.125	116.00	24.55	2.814	2.225	56.75	1773.29

TABLE 8.18 Continued

Nominal Pipe Size	Schedule No. Carbon & Alloy Steels	Stainless Steels	Weight	Outside Dia. Inches	Inside Dia. Inches	Wall Thickness Inches	Weight per Ft. lb./ft.	Wt. of Water Per Ft. lb./ft.	Sq.Ft. Outside Surface per Ft.	Sq.Ft. Inside Surface per Ft.	Transverse Area Sq.Inches	Flow GPM Velocity of 10FPS
		5S		12.750	12.438	0.156	21.02	52.58	3.338	3.256	121.50	3797.01
		10S		12.750	12.390	0.180	24.16	52.17	3.338	3.244	120.57	3767.76
				12.750	12.344	0.203	27.20	51.78	3.338	3.232	119.67	3739.84
				12.750	12.312	0.219	29.30	51.52	3.338	3.229	119.06	3720.47
				12.750	12.274	0.238	31.80	51.20	3.338	3.213	118.32	3697.84
	20			12.750	12.250	0.250	33.40	51.00	3.338	3.207	117.86	3683.10
				12.750	12.192	0.279	37.20	50.52	3.338	3.192	116.75	3648.30
				12.750	12.150	0.300	40.00	50.17	3.338	3.181	115.94	3623.21
	30			12.750	12.090	0.330	43.80	49.68	3.338	3.165	114.80	3587.51
				12.750	12.062	0.344	45.50	49.45	3.338	3.158	114.27	3570.92
12 inch	40	40S	Std.	12.750	12.000	0.375	49.60	48.94	3.338	3.142	113.10	3534.80
				12.750	11.938	0.406	53.60	48.43	3.338	3.125	111.93	3497.87
				12.750	11.874	0.438	57.50	47.92	3.338	3.109	110.74	3460.47
		80S	X-Strong	12.750	11.750	0.500	65.40	46.92	3.338	3.076	108.43	3388.67
	60			12.750	11.626	0.562	73.20	45.94	3.338	3.044	106.16	3317.43
				12.750	11.500	0.625	80.90	44.95	3.338	3.011	103.87	3245.91
	80			12.750	11.376	0.687	88.60	43.98	3.338	2.978	101.64	3176.29
	100			12.750	11.064	0.843	108.00	41.60	3.338	2.897	96.14	3004.46
				12.750	11.000	0.875	110.90	41.12	3.338	2.880	95.03	2969.79
	120			12.750	10.750	1.000	125.50	39.27	3.338	2.814	90.76	2836.34
	140			12.750	10.500	1.125	140.00	37.47	3.338	2.749	86.59	2705.96
				12.750	10.312	1.219	150.10	36.14	3.338	2.700	83.52	2600.92
	160			12.750	10.126	1.312	161.00	34.85	3.338	2.651	80.53	2516.61
		5S		14.000	13.688	0.156	23.11	63.67	3.665	3.584	147.15	4598.65
		10S		14.000	13.624	0.188	28.00	63.08	3.665	3.567	145.78	4555.66
				14.000	13.560	0.220	32.00	62.49	3.665	3.550	144.41	4512.95
				14.000	13.624	0.298	36.00	62.16	3.665	3.541	143.65	4480.02
	10			14.000	13.500	0.250	37.00	61.94	3.665	3.534	143.14	4473.10
	20			14.000	13.376	0.312	46.00	60.80	3.665	3.502	140.52	4391.30
	30		Std.	14.000	13.250	0.375	55.00	59.68	3.665	3.469	137.89	4308.96
				14.000	13.188	0.406	63.00	59.11	3.665	3.453	136.60	4268.73
	40			14.000	13.124	0.438	63.00	58.54	3.665	3.436	135.26	4227.40
				14.000	13.062	0.469	68.00	57.98	3.665	3.420	134.00	4187.55
14 inch			X-Strong	14.000	13.000	0.500	72.00	57.43	3.665	3.403	132.73	4147.89
	60			14.000	12.814	0.593	85.00	55.80	3.665	3.355	128.96	4030.05
				14.000	12.750	0.625	89.00	55.25	3.665	3.338	127.68	3989.89
	80			14.000	12.688	0.656	94.00	54.71	3.665	3.322	126.44	3951.16
	100			14.000	12.500	0.750	107.00	53.10	3.665	3.273	122.72	3834.96
	120			14.000	12.126	0.937	131.00	49.97	3.665	3.175	115.49	3608.91
	140			14.000	11.814	1.093	151.00	47.43	3.665	3.093	109.62	3425.59
				14.000	11.500	1.250	171.00	44.95	3.665	3.011	103.87	3245.91
	160			14.000	11.312	1.344	182.00	43.49	3.665	2.961	100.50	3140.65
				14.000	11.188	1.406	190.00	42.54	3.665	2.929	98.31	3072.17
		5S		16.000	15.670	0.165	28.43	83.45	4.189	4.102	192.95	6026.09
		10S		16.000	15.624	0.188	32.00	82.96	4.189	4.090	191.72	5991.36
				16.000	15.524	0.238	40.00	81.90	4.189	4.064	189.28	5914.91
	10			16.000	15.500	0.250	42.00	81.65	4.189	4.058	188.69	5896.64
				16.000	15.438	0.281	47.00	81.00	4.189	4.042	187.19	5849.66
	20			16.000	15.376	0.312	52.00	80.35	4.189	4.025	185.69	5802.67
				16.000	15.312	0.344	57.00	79.68	4.189	4.009	184.14	5754.46
	30		Std	16.000	15.250	0.375	63.00	79.04	4.189	3.992	182.65	5707.96
				16.000	15.188	0.406	68.00	78.39	4.189	3.976	181.17	5661.64
				16.000	15.124	0.438	73.00	77.74	4.189	3.959	179.65	5614.02
				16.000	15.062	0.469	78.00	77.10	4.189	3.943	178.18	5568.09
16 inch	40		X-Strong	16.000	15.000	0.500	83.00	76.47	4.189	3.927	176.72	5522.34
				16.000	14.938	0.531	98.00	75.84	4.189	3.911	175.26	5476.79
	60			16.000	14.688	0.656	108.00	73.32	4.189	3.845	169.44	5295.00
				16.000	14.626	0.687	112.00	72.70	4.189	3.829	168.01	5250.40
				16.000	14.500	0.750	122.00	71.46	4.189	3.796	165.13	5160.32
	80			16.000	14.314	0.843	137.00	69.63	4.189	3.747	160.92	5028.78
	100			16.000	13.938	1.031	166.00	66.02	4.189	3.649	152.58	4768.08
	120			16.000	13.564	1.218	193.00	62.53	4.189	3.551	144.50	4515.81
	140			16.000	13.124	1.438	224.00	58.54	4.189	3.436	135.26	4227.40
				16.000	13.000	1.500	232.00	57.43	4.189	3.403	132.73	4147.89
	160			16.000	12.814	1.593	245.00	55.80	4.189	3.355	128.96	4030.05

Table of PROPERTIES of PIPE, continued:

Ralph L. Vandagriff

Date: 9/24/96

Boiler House Notes:

Page No:

TABLE 8.18 Continued

| Table of PROPERTIES of PIPE, continued: | | | | | | | | | | | |

Ralph L. Vandagriff   Boiler House Notes:

Date: 9/24/96   Page No:

Nominal Pipe Size	Schedule No. Carbon & Alloy Steels	Schedule No. Stainless Steels	Weight	Outside Dia. Inches	Inside Dia. Inches	Wall Thickness Inches	Weight per Ft., lb./ft.	Wt. of Water Per Ft. lb./ft.	Sq.Ft. Outside Surface per Ft.	Sq.Ft. Inside Surface per Ft.	Transverse Area Sq.Inches	Flow GPM Velocity of 10FPS
		5S		18.000	17.670	0.165	31.14	106.11	4.712	4.626	245.22	7663.27
		10S		18.000	17.624	0.188	35.87	105.56	4.712	4.614	243.95	7623.42
	10			18.000	17.500	0.250	47.00	104.09	4.712	4.582	240.53	7516.52
	20			18.000	17.376	0.312	59.00	102.61	4.712	4.549	237.13	7410.38
			Std.	18.000	17.250	0.375	71.00	101.13	4.712	4.516	233.71	7303.30
	30			18.000	17.124	0.438	82.00	99.65	4.712	4.483	230.30	7197.00
			X-Strong	18.000	17.000	0.500	93.00	98.22	4.712	4.451	226.98	7093.14
	40			18.000	16.876	0.562	105.00	96.79	4.712	4.418	223.68	6990.04
				18.000	16.812	0.594	110.00	96.06	4.712	4.401	221.99	6937.13
18 Inch				18.000	16.750	0.625	116.00	95.35	4.712	4.385	220.35	6886.06
	60			18.000	16.500	0.750	138.00	92.62	4.712	4.320	213.83	6682.04
				18.000	16.376	0.812	149.00	91.14	4.712	4.287	210.62	6581.98
	80			18.000	16.126	0.937	171.00	88.38	4.712	4.222	204.24	6382.55
	100			18.000	15.688	1.156	208.00	83.64	4.712	4.107	193.30	6040.54
	120			18.000	15.250	1.375	244.00	79.04	4.712	3.992	182.65	5707.98
	140			18.000	14.876	1.562	275.00	75.21	4.712	3.895	173.81	5431.42
				18.000	14.626	1.687	294.00	72.70	4.712	3.829	168.01	5250.40
	160			18.000	14.438	1.781	309.00	70.84	4.712	3.780	163.72	5116.29
		5S		20.000	19.624	0.188	39.68	130.88	5.236	5.138	302.46	9451.83
		10S		20.000	19.564	0.218	46.22	130.08	5.236	5.122	300.61	9394.12
	10			20.000	19.500	0.250	53.00	129.23	5.236	5.105	298.65	9332.78
				20.000	19.374	0.313	66.00	127.56	5.236	5.072	294.90	9212.54
	20		Std.	20.000	19.250	0.375	79.00	125.94	5.236	5.040	291.04	9094.99
				20.000	19.124	0.438	92.00	124.29	5.236	5.007	287.24	8976.32
	30		X-Strong	20.000	19.000	0.500	105.00	122.69	5.236	4.974	283.53	8860.29
				20.000	18.876	0.562	117.00	121.09	5.236	4.942	279.84	8745.02
	40			20.000	18.814	0.593	123.00	120.30	5.236	4.926	278.01	8687.67
20 Inch				20.000	18.750	0.625	129.00	119.48	5.236	4.909	276.12	8628.66
	60			20.000	18.376	0.812	167.00	114.76	5.236	4.811	265.21	8287.97
				20.000	18.250	0.875	179.00	113.19	5.236	4.778	261.59	8174.80
				20.000	18.188	0.906	185.00	112.42	5.236	4.762	259.81	8119.15
	80			20.000	17.938	1.031	209.00	109.35	5.236	4.696	252.72	7897.49
	100			20.000	17.438	1.281	256.00	103.34	5.236	4.565	238.83	7463.36
	120			20.000	17.000	1.500	297.00	98.22	5.236	4.451	226.98	7093.14
	140			20.000	16.500	1.750	342.00	92.52	5.236	4.320	213.83	6682.04
				20.000	16.312	1.844	357.00	90.43	5.236	4.270	208.96	6530.63
	160			20.000	16.064	1.968	379.00	87.70	5.236	4.205	202.67	6333.57
		5S		22.000	21.624	0.188	43.66	158.91	5.760	5.661	367.25	11476.59
		10S		22.000	21.564	0.218	50.90	158.03	5.760	5.645	365.22	11412.99
	10			22.000	21.500	0.250	58.00	157.10	5.760	5.629	363.05	11345.35
				22.000	21.376	0.312	72.00	155.29	5.760	5.596	358.88	11214.86
	20		Std.	22.000	21.250	0.375	87.00	153.46	5.760	5.563	354.66	11083.04
				22.000	21.126	0.437	103.00	151.66	5.760	5.531	350.53	10954.07
	30		X-Strong	22.000	21.000	0.500	115.00	149.87	5.760	5.498	346.36	10823.79
				22.000	20.876	0.562	129.00	148.11	5.760	5.465	342.28	10696.35
22 Inch				22.000	20.750	0.625	143.00	146.33	5.760	5.432	338.16	10567.62
				22.000	20.624	0.688	157.00	144.56	5.760	5.399	334.07	10439.67
				22.000	20.500	0.750	170.00	142.82	5.760	5.367	330.06	10314.51
	60			22.000	20.250	0.875	197.44	139.36	5.760	5.301	322.06	10064.47
	80			22.000	19.750	1.125	250.85	132.56	5.760	5.171	306.36	9573.60
	100			22.000	19.250	1.375	302.92	125.94	5.760	5.040	291.04	9094.99
	120			22.000	18.750	1.625	353.55	119.48	5.760	4.909	276.12	8628.66
	140			22.000	18.250	1.875	403.05	113.19	5.760	4.778	261.59	8174.80
	160			22.000	17.750	2.125	451.12	107.07	5.760	4.647	247.45	7732.92

TABLE 8.18 Continued

	Table of PROPERTIES of PIPE, continued:

Ralph L. Vandagriff

Date: 9/24/96

Nominal Pipe Size	Schedule No. Carbon & Alloy Steels	Schedule No. Stainless Steels	Weight	Outside Dia. Inches	Inside Dia. Inches	Wall Thickness Inches	Weight per Ft. lb./ft.	Wt. of Water Per Ft. lb./ft.	Sq.Ft. Outside Surface per Ft.	Sq.Ft. Inside Surface per Ft.	Transverse Area Sq.Inches	Flow GPM Velocity of 10FPS
		5S		24.000	23.564	0.218	55.38	188.71	6.283	6.169	436.10	13629.2
	10	10S		24.000	23.500	0.250	63.00	187.88	6.283	6.152	433.74	13554.3
				24.000	23.376	0.312	79.00	185.71	6.283	6.120	429.17	13411.6
	20		Std.	24.000	23.250	0.375	95.00	183.71	6.283	6.087	424.56	13267.4
				24.000	23.126	0.437	110.00	181.76	6.283	6.054	420.04	13126.3
			X-Strong	24.000	23.000	0.500	125.00	179.78	6.283	6.021	415.48	12983.6
	30			24.000	22.876	0.562	141.00	177.85	6.283	5.989	411.01	12844.0
				24.000	22.750	0.625	156.00	175.89	6.283	5.956	406.49	12702.9
	40			24.000	22.626	0.687	171.00	173.98	6.283	5.923	402.07	12564.8
24 inch	60			24.000	22.500	0.750	186.00	172.05	6.283	5.891	397.61	12425.3
				24.000	22.064	0.968	238.00	165.45	6.283	5.776	382.35	11948.4
				24.000	21.938	1.031	253.00	163.56	6.283	5.743	377.99	11812.3
	80			24.000	21.564	1.218	297.00	158.03	6.283	5.645	365.22	11413.0
	100			24.000	20.938	1.531	367.00	148.99	6.283	5.482	344.32	10760.0
	120			24.000	20.376	1.812	429.00	141.10	6.283	5.334	326.08	10190.1
	140			24.000	19.876	2.062	484.00	134.26	6.283	5.204	310.28	9696.1
				24.000	19.626	2.187	510.00	130.90	6.283	5.138	302.52	9453.8
	160			24.000	19.314	2.343	542.00	126.77	6.283	5.056	292.98	9155.6
	10			26.000	25.500	0.250	67.00	220.99	6.807	6.676	510.71	15959.6
				26.000	25.376	0.312	84.00	218.84	6.807	6.643	505.75	15804.7
			Std.	26.000	25.250	0.375	103.00	216.68	6.807	6.610	500.74	15648.2
				26.000	25.126	0.437	119.00	214.55	6.807	6.578	495.84	15494.9
26 inch	20		X-Strong	26.000	25.000	0.500	136.00	212.41	6.807	6.545	490.88	15339.8
				26.000	24.876	0.562	153.00	210.30	6.807	6.513	486.02	15188.0
				26.000	24.750	0.625	169.00	208.16	6.807	6.480	481.11	15034.6
				26.000	24.624	0.688	186.00	206.06	6.807	6.447	476.22	14881.9
				26.000	24.500	0.750	202.00	203.99	6.807	6.414	471.44	14732.4
	10			28.000	27.376	0.312	92.42	254.70	7.330	7.167	588.61	18394.2
28 inch			Std.	28.000	27.250	0.375	110.65	252.36	7.330	7.134	583.21	18225.3
	20		X-Strong	28.000	27.000	0.500	146.87	247.75	7.330	7.069	572.56	17892.4
	30			28.000	26.750	0.625	182.75	243.18	7.330	7.003	562.00	17562.6
			5S		30.000	29.500	0.250	79.44	295.75	7.854	7.723	683.49	21359.2
	10	10S		30.000	29.376	0.312	99.00	293.27	7.854	7.691	677.76	21180.0
			Std.	30.000	29.250	0.375	119.00	290.76	7.854	7.658	671.96	20998.7
30 inch				30.000	29.126	0.437	138.00	288.30	7.854	7.625	666.27	20821.0
	20		X-Strong	30.000	29.000	0.500	158.00	285.81	7.854	7.592	660.52	20641.3
				30.000	28.876	0.562	177.00	283.37	7.854	7.560	654.88	20465.2
	30			30.000	28.750	0.625	196.00	280.91	7.854	7.527	649.18	20286.9
	10			32.000	31.376	0.312	105.77	334.57	8.378	8.214	773.19	24162.2
			Std.	32.000	31.250	0.375	126.66	331.88	8.378	8.181	766.99	23968.5
32 inch	20		X-Strong	32.000	31.000	0.500	168.23	326.60	8.378	8.116	754.77	23586.5
	30			32.000	30.750	0.625	209.46	321.35	8.378	8.050	742.64	23207.6
	40			32.000	30.624	0.688	229.94	318.72	8.378	8.017	736.57	23017.8
	10			34.000	33.376	0.312	112.45	378.58	8.901	8.738	874.90	27340.7
			Std.	34.000	33.250	0.375	134.69	375.72	8.901	8.705	868.31	27134.6
34 inch	20		X-Strong	34.000	33.000	0.500	178.91	370.10	8.901	8.639	855.30	26728.1
	30			34.000	32.750	0.625	222.81	364.51	8.901	8.574	842.39	26324.7
	40			34.000	32.624	0.688	244.63	361.71	8.901	8.541	835.92	26122.5
	10			36.000	35.376	0.312	119.12	425.31	9.425	9.261	982.90	30715.6
			Std.	36.000	35.250	0.375	142.70	422.28	9.425	9.228	975.91	30497.1
36 inch	20		X-Strong	36.000	35.000	0.500	189.60	416.32	9.425	9.163	962.12	30066.1
	30			36.000	34.750	0.625	236.16	410.39	9.425	9.098	948.42	29638.1
	40			36.000	34.500	0.750	270.62	404.51	9.425	9.032	934.82	29213.2

TABLE 8.19

(All Dimensions in Inches)

Nom. Pipe Size	A	B	C	D	E	F	L USA	L Short	G
1/2	1 1/2	5/8	1 7/8	......	......	1	3	......	1 3/8
3/4	1 1/8	7/16	1 11/16	......	......	1 1/2	3	2	1 11/16
1	1 1/2	7/8	2 3/16	1	1 5/8	1 1/2	4	2	2
1 1/4	1 7/8	1	2 3/4	1 1/4	2 1/16	1 1/2	4	2	2 1/2
1 1/2	2 1/4	1 1/8	3 1/4	1 1/2	2 7/16	1 1/2	4	2	2 7/8
2	3	1 3/8	4 3/16	2	3 3/16	1 1/2	6	2 1/2	3 5/8
2 1/2	3 3/4	1 3/4	5 3/16	2 1/2	3 15/16	1 1/2	6	2 1/2	4 1/8
3	4 1/2	2	6 1/4	3	4 3/4	2	6	2 1/2	5
3 1/2	5 1/4	2 1/4	7 1/4	3 1/2	5 1/2	2 1/2	6	3	5 1/2
4	6	2 1/2	8 1/4	4	6 1/4	2 1/2	6	3	6 3/16
5	7 1/2	3 1/8	10 5/16	5	7 3/4	3	8	3	7 5/16
6	9	3 3/4	12 5/16	6	9 5/16	3 1/2	8	3 1/2	8 1/2
8	12	5	16 5/16	8	12 5/16	4	8	4	10 5/8
10	15	6 1/4	20 3/8	10	15 3/8	5	10	5	12 3/4
12	18	7 1/2	24 3/8	12	18 3/8	6	10	6	15
14	21	8 3/4	28	14	21	6 1/2	12	......	16 1/4
16	24	10	32	16	24	7	12	......	18 1/2
18	27	11 1/4	36	18	27	8	12	......	21
20	30	12 1/2	40	20	30	9	12	......	23
24	36	15	48	24	36	10 1/2	12	......	27 1/4
30	45	18 1/2	60	30	45	10 1/2	.....	......	......

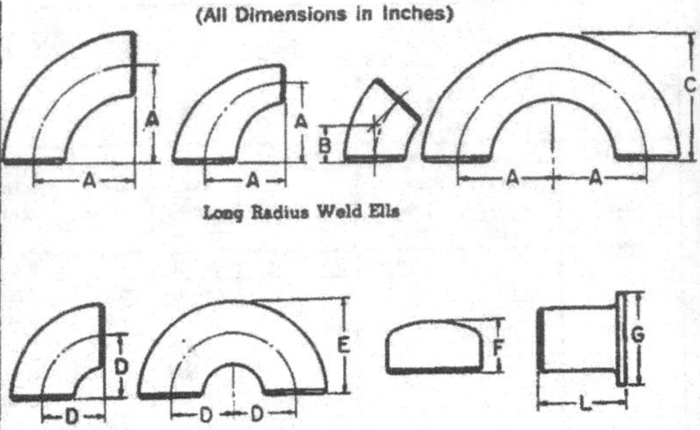

Long Radius Weld Ells

Short Radius Weld Ells

Caps

Stub Ends

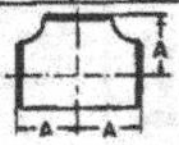

Straight Tees

Nom. Pipe Size	Outlet	A	D	L
1	1	1 1/2	......	......
1	3/4	1 1/2	1 1/2	2
1	1/2	1 1/2	1 1/2	2
1 1/4	1 1/4	1 7/8	......	......
1 1/4	1	1 7/8	1 7/8	2
1 1/4	3/4	1 7/8	1 7/8	2
1 1/4	1/2	1 7/8	1 7/8	2
1 1/2	1 1/2	2 1/4	......	......
1 1/2	1 1/4	2 1/4	2 1/4	2 1/2
1 1/2	1	2 1/4	2 1/4	2 1/2
1 1/2	3/4	2 1/4	2 1/4	2 1/2
1 1/2	1/2	2 1/4	2 1/4	2 1/2
2	2	2 1/2	......	......
2	1 1/2	2 1/2	2 3/8	3
2	1 1/4	2 1/2	2 1/4	3
2	1	2 1/2	2	3
2	3/4	2 1/2	1 3/4	3
2 1/2	2 1/2	3	......	......
2 1/2	2	3	2 3/4	3 1/2
2 1/2	1 1/2	3	2 5/8	3 1/2
2 1/2	1 1/4	3	2 1/2	3 1/2
2 1/2	1	3	2 1/4	3 1/2
3	3	3 3/8	......	......
3	2 1/2	3 3/8	3 1/4	3 1/2
3	2	3 3/8	3	3 1/2
3	1 1/2	3 3/8	2 7/8	3 1/2
3	1 1/4	3 3/8	2 3/4	3 1/2
3 1/2	3 1/2	3 3/4	......	......
3 1/2	3	3 3/4	3 5/8	4
3 1/2	2 1/2	3 3/4	3 1/2	4
3 1/2	2	3 3/4	3 1/4	4
3 1/2	1 1/2	3 3/4	3 1/8	4

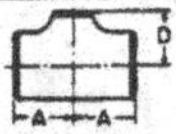

Reducing Tees

Nom. Pipe Size	Outlet	A	D	L
4	4	4 1/8	......	......
4	3 1/2	4 1/8	4	4
4	3	4 1/8	3 7/8	4
4	2 1/2	4 1/8	3 3/4	4
4	2	4 1/8	3 1/2	4
4	1 1/2	4 1/8	3 3/8	4
5	5	4 7/8	......	......
5	4	4 7/8	4 5/8	5
5	3 1/2	4 7/8	4 1/2	5
5	3	4 7/8	4 3/8	5
5	2 1/2	4 7/8	4 1/4	5
5	2	4 7/8	4 1/8	5
6	6	5 5/8	......	......
6	5	5 5/8	5 3/8	5 1/2
6	4	5 5/8	5 1/8	5 1/2
6	3 1/2	5 5/8	5	5 1/2
6	3	5 5/8	4 7/8	5 1/2
6	2 1/2	5 5/8	4 3/4	5 1/2
8	8	7	......	......
8	6	7	6 5/8	6
8	5	7	6 3/8	6
8	4	7	6 1/8	6
8	3 1/2	7	6	6
10	10	8 1/2	......	......
10	8	8 1/2	8	7
10	6	8 1/2	7 5/8	7
10	5	8 1/2	7 1/2	7
10	4	8 1/2	7 1/4	7
12	12	10	......	......
12	10	10	9 1/2	8
12	8	10	9	8
12	6	10	8 5/8	8
12	5	10	8 1/2	8

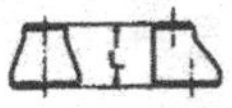

Con. & Ecc. Reducers

Nom. Pipe Size	Outlet	A	D	L
14	14	11	......	......
14	12	11	10 5/8	13
14	10	11	10 1/8	13
14	8	11	9 3/4	13
14	6	11	9 3/8	13
16	16	12	......	......
16	14	12	12	14
16	12	12	11 5/8	14
16	10	12	11 1/8	14
16	8	12	10 3/4	14
16	6	12	10 3/8	14
18	18	13 1/2	......	......
18	16	13 1/2	13	15
18	14	13 1/2	13	15
18	12	13 1/2	12 5/8	15
18	10	13 1/2	12 1/8	15
18	8	13 1/2	11 3/4	15
20	20	15	......	......
20	18	15	14 1/2	20
20	16	15	14	20
20	14	15	14	20
20	12	15	13 5/8	20
20	10	15	13 1/8	20
20	8	15	12 3/4	20
24	24	17	......	......
24	20	17	17	20
24	18	17	16 1/2	20
24	16	17	16	20
24	14	17	16	20
24	12	17	15 5/8	20
24	10	17	15 1/8	20

Reprinted by permission of Taylor Forge & Pipe Works

Table 8.20

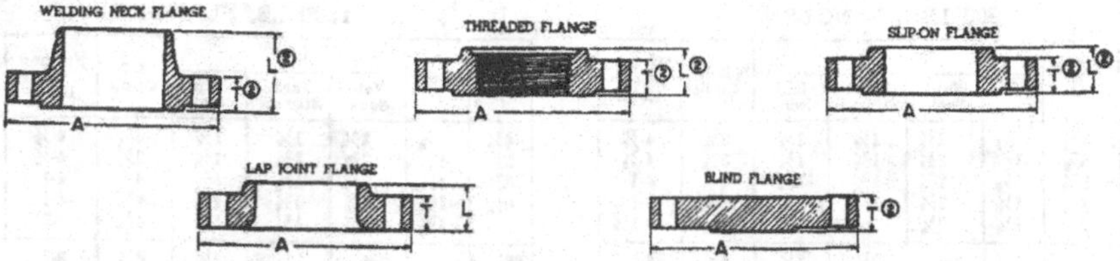
WELDING NECK FLANGE
THREADED FLANGE
SLIP-ON FLANGE
LAP JOINT FLANGE
BLIND FLANGE
A
T
L

150 LB. FLANGES

Nom. Pipe Size	A	T①	L① Weld Neck	L① Thrd. Slip on	L① Lap Joint	Bolt Circle	No. and Sizes of Holes
1/2	3 1/2	7/16	1 7/8	5/8	5/8	2 3/8	4-5/8
3/4	3 7/8	1/2	2 1/16	5/8	5/8	2 3/4	4-5/8
1	4 1/4	9/16	2 3/16	11/16	11/16	3 1/8	4-5/8
1 1/4	4 5/8	5/8	2 1/4	13/16	13/16	3 1/2	4-5/8
1 1/2	5	11/16	2 7/16	7/8	7/8	3 7/8	4-5/8
2	6	3/4	2 1/2	1	1	4 3/4	4-3/4
2 1/2	7	7/8	2 3/4	1 1/8	1 1/8	5 1/2	4-3/4
3	7 1/2	15/16	2 3/4	1 3/16	1 3/16	6	4-3/4
3 1/2	8 1/2	15/16	2 13/16	1 1/4	1 1/4	7	8-3/4
4	9	15/16	3	1 5/16	1 5/16	7 1/2	8-3/4
5	10	15/16	3 1/2	1 7/16	1 7/16	8 1/2	8-7/8
6	11	1	3 1/2	1 9/16	1 9/16	9 1/2	8-7/8
8	13 1/2	1 1/8	4	1 3/4	1 3/4	11 3/4	8-7/8
10	16	1 3/16	4	1 15/16	1 15/16	14 1/4	12-1
12	19	1 1/4	4 1/2	2 3/16	2 3/16	17	12-1
14	21	1 3/8	5	2 1/4	3 1/8	18 3/4	12-1 1/8
16	23 1/2	1 7/16	5	2 1/2	3 7/16	21 1/4	16-1 1/8
18	25	1 9/16	5 1/2	2 11/16	3 13/16	22 3/4	16-1 1/4
20	27 1/2	1 11/16	5 11/16	2 7/8	4 1/16	25	20-1 1/4
24	32	1 7/8	6	3 1/4	4 3/8	29 1/2	20-1 3/8

300 LB. FLANGES

A	T①	L③ Weld Neck	L③ Thrd. Slip on	L③ Lap Joint	Bolt Circle	No. and Size of Holes	Nom. Pipe Size
3 3/4	9/16	2 1/16	7/8	7/8	2 5/8	4-5/8	1/2
4 5/8	5/8	2 1/4	1	1	3 1/4	4-3/4	3/4
4 7/8	11/16	2 7/16	1 1/16	1 1/16	3 1/2	4-3/4	1
5 1/4	3/4	2 9/16	1 1/16	1 1/16	3 7/8	4-3/4	1 1/4
6 1/8	13/16	2 11/16	1 3/16	1 3/16	4 1/2	4-7/8	1 1/2
6 1/2	7/8	2 3/4	1 5/16	1 5/16	5	8-3/4	2
7 1/2	1	3	1 1/2	1 1/2	5 7/8	8-7/8	2 1/2
8 1/4	1 1/8	3 1/8	1 11/16	1 11/16	6 5/8	8-7/8	3
9	1 3/16	3 3/16	1 3/4	1 3/4	7 1/4	8-7/8	3 1/2
10	1 1/4	3 3/8	1 7/8	1 7/8	7 7/8	8-7/8	4
11	1 3/8	3 7/8	2	2	9 1/4	8-7/8	5
12 1/2	1 7/16	3 7/8	2 1/16	2 1/16	10 5/8	12-7/8	6
15	1 5/8	4 3/8	2 7/16	2 7/16	13	12-1	8
17 1/2	1 7/8	4 5/8	2 5/8	3 3/4	15 1/4	16-1 1/8	10
20 1/2	2	5 1/8	2 7/8	4	17 3/4	16-1 1/4	12
23	2 1/8	5 5/8	3	4 3/8	20 1/4	20-1 1/4	14
25 1/2	2 1/4	5 3/4	3 1/4	4 3/4	22 1/2	20-1 3/8	16
28	2 3/8	6 1/4	3 1/2	5 1/8	24 3/4	24-1 3/8	18
30 1/2	2 1/2	6 3/8	3 3/4	5 1/2	27	24-1 3/8	20
36	2 3/4	6 5/8	4 3/16	6	32	24-1 5/8	24

400 LB. FLANGES

Nom. Pipe Size	A	T②	L③ Weld Neck	L③ Thrd. Slip on	L③ Lap Joint	Bolt Circle	No. and Size of Holes
1/2	3 3/4	9/16	2 1/16	7/8	7/8	2 5/8	4-5/8
3/4	4 5/8	5/8	2 1/4	1	1	3 1/4	4-3/4
1	4 7/8	11/16	2 7/16	1 1/16	1 1/16	3 1/2	4-3/4
1 1/4	5 1/4	13/16	2 5/8	1 1/8	1 1/8	3 7/8	4-3/4
1 1/2	6 1/8	7/8	2 3/4	1 1/4	1 1/4	4 1/2	4-7/8
2	6 1/2	1	2 7/8	1 7/16	1 7/16	5	8-3/4
2 1/2	7 1/2	1 1/8	3 1/8	1 5/8	1 5/8	5 7/8	8-7/8
3	8 1/4	1 1/4	3 1/4	1 13/16	1 13/16	6 5/8	8-7/8
3 1/2	9	1 3/8	3 3/8	1 15/16	1 15/16	7 1/4	8-1
4	10	1 3/8	3 1/2	2	2	7 7/8	8-1
5	11	1 1/2	4	2 1/8	2 1/8	9 1/4	8-1
6	12 1/2	1 5/8	4 1/16	2 1/4	2 1/4	10 5/8	12-1
8	15	1 7/8	4 5/8	2 11/16	2 11/16	13	12-1 1/8
10	17 1/2	2 1/8	4 7/8	2 7/8	4	15 1/4	16-1 1/4
12	20 1/2	2 1/4	5 3/8	3 1/8	4 1/4	17 3/4	16-1 3/8
14	23	2 3/8	5 7/8	3 5/16	4 5/8	20 1/4	20-1 3/8
16	25 1/2	2 1/2	6	3 11/16	5	22 1/2	20-1 1/2
18	28	2 5/8	6 1/2	3 7/8	5 3/8	24 3/4	24-1 1/2
20	30 1/2	2 3/4	6 5/8	4	5 3/4	27	24-1 5/8
24	36	3	6 7/8	4 1/2	6 1/4	32	24-1 7/8

600 LB. FLANGES

A	T①	L② Weld Neck	L② Thrd. Slip on	L② Lap Joint	Bolt Circle	No. and Size of Holes	Nom. Pipe Size
3 3/4	9/16	2 1/16	7/8	7/8	2 5/8	4-5/8	1/2
4 5/8	5/8	2 1/4	1	1	3 1/4	4-3/4	3/4
4 7/8	11/16	2 7/16	1 1/16	1 1/16	3 1/2	4-3/4	1
5 1/4	13/16	2 5/8	1 1/8	1 1/8	3 7/8	4-3/4	1 1/4
6 1/8	7/8	2 3/4	1 1/4	1 1/4	4 1/2	4-7/8	1 1/2
6 1/2	1	2 7/8	1 7/16	1 7/16	5	8-3/4	2
7 1/2	1 1/8	3 1/8	1 5/8	1 5/8	5 7/8	8-7/8	2 1/2
8 1/4	1 1/4	3 1/4	1 13/16	1 13/16	6 5/8	8-7/8	3
9	1 3/8	3 3/8	1 15/16	1 15/16	7 1/4	8-1	3 1/2
10 3/4	1 1/2	4	2 1/8	2 1/8	8 1/2	8-1	4
13	1 3/4	4 1/2	2 3/8	2 3/8	10 1/2	8-1 1/8	5
14	1 7/8	4 5/8	2 5/8	2 5/8	11 1/2	12 1 1/8	6
16 1/2	2 3/16	5 1/4	3	3	13 3/4	12-1 1/4	8
20	2 1/2	6	3 3/8	4 3/8	17	16-1 3/8	10
22	2 5/8	6 1/8	3 5/8	4 5/8	19 1/4	20-1 3/8	12
23 3/4	2 3/4	6 1/2	3 11/16	5	20 3/4	20-1 1/2	14
27	3	7	4 3/16	5 1/2	23 3/4	20-1 5/8	16
29 1/4	3 1/4	7 1/4	4 5/8	6	25 3/4	20-1 3/4	18
32	3 1/2	7 1/2	5	6 1/2	28 1/2	24-1 3/4	20
37	4	8	5 1/2	7 1/4	33	24-2	24

Reprinted by permission of Taylor Forge & Pipe Works

Table 8.20 Continued

900 LB. FLANGES

Nom. Pipe Size	A	T(2)	L(2) Weld Neck	L(2) Thrd. Slip on	L(2) Lap Joint	Bolt Circle	No. and Size of Holes
1/2	4 3/4	7/8	2 3/8	1 1/4	1 1/4	3 1/4	4-7/8
3/4	5 1/8	1	2 3/4	1 3/8	1 3/8	3 1/2	4-7/8
1	5 7/8	1 1/8	2 7/8	1 5/8	1 5/8	4	4-1
1 1/4	6 1/4	1 1/8	2 7/8	1 5/8	1 5/8	4 3/8	4-1
1 1/2	7	1 1/4	3 1/4	1 3/4	1 3/4	4 7/8	4-1 1/8
2	8 1/2	1 1/2	4	2 1/4	2 1/4	6 1/2	8-1
2 1/2	9 5/8	1 5/8	4 1/8	2 1/2	2 1/2	7 1/2	8-1 1/8
3	9 1/2	1 1/2	4	2 1/8	2 1/8	7 1/2	8-1
3 1/2	......	......	......	......	......	......	......
4	11 1/2	1 3/4	4 1/2	2 3/4	2 3/4	9 1/4	8-1 1/4
5	13 3/4	2	5	3 1/8	3 1/8	11	8-1 3/8
6	15	2 3/16	5 1/2	3 3/8	3 3/8	12 1/2	12-1 1/4
8	18 1/2	2 1/2	6 3/8	4	4 1/2	15 1/2	12-1 1/2
10	21 1/2	2 3/4	7 1/4	4 1/4	5	18 1/2	16-1 1/2
12	24	3 1/8	7 7/8	4 5/8	5 5/8	21	20-1 1/2
14	25 1/4	3 3/8	8 3/8	5 1/8	6 1/8	22	20-1 5/8
16	27 3/4	3 1/2	8 1/2	5 1/4	6 1/2	24 1/4	20-1 3/4
18	31	4	9	6	7 1/2	27	20-2
20	33 3/4	4 1/4	9 3/4	6 1/4	8 1/4	29 1/2	20-2 1/8
24	41	5 1/2	11 1/2	8	10 1/2	35 1/2	20-2 5/8

1500 LB. FLANGES

A	T(2)	L(2) Weld Neck	L(2) Thrd. Slip on	L(2) Lap Joint	Bolt Circle	No. and Size of Holes	Nom. Pipe Size
4 3/4	7/8	2 3/8	1 1/4	1 1/4	3 1/4	4-7/8	1/2
5 1/8	1	2 3/4	1 3/8	1 3/8	3 1/2	4-7/8	3/4
5 7/8	1 1/8	2 7/8	1 5/8	1 5/8	4	4-1	1
6 1/4	1 1/8	2 7/8	1 5/8	1 5/8	4 3/8	4-1	1 1/4
7	1 1/4	3 1/4	1 3/4	1 3/4	4 7/8	4-1 1/8	1 1/2
8 1/2	1 1/2	4	2 1/4	2 1/4	6 1/2	8-1	2
9 5/8	1 5/8	4 1/8	2 1/2	2 1/2	7 1/2	8-1 1/8	2 1/2
10 1/2	1 7/8	4 5/8	2 7/8	2 7/8	8	8-1 1/4	3
......	......	......	......	......	......	......	3 1/2
12 1/4	2 1/8	4 7/8	3 9/16	3 9/16	9 1/2	8-1 3/8	4
14 3/4	2 7/8	6 1/8	4 1/8	4 1/8	11 1/2	8 1 5/8	5
15 1/2	3 1/4	6 3/4	4 11/16	4 11/16	12 1/2	12-1 1/2	6
19	3 5/8	8 3/8	5 5/8	5 5/8	15 1/2	12-1 3/4	8
23	4 1/4	10	6 1/4	7	19	12-2	10
26 1/2	4 7/8	11 1/8	7 1/8	8 5/8	22 1/2	16-2 1/8	12
29 1/2	5 1/4	11 3/4	......	9 1/2	25	16-2 3/8	14
32 1/2	5 3/4	12 1/4	......	10 1/4	27 1/4	16-2 5/8	16
36	6 3/8	12 7/8	......	10 7/8	30 1/2	16-2 7/8	18
38 3/4	7	14	......	11 1/2	32 3/4	16-3 1/8	20
46	8	16	......	13	39	16-3 5/8	24

2500 LB. FLANGES

Nom. Pipe Size	A	T(2)	L(2) Weld Neck	L(2) Thrd.	L(2) Lap Joint	Bolt Circle	No. and Size of Holes
1/2	5 1/4	1 3/16	2 7/8	1 9/16	1 9/16	3 1/2	4- 7/8
3/4	5 1/2	1 1/4	3 1/8	1 11/16	1 11/16	3 3/4	4- 7/8
1	6 1/4	1 3/8	3 1/2	1 7/8	1 7/8	4 1/4	4-1
1 1/4	7 1/4	1 1/2	3 3/4	2 1/16	2 1/16	5 1/8	4-1 1/8
1 1/2	8	1 3/4	4 3/8	2 3/8	2 3/8	5 3/4	4-1 1/4
2	9 1/4	2	5	2 3/4	2 3/4	6 3/4	8-1 1/8
2 1/2	10 1/2	2 1/4	5 5/8	3 1/8	3 1/8	7 3/4	8-1 1/4
3	12	2 5/8	6 5/8	3 5/8	3 5/8	9	8-1 3/8
4	14	3	7 1/2	4 1/4	4 1/4	10 3/4	8-1 5/8
5	16 1/2	3 5/8	9	5 1/8	5 1/8	12 3/4	8-1 7/8
6	19	4 1/4	10 3/4	6	6	14 1/2	8-2 1/8
8	21 3/4	5	12 1/2	7	7	17 1/4	12-2 1/8
10	26 1/2	6 1/2	16 1/2	9	9	21 1/4	12-2 5/8
12	30	7 1/4	18 1/4	10	10	24 3/8	12-2 7/8

(1) Bore to match schedule of attached pipe.

(2) Includes 1/16" raised face in 150 pound and 300 pound standard. Does *not* include raised face in 400, 600, 900, 1500 and 2500 pound standard.

(3) Inside pipe diameters are also provided by this table.

Reprinted by permission of Taylor Forge & Pipe Works

WELDING NECK FLANGE BORES(1)(3)

Nom. Pipe Size	Outside Diameter	Sched. 10	Sched. 20	Sched. 30	Standard Wall	Sched. 40	Sched. 60	Extra Strong	Sched. 80	Sched. 100	Sched. 120	Sched. 140	Sched. 160	Double Extra Strong	Nom. Pipe Size
1/2	0.840	....	....	....	0.622	0.622	....	0.546	0.546	....	....	....	0.466	0.252	1/2
3/4	1.050	....	....	....	0.824	0.824	....	0.742	0.742	....	....	....	0.614	0.434	3/4
1	1.315	....	....	....	1.049	1.049	....	0.957	0.957	....	....	....	0.815	0.599	1
1 1/4	1.660	....	....	....	1.380	1.380	....	1.278	1.278	....	....	....	1.160	0.896	1 1/4
1 1/2	1.900	....	....	....	1.610	1.610	....	1.500	1.500	....	....	....	1.338	1.100	1 1/2
2	2.375	....	....	....	2.067	2.067	....	1.939	1.939	....	....	....	1.689	1.503	2
2 1/2	2.875	....	....	....	2.469	2.469	....	2.323	2.323	....	....	....	2.125	1.771	2 1/2
3	3.500	....	....	....	3.068	3.068	....	2.900	2.900	....	....	....	2.624	2.300	3
3 1/2	4.000	....	....	....	3.548	3.548	....	3.364	3.364	....	....	....	....	....	3 1/2
4	4.500	....	....	....	4.026	4.026	....	3.826	3.826	....	3.624	....	3.438	3.152	4
5	5.563	....	....	....	5.047	5.047	....	4.813	4.813	....	4.563	....	4.313	4.063	5
6	6.625	....	....	....	6.065	6.065	....	5.761	5.761	....	5.501	....	5.189	4.897	6
8	8.625	....	8.125	8.071	7.981	7.981	7.813	7.625	7.625	7.439	7.189	7.001	6.813	6.875	8
10	10.750	....	10.250	10.136	10.020	10.020	9.750	9.750	9.564	9.314	9.064	8.750	8.500	....	10
12	12.750	....	12.250	12.090	12.000	11.938	11.626	11.750	11.376	11.064	10.750	10.500	10.126	....	12
14	14.000	13.500	13.375	13.250	13.250	13.124	12.814	13.000	12.500	12.126	11.814	11.500	11.188	....	14
16	16.000	15.500	15.375	15.250	15.250	15.000	14.688	15.000	14.314	13.938	13.564	13.124	12.814	....	16
18	18.000	17.500	17.375	17.124	17.250	16.876	16.500	17.000	16.126	15.688	15.250	14.876	14.438	....	18
20	20.000	19.500	19.250	19.000	19.250	18.814	18.376	19.000	17.938	17.438	17.000	16.500	16.064	....	20
24	24.000	23.500	23.250	22.876	23.250	22.626	22.064	23.000	21.564	20.938	20.376	19.876	19.314	....	24
30	30.000	29.376	29.000	28.750	29.250	....	....	29.000	....	....	....	....	....	....	30

TABLE 8.21

LENGTH OF ALLOY STEEL STUD BOLTS

Standard ANSI B16.5
Note: All dimensions in inches.
The length of stud bolts do not include the height of crown.
Bolt holes are 1/8 inch larger than bolt diameters.
Sizes 22, 26, 28 and 30 are not covered by ANSI B16.5.

Nominal Pipe Size	150 lb. USA Standard Flanges				300 lb. USA Standard Flanges				400 lb. USA Standard Flanges				600 lb. USA Standard Flanges			
	Number of Bolts	Diameter of Bolts	Length of Stud Bolts 1/16 Raised Face	Ring Joint	Number of Bolts	Diameter of Bolts	Length of Stud Bolts 1/16 Raised Face	Ring Joint	Number of Bolts	Diameter of Bolts	Length of Stud Bolts 1/4 Raised Face	Ring Joint	Number of Bolts	Diameter of Bolts	Length of Stud Bolts 1/4 Raised Face	Ring Joint
1/2	4	1/2	2 1/2		4	1/2	2 3/4	3	4	1/2	3 1/4	3	4	1/2	3 1/4	3
3/4	4	1/2	2 1/2		4	5/8	3	3 1/2	4	5/8	3 1/2	3 1/2	4	5/8	3 1/2	3 1/2
1	4	1/2	2 3/4	3 1/4	4	5/8	3 1/4	3 3/4	4	5/8	3 3/4	3 3/4	4	5/8	3 3/4	3 3/4
1 1/4	4	1/2	2 3/4	3 1/4	4	5/8	3 1/4	3 3/4	4	5/8	4	4	4	5/8	4	4
1 1/2	4	1/2	3	3 1/2	4	3/4	3 3/4	4 1/4	4	3/4	4 1/4	4 1/4	4	3/4	4 1/4	4 1/4
2	4	5/8	3 1/4	3 3/4	8	5/8	3 1/2	4 1/4	8	5/8	4 1/4	4 1/2	8	5/8	4 1/4	4 1/2
2 1/2	4	5/8	3 1/2	4	8	3/4	4	4 3/4	8	3/4	4 3/4	5	8	3/4	4 3/4	5
3	4	5/8	3 3/4	4 1/4	8	3/4	4 1/4	5	8	3/4	5	5 1/4	8	3/4	5	5 1/4
3 1/2	8	5/8	3 3/4	4 1/4	8	3/4	4 1/2	5 1/4	8	7/8	5 1/2	5 3/4	8	7/8	5 1/2	5 3/4
4	8	5/8	3 3/4	4 1/4	8	3/4	4 1/2	5 1/4	8	7/8	5 1/2	5 3/4	8	7/8	5 3/4	6
5	8	3/4	4	4 1/2	8	3/4	4 3/4	5 1/2	8	7/8	5 3/4	6	8	1	6 1/2	6 3/4
6	8	3/4	4	4 1/2	12	3/4	5	5 3/4	12	7/8	6	6 1/4	12	1	6 3/4	7
8	8	3/4	4 1/4	4 3/4	12	7/8	5 1/2	6 1/4	12	1	6 3/4	7	12	1 1/8	7 3/4	7 3/4
10	12	7/8	4 3/4	5 1/4	16	1	6 1/4	7	16	1 1/8	7 1/2	7 3/4	16	1 1/4	8 1/2	8 3/4
12	12	7/8	4 3/4	5 1/4	16	1 1/8	6 3/4	7 1/2	16	1 1/4	8	8 1/4	20	1 1/4	8 3/4	9
14	12	1	5 1/4	5 3/4	20	1 1/8	7	7 3/4	20	1 1/4	8 1/4	8 1/2	20	1 3/8	9 1/4	9 1/2
16	16	1	5 1/2	6	20	1 1/4	7 1/2	8 1/4	20	1 3/8	8 3/4	9	20	1 1/2	10	10 1/4
18	16	1 1/8	6	6 1/2	24	1 1/4	7 3/4	8 1/2	24	1 3/8	9	9 1/4	20	1 5/8	10 3/4	11
20	20	1 1/8	6 1/4	6 3/4	24	1 1/4	8 1/4	9	24	1 1/2	9 3/4	10	24	1 5/8	11 1/2	11 3/4
22	20	1 1/4	6 1/2	7	24	1 1/2	8 3/4	9 3/4	24	1 5/8	10	10 1/2	24	1 3/4	12	12 1/2
24	20	1 1/4	7	7 1/2	24	1 1/2	9 1/4	10 1/4	24	1 3/4	10 3/4	11 1/4	24	1 7/8	13	13 1/4
26	24	1 1/4	7		28	1 5/8	10	11	28	1 3/4	11 1/2	12	28	1 7/8	13 1/4	13 3/4
28	28	1 1/4	7		28	1 5/8	10 1/2	11 1/2	28	1 7/8	12 1/4	12 3/4	28	2	13 3/4	14 1/4
30	28	1 1/4	7 1/4		28	1 3/4	11 1/4	12 1/4	28	2	13	13 1/2	28	2	14	14 1/2

TABLE 8.21

LENGTH OF ALLOY STEEL STUD BOLTS, continued

Standard ANSI B16.5. The length of stud bolts do not include the height of crown. Note: All dimensions in inches.

Bolt holes are 1/8 inch larger than bolt diameters. Sizes 22, 26, 28 and 30 are not covered by ANSI B16.5.

Nominal Pipe Size	900 lb. USA Standard Flanges				1500 lb. USA Standard Flanges				2500 lb. USA Standard Flanges			
	Number of Bolts	Diameter of Bolts	Length of Stud Bolts		Number of Bolts	Diameter of Bolts	Length of Stud Bolts		Number of Bolts	Diameter of Bolts	Length of Stud Bolts	
			1/4 Raised Face	Ring Joint			1/4 Raised Face	Ring Joint			1/4 Raised Face	Ring Joint
1/2	4	3/4	4 1/4	4 1/4	4	3/4	4 1/4	4 1/4	4	3/4	5 1/4	5 1/4
3/4	4	3/4	4 1/2	4 1/2	4	3/4	4 1/2	4 1/2	4	3/4	5 1/4	5 1/4
1	4	7/8	5	5	4	7/8	5	5	4	7/8	5 3/4	5 3/4
1 1/4	4	7/8	5	5	4	7/8	5	5	4	1	6 1/4	6 1/2
1 1/2	4	1	5 1/2	5 1/2	4	1	5 1/2	5 1/2	4	1 1/8	7	7 1/4
2	8	7/8	5 3/4	5 3/4	8	7/8	5 3/4	5 3/4	8	1	7 1/4	7 1/2
2 1/2	8	1	6 1/4	6 1/4	8	1	6 1/4	6 1/4	8	1 1/8	8	8 1/4
3	8	7/8	5 3/4	6	8	1 1/8	7	7	8	1 1/4	9	9 1/4
4	8	1 1/8	6 3/4	7	8	1 1/4	7 3/4	7 3/4	8	1 1/2	10 1/4	10 3/4
5	8	1 1/4	7 1/2	7 3/4	8	1 1/2	9 3/4	9 3/4	8	1 3/4	12	12 3/4
6	12	1 1/8	7 3/4	7 3/4	12	1 3/8	10 1/4	10 1/2	8	2	13 3/4	14 1/2
8	12	1 3/8	8 3/4	9	12	1 5/8	11 1/2	12	12	2	15 1/4	16
10	16	1 3/8	9 1/4	9 1/2	12	1 7/8	13 1/4	13 3/4	12	2 1/2	19 1/2	20 1/2
12	20	1 3/8	10	10 1/4	16	2	14 3/4	15 1/2	12	2 3/4	21 1/2	22 1/2
14	20	1 1/2	10 3/4	11 1/4	16	2 1/4	16	17				
16	20	1 5/8	11 1/4	11 3/4	16	2 1/2	17 1/2	18 1/2				
18	20	1 7/8	12 3/4	13 1/2	16	2 3/4	19 1/2	20 1/2				
20	20	2	13 1/2	14 1/4	16	3	21 1/2	22 1/2				
24	20	2 1/2	17 1/4	17 3/4	16	3 1/2	24 1/2	25 3/4				
26	20	2 3/4	17 1/2	18 3/4								
28	20	3	18 1/4	19 1/2								
30	20	3	18 3/4	20								

TABLE 8.22

Boiler House Notes:			Page:

American Standard Flange Facings

re: ASA Standard B16.5 - 1953

Nominal Pipe Size (inches)	Raised Face Dimensions		
	Outside Diameter (inches)	Height (inches)	
		150# & 300# series	400# thru 2500# series
1/2	1.3750	0.0625	0.25
3/4	1.6875	0.0625	0.25
1	2.0000	0.0625	0.25
1 1/4	2.5000	0.0625	0.25
1 1/2	2.8750	0.0625	0.25
2	3.6250	0.0625	0.25
2 1/2	4.1250	0.0625	0.25
3	5.0000	0.0625	0.25
3 1/2	5.5000	0.0625	0.25
4	6.1875	0.0625	0.25
5	7.3125	0.0625	0.25
6	8.5000	0.0625	0.25
8	10.6250	0.0625	0.25
10	12.7500	0.0625	0.25
12	15.0000	0.0625	0.25
14	16.2500	0.0625	0.25
16	18.5000	0.0625	0.25
18	21.0000	0.0625	0.25
20	23.0000	0.0625	0.25
24	27.2500	0.0625	0.25

Note: A tolerance of +/- 1/64" is allowed on the inside and outside diameter of all facings.

Large Diameter Flanges

150 lb. W.S.P. @ 750 deg. F.

Forged & rolled steel, ASTM A105 - Gr. 1

26	29.000	0.0625	
28	31.000	0.0625	
30	33.250	0.0625	
32	35.250	0.0625	
34	37.375	0.0625	
36	39.375	0.0625	

re: Ladish Company, parts # 505 and 506	Catalog # 55	
Ralph L. Vandagriff Date: 8/3/98		

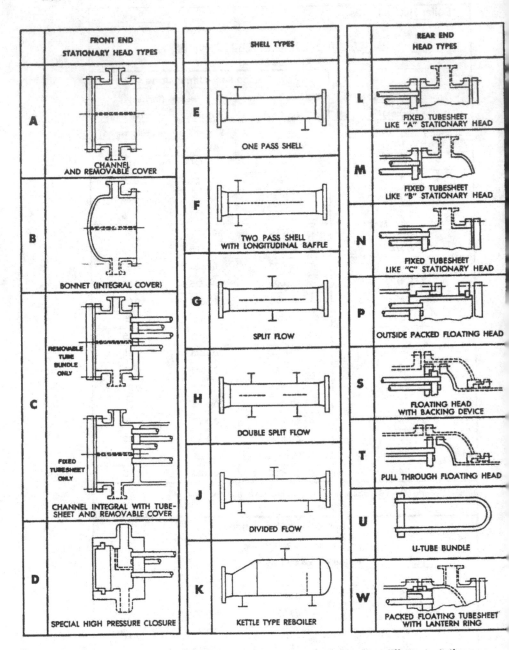

FRONT END
STATIONARY HEAD TYPES

SHELL TYPES

REAR END
HEAD TYPES

A — CHANNEL AND REMOVABLE COVER

B — BONNET (INTEGRAL COVER)

C — REMOVABLE TUBE BUNDLE ONLY / FIXED TUBESHEET ONLY / CHANNEL INTEGRAL WITH TUBE-SHEET AND REMOVABLE COVER

D — SPECIAL HIGH PRESSURE CLOSURE

E — ONE PASS SHELL

F — TWO PASS SHELL WITH LONGITUDINAL BAFFLE

G — SPLIT FLOW

H — DOUBLE SPLIT FLOW

J — DIVIDED FLOW

K — KETTLE TYPE REBOILER

L — FIXED TUBESHEET LIKE "A" STATIONARY HEAD

M — FIXED TUBESHEET LIKE "B" STATIONARY HEAD

N — FIXED TUBESHEET LIKE "C" STATIONARY HEAD

P — OUTSIDE PACKED FLOATING HEAD

S — FLOATING HEAD WITH BACKING DEVICE

T — PULL THROUGH FLOATING HEAD

U — U-TUBE BUNDLE

W — PACKED FLOATING TUBESHEET WITH LANTERN RING

FIGURE 8.2   Nomenclature of heat exchangers depicting the different stationary head types (front end and near end) and shell types.

1. Stationary Head—Channel
2. Stationary Head—Bonnet
3. Stationary Head Flange—Channel or Bonnet
4. Channel Cover
5. Stationary Head Nozzle
6. Stationary Tubesheet
7. Tubes
8. Shell
9. Shell Cover
10. Shell Flange—Stationary Head End
11. Shell Flange—Rear Head End
12. Shell Nozzle
13. Shell Cover Flange
14. Expansion Joint
15. Floating Tubesheet
16. Floating Head Cover
17. Floating Head Flange
18. Floating Head Backing Device
19. Split Shear Ring

20. Slip-on Backing Flange
21. Floating Head Cover—External
22. Floating Tubesheet Skirt
23. Packing Box Flange
24. Packing
25. Packing Follower Ring
26. Lantern Ring
27. Tie Rods and Spacers
28. Transverse Baffles or Support Plates
29. Impingement Baffle
30. Longitudinal Baffle
31. Pass Partition
32. Vent Connection
33. Drain Connection
34. Instrument Connection
35. Support Saddle
36. Lifting Lug
37. Support Bracket
38. Weir
39. Liquid Level Connection

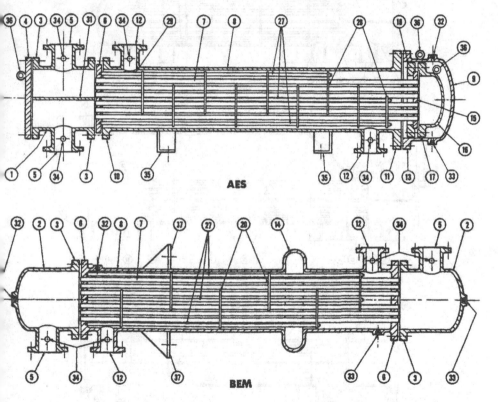

FIGURE 8.3   To establish a standard terminology various types of heat exchangers are illustrated with their typical pipes and connections numbered for identification.

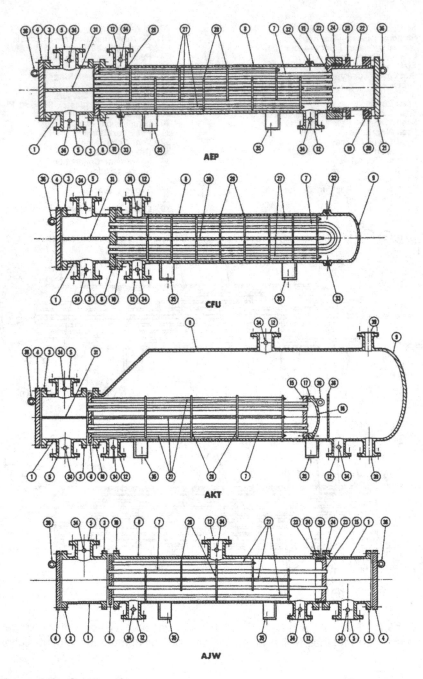

AEP

CFU

AKT

AJW

Figure 8.3   Continued

9

Boiler Energy Conservation

Energy Conservation Measures; Economizer Design; Deep Economizers; Economizer Steaming in HRSGs; Excess Combustion Air; Controlling Excess Combustion Air; Preheating Feedwater; Condensate and Blowdown; Flash Steam Heat Recovery; Steam Generator Overall Efficiency.

I. ENERGY CONSERVATION MEASURES [1,6,7,17]

A. Heat Recovery

One of the most feasible methods of conserving energy in a steam generator is the application of heat recovery equipment. This can be accomplished by the utilization of the combustion flue gases to increase the incoming feedwater temperature or, if needed, increase the combustion air temperature. The most frequently used method on packaged boilers is the economizer that increases the feedwater temperature. This is preferred because the capital investment is less than that of an air preheater; there is lower draft loss, thus lower fan horsepower required; and, finally, reduced furnace heat absorption. See Figure 9.1 on boiler house recoverable losses.

An economizer absorbs heat from the flue gases and adds it to the feedwater as sensible heat before the feedwater enters the boiler. This cools the combustion flue gases and increases overall efficiency of the unit. For every 40°F that the flue gas is cooled by an economizer, the overall boiler efficiency increases by approximately 1.0%.

B. Economizer Design and Construction

The design and use of economizers naturally paralleled the development of boilers. Economizers in large field-erected boilers are usually arranged for downward flow of gas and upward flow of water. Economizers for packaged boilers usually have the water flow down and the gas up. Economizer design is basically very simple. Tubes are continuous horizontal J-bend from inlet to outlet headers, with

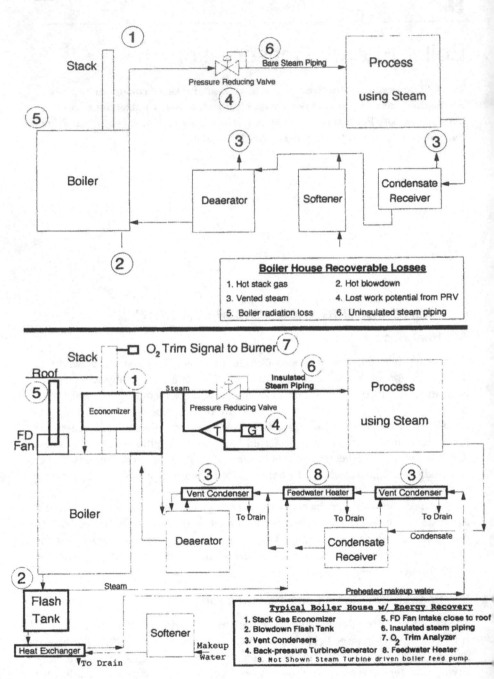

FIGURE 9.1   Boiler house recoverable energy losses.

welded terminals to eliminate seat leakage. The square pitch or in-line spacing is arranged for uniform heat absorption, good external cleaning, and minimum draft loss. The economizer has had an interesting transition period in recent years. Initially, these units were designed as bare tube surfaces: water inside—gas outside. Adding extended surface or fins to these tubes made the unit much more efficient in heat transfer and reduced the physical size. The same performance of the bare tube economizer was now possible with a more compact fin tube design. Early designs utilized large tubes and cast iron construction. Gradually, this design was improved with the introduction of 2-in. O.D. tubes and welded steel fabrication. The enclosure is double cased and insulated complete with structural reinforcing members. Normal accessories consist of support steel, interconnecting gas ducts, and feedwater piping. Frequently, a feedwater bypass line is provided for operation of the boiler with the economizer out of service for maintenance or for operation at minimum steaming conditions.

A fin spacing of five per inch was first used on natural gas firing. Gradually, distillate fuel oil firing was applied with a fin spacing of four per inch. In recent years, heavy fuel oils have provided very successful results at a fin spacing of 2–2.5 per inch. Fin thickness is also an important design consideration. An economizer fired with a clean gaseous fuel is designed with a fin thickness of 0.060 in. On oil firing, a heavier fin, usually 0.105 in. is required owing to the erosive atmosphere and higher fouling characteristics of this fuel. In both cases, the fin height or distance from the tube is 0.75 in. These fins are continuously welded to the tube, forming a sealed bond. Sufficient heating surface is installed in the economizer to absorb enough heat to give the desired gas exit temperature. Economizers are generally designed to reduce gas temperatures by 200°–300°F. The rise in water temperature varies between 70° and 100°F. In terms of efficiency, the increase is 5–6%. The system resistance of the economizer is generally about 5 lb pressure drop on the water side and 1.0–1.5 in. static pressure loss on the gas side. Location of the economizer will vary with the overall design of the boiler unit and surrounding space limitations. One of the most common applications is the location of the economizer in ductwork beneath the stack. The economizer is arranged above the boiler on structural support steel with minimum interconnecting gas ducts and stack length. This arrangement also conserves space at the grade level area. Equipment is offset from the boiler gas outlet for water wash or compressed air cleaning with drain provisions.

Designing the economizer for counterflow of gas and water results in a maximum mean temperature difference for heat transfer. To avoid generating steam in the economizer, the design ordinarily provides exiting water temperatures below that of saturated steam during normal operations. Under certain operating conditions, the economizer may be designed for upward flow of water and downward flow of gas, to avoid water hammer. (*Note*: Sometimes steaming economizers are designed into large supercritical boiler systems used by utilities.)

Because steel is subject to corrosion even in the presence of extremely low concentrations of oxygen, it is necessary to provide water that is practically 100% oxygen-free. It is common practice to use deaerators for oxygen removal.

When selecting a new boiler with an economizer, the boiler surface is reduced for a fixed output, and it is necessary to design equipment proportions carefully. Because of the lowered flue gas temperatures provided by an economizer, economical reductions in induced draft fan size and horsepower plus smaller dust collection equipment (if required) can be used.

The temperature of the flue gas entering the economizer will vary with different types of boilers, operating load conditions, fuel characteristics, and combustion conditions. The temperature to which the gases can be cooled while passing through the economizer is determined by the following:

1. Amount of heat that can be absorbed
2. Temperature of the entering feedwater or combustion air
3. Dew point of the flue gases
4. Economical exit temperature, below which any gain in efficiency is offset by increased costs

Some design criteria are

1. Low-load operation creates design complications.
2. If combustion becomes dirty, the unit will clog.
3. Draft fans are required to overcome the resistance imposed by the economizer. Bare tubes have less resistance than finned tubes.
4. To minimize cold-end corrosion, particularly at low load, it is necessary to bypass all or part of the flue gases around the economizer, permitting base plant operation, and also to recirculate a portion of the feedwater. The alternative to this is use of corrosion-resistant alloy metals in the cold end.

C. Economizer Velocity Limits

The ultimate goal of economizer design is to achieve the necessary heat transfer at minimum cost. A key design criterion for economizers is the maximum allowable gas velocity (defined at the minimum cross-sectional free-flow area in the tube bundle). Higher velocities provide better heat transfer and reduce capital cost. For clean-burning fuels, such as gas and low-ash oil, velocities are typically set by the maximum economical pressure loss. For high-ash oil and coal, gas side velocities are limited by the erosion potential of the fly ash. This erosion potential is primarily determined by the percentage of Al_2O_3 and SiO_2 in the ash, the total ash in the fuel, and the gas maximum velocity. Experience dictates acceptable velocities.

D. Economizer Cleaning and Corrosion Protection

Cleanliness is important to keep the tube and fin surfaces free of deposits for maximum heat transfer. A soot-blowing system is employed for this purpose. Economizers are designed with tube spacing and tube bank depths best suited for this external cleaning.

Feedwater to the economizer should be deaerated and heated to a temperature, preferable about 220°F or higher, to minimize internal tube corrosion from dissolved oxygen and external metal corrosion from the formation of condensation. The possibility of sulfuric acid corrosion of economizer tubes does exist. The tube metal temperature of economizers is essentially the same as that of the water in the tube, because the temperature drop through the tube wall is minimal. The fin temperature of extended surfaces remains considerably higher and is not subject to moisture condensation. External corrosion of economizers may occur when the water vapor in the flue gas condenses on the surfaces of the tubes, and corrosion is accelerated when this happens in the presence of the products of combustion of sulfur. The rate of corrosion increases as the metal temperature is reduced. As the amount of sulfur increases, the dew point increases and so does the potential rate of corrosion. The basic cause of low-temperature corrosion is well known. During combustion most of the sulfur in the fuel burns to sulfur dioxide (SO_2), with a small part (1–3%) forming sulfur trioxide (SO_3). Passing through the boiler, the SO_3 combines with water to form sulfuric acid vapor (H_2SO_4). In most cases, this sulfuric acid vapor makes up about 10–50 ppm of the flue gas composition. The presence of sulfuric acid raises the dew point of the flue gas above that of water. The higher the sulfur content, the lower the acid dew point. If metal temperatures within the boiler fall below the acid dew point, sulfuric acid condenses and acid corrosion results. However, with a sufficiently high incoming feedwater temperature, tube metal wastage can be eliminated. Also, some efficiency can be given up to keep the gas exit temperature out of the economizer higher. At full steam load, 350°F is typical for no. 6 fuel oil firing, whereas 300°F or less might be practical when a sulfur-free fuel is fired.

II. ECONOMIZER DESIGN [57]

A. Relation Between Boiler Size and Economizer Size

The question is sometimes asked; "If there is sufficient heat left in the exhaust for an economizer, why not make the boiler bigger and do away with the need for an economizer?" In other words, as efficiency is increased by reducing the exit temperature, why not install more boiler surface to extract more heat? Once again the question is one of practical economics.

A typical industrial boiler producing saturated steam operates at 10 bar. At this pressure the steam temperature leaving the boiler is 186°C (Fig. 9.2a). For

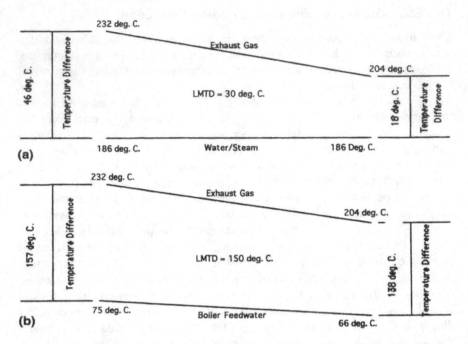

FIGURE 9.2 (a) Effective temperature difference when all heat transferred in the boiler is 30°C. (b) The same cooling effect is achieved in the exhaust gas by an economizer rather than in the boiler.

heat to flow from the products of combustion to the water/steam, there must be a finite temperature difference between the two fluid streams. Therefore, the limiting case is when exhaust gas temperature and steam temperature are equal. The overall heat transfer is in proportion to the temperature difference between the two streams.

It can be seen from Figure 9.2a that the effective temperature difference (LMTD) in the case where all the heat is transferred in the boiler is approximately 30°C. This effective temperature difference may be seen as the driving force behind the transfer of the heat. The smaller this driving force is, the greater will be the area of surface required for the transfer of a given quantity of heat.

In Figure 9.2b, the same cooling effect is being achieved in the exhaust gas by an economizer, rather than in the boiler. In this case, the effective temperature difference is approximately 150°C so the driving force available to transfer the same amount of heat from the gases is five times greater than in the previous case. The surface area required to effect this transfer is, therefore, reduced by a similar factor.

Additional surface means additional cost, as well as greater space require-ments and floor loading. It is clearly far more economical to install an economizer than to increase the size of the boiler by putting in five times the extra surface.

B. Extended Surfaces [57]

1. Extended Surfaces: the Types and Reasons Why

In addition to the reduction in the amount of surface required by the correct specification of an economizer in relation to its boiler, the physical size of the unit can be made much smaller by the introduction of an extended surface for heat transfer.

Example. Heat Transfer Across an Economizer Tube. The flow of heat from gas to liquid has to overcome a series of resistances as follows:

1. Boundary layer between gas and tube
2. Tube wall
3. Boundary layer between tube and water

In addition, there are resistances because of fouling, at both the gas and water interfaces, which vary with conditions.

The waterside heat-transfer coefficient is always far in excess of that for the gas side and the resistance through the metal itself is relatively insignificant. The area that controls the rate of heat transfer is, therefore, the gas side, and anything that can be done to improve the flow of heat in this region will improve the performance of a given length of tube. The addition of fins or gills to the gas-swept side of the tube increases the area available to transfer the heat, thereby reducing the total length of tubing required and, hence, the size of the unit.

Table 9.1 shows a comparison between plain-tube and extended-surface designs for the same duty in a power station boiler. It is evident that consideration of first cost, pressure drops both on the gas and water side and, hence, operating costs (fan and pumping loads) as well as the clear advantage of size reduction

TABLE 9.1 The Effect of Extended Surface in Economizers

	No. of tubes	Rows wide	Rows high	Center distance (mm)	Height (mm)
Plain tube economizer	11,424	272	42	76	3,192
Extended-surface economizer	1,968	164	12	127	1,524

show a convincing case for the use of extended surfaces. The same is true for smaller general industrial-type economizers.

The methods of achieving the extended surface can be broadly classified into three types:

1. Integral: cast, rolled, or extruded
2. Metallurgical bond: welded or brazed fins
3. Mechanical bond: crimped or wrapped-on fins

Owing to the relatively high temperatures involved, it is generally considered inadvisable to employ mechanically attached fins, as differential expansion between fin and tube can cause separation of the base of the fin from the tube. This greatly impairs the heat transfer and introduces the risk of particle accumulation under the fin in the form of grit or soot.

The principal forms of fin in use are as follows:

1. A helically wound version, with the base of the fin continuously welded to the base tube: steel fin on steel tube
2. A parallel fin arrangement attached by high-frequency welding: steel fin on steel tube
3. A parallel fin, cast-iron sleeve on a steel tube
4. An all cast-iron parallel-finned tube

The choice of surface is determined largely by the type of fuel and the quality of the feedwater. For very clean gas, such as natural gas, the helical form of fin is increasingly being employed and is a compact surface. For gas-fired boilers and when adequate draught is available, this, therefore, is the preferred form.

The other three types of fin all have in common the parallel-fin configuration, which, although providing a smaller heat transfer for a given length of tube, has advantages over the helical type in terms of draught loss across the tube and, more significantly, the tendency to fouling is much reduced. The straight gas passages, afforded by the parallel fins, allow solids in the gas stream to be carried through the economizer tube banks, thus minimizing the deposits on the heat-transfer surface. Such deposition as does occur can effectively be dealt with by steam or compressed air blowing. Here again, the penetration of the blowers is enhanced by the straight passage offered by the parallel fins.

Apart from solid deposition, there is the important question of sulfatic compounds in the gases and the consequent danger of acid formation on the finned tube surfaces. Where either the fuel is sulfur-free or the metal temperature is sufficiently above the acid dew point, to rule out any possibility of acid formation, the all-steel, welded fin type is preferred. If there is the possibility that, under conditions of partial load or intermittent use, acid formation may occur, the adoption of a cast-iron finned surface is recommended. As mentioned earlier, the lower corrosion rate and the ease with which a substantial section can be economically produced make cast iron a suitable material for the purpose.

The increasing use of poorer quality fuel oils and the variations that are being experienced in their composition, call for the use of a cast-iron finned economizer. In addition, the return to coal firing is opening up renewed outlets for this type of surface. The cast-iron–protected steel tube type is used wherever possible on heavy oil and coal-fired boilers when the water quality is sufficiently high to rule out oxidization attack inside the tube. If this is not so, as sometimes occurs on medium-pressure shell boilers, the solid cast-iron finned tube is employed. This provides an effective long-life solution to attack from both without and within.

Nowadays, the all-steel forms of finned tube tend to be cheaper to produce and an economical form of economizer employed with sulfur-bearing fuels combines the cast-iron–protected type at the cold end, with an all-steel main section in the area where metal temperatures are safely above the dew point level. This particular principal is widely employed in marine steam boilers, as well as on land, among the many vessels incorporating such a combined economizer is the QE2.

As a footnote to this section, I would like to refer briefly to the question of how the deposition of acids on economizer tubes is caused. The precise mechanism of its formation has been the subject of several studies over the years and despite this work, the prediction of the manner and intensity of acid formation is still a somewhat imprecise discipline when related to the varying conditions that apply in practice. Acid is formed when the SO_3, present in the products of combustion of heavy oils and certain coals, combines with water and condenses on the cooled metal surfaces of the economizer tubes. This condensation is a local phenomenon related to the temperature of the metal. As the metal is generally at a temperature similar to that of the water inside the tubes, it is the water temperature that exerts the major influence on the condensation and not, as is often thought, the temperature of the exhaust gases. The phenomenon can completely be avoided by the maintenance of the feedwater temperature at a sufficiently high level to keep the whole of the economizer well out of the dew point range. This can be done by several methods, including electrical preheating of the feedwater to the economizer, steam injection into the feedwater, and recirculation of part of the water in the economizer around the cold end. In this way, the need for cast iron, as described in the foregoing, can be eliminated.

It must be appreciated, however, that this may be done at the expense of efficiency gain. Clearly, the lower the water temperature at the inlet to the economizer, the greater is the potential for extraction of heat from the exhaust gases and the consequent efficiency gain and fuel saving. The skill of the economizer designer is to balance the economics of the installation and achieve a situation in which the maximum heat is recovered consistent with an acceptable rate of corrosion of the economizer. At the same time attention must be paid to the downstream back-end equipment (ductwork, chimney, and such), where the cooling effect of the ambient air is more noticeable and the gas temperature might reach dangerously low levels without proper precautions. The balance of such considerations becomes more delicate with every increase in the price of the fuel being burned. In short, the aim is not necessarily to eliminate corrosion completely, but to hold it at a level at which the rate of deterioration of the surfaces is consistent with a sound return on investment. As previously mentioned, a lifetime of 10 years for the critical parts of an economizer is not an unreasonable expectation. Within this time the savings effected are sufficient to justify the expenditure on the "extra" heat recovery.

See Table 9.2 on strengths and weaknesses of various economizers designs.

TABLE 9.2

BOILER HOUSE NOTES:							Page No:	
Ralph L. Vandagriff Date: 6/29/98	Strengths and Weaknesses of Various Economizer Designs							
Tube Arrangements	IN - LINE				STAGGERED			
Tube Surface:	Spiral Fin	Stud	Bare	Steel H Fin	Continuous Fin	Spiral Fin	Stud	Bare
Price:	Low	Moderate	Very High	Moderate	High	Low	Moderate	High
Space Required:	Small	Medium	Large	Medium	Medium	Small	Medium	Large
Gas Pressure Drop.	Medium	Medium	High	Low	High	High	High	High
Weight	Medium	Medium	High	Medium	Medium	Low	Medium	High
Number of welds	Low	Medium	High	Low	Medium	Low	Medium	High
Average Life Years	10 to 15	20	30 plus	30 plus	15 to 20	5 to 10	10	15 to 20
Operational Availability w/ following maximum gas velocities, FPS Coal - HGI 50 or Oil - No. 6	Low 40	Fair 40	Good 50	Good 50	Moderate 50	Low 40	Moderate 40	Good 50
Operational Availability w/ following maximum gas velocities, FPS Coal - HGI 50	Poor 75	Fair 75	Moderate 75	Would not install 75	Fair 75	Very Poor 75	Poor 75	Fair 75
Note: Coal listed is Hardgrove Index 50, Average ash approximately 15/20% and ash SiO2 of 45/50%.								
Above Information courtesy of : Boiler Tube of America, div. Senior Engineering.								

III. DEEP ECONOMIZERS [18]

Deep economizers are economizers that are designed to handle the acidic condensate that results from cooling a flue gas below 270°F. The primary design variable
is the material of construction of the tubes at the cold end of the device. Typical
systems design are the following:

1. Carbon steel tubes with a throwaway section at the cold end: These
   systems are designed with modular sections at the cold end that are
   easily removed and replaced on a periodical basis.
2. Stainless steel tubes that withstand the corrosive environment: These
   are standard economizers with stainless steel tubes.
3. Carbon steel tubes for the bulk of the exchanger and stainless steel tubes
   for the cold end: These systems have carbon steel tubes for the main section
   of the economizer with stainless steel only for the cold end section.
4. Glass-tubed heat exchangers: These systems use glass tubes. They have
   been applied most extensively in gas–gas service as air preheaters.
   Applications with gas–liquid systems are under development.
5. Teflon tubes: These systems use Teflon tubes to withstand the corrosive
   environment. CHX and duPont have developed the unit and have
   solved the critical-sealing problems that usually result in applying
   Teflon tubes in heat exchangers. Exhaust gas temperatures higher than
   500°F require two-stage systems.

Although Teflon or glass would be most suitable for applications with very
corrosive gases containing high concentrations of sulfuric and hydrofluoric acids,
Inconel or other high alloys can be substituted for the stainless steel sections, or
a shorter replacement period can be considered for the throwaway-type units.

A critical feature for the use of deep economizers is a suitable heat sink
to cool the exhaust gas to the 150°–160°F. range. Typical heating plant condensate return systems operate at 150°–160°F, and usually the combined flow of
cold makeup and condensate return provides for a suitable cold inlet temperature
of 130°–140°F.

IV. ECONOMIZER STEAMING IN HRSGs [58]

When the economizer in a boiler of HRSG starts generating steam, particularly
with downward flow of water, problems can arise in the form of water hammer,
vibration, and so on. With upward water flow design, a certain amount of steaming, 3–5%, can be tolerated as the bubbles have a natural tendency to go up along
with the water. However, steaming should generally be avoided. To understand
why the economizer is likely to steam, we should first look at the characteristics
of a gas turbine as a function of ambient temperature and load.

In single-shaft machines, which are widely used, as the ambient temperature or load decreases, the exhaust gas temperature decreases. The variation in mass flow is marginal compared with fossil fuel-fired boilers, while the steam or water flow drops off significantly. (The effect of mass flow increase usually does not offset the effect of lower exhaust gas temperature.) The energy-transferring ability of the economizer, which is governed by the gas side heat-transfer coefficient, does not change much with gas turbine load or ambient temperature; hence, nearly the same duty is transferred with a smaller water flow through the economizer, which results in a water exit temperature approaching that of saturation. Consequently, we should design the economizer such that it does not steam in the lowest unfired ambient case, which will ensure that steaming does not occur at other ambient conditions. A few other steps may also be taken, such as designing the economizer with a horizontal gas flow with horizontal tubes. This ensures that the last few rows of the economizer, which are likely to steam, have a vertical flow of the steam–water mixture.

In conventional fossil fuel-fired boilers the gas flow decreases in proportion to the water flow, and the energy-transferring ability of the economizer is also lower at lower loads. Hence, steaming is not a concern in these boilers; usually the approach point increases at lower loads in fired boilers, whereas it is a concern in HRSGs.

A. Options [59]

1. Reverse the flow direction of water using valves, which is cumbersome; in nonsteaming mode the economizer operates in counterflow configuration, whereas in the steaming mode, it operates in parallel flow configuration.
2. The exhaust gas may be bypassed around the economizer to decrease its duty and thus prevent its steaming. This is a loss of energy.
3. Some boilers are designed so that the gas flow to the boiler itself is bypassed during steaming conditions; this is not recommended, as it results in a significant loss of energy by virtue of the evaporator not handling the entire gas stream.
4. Bypass a portion of the economizer surface on the water side so that the surface area participating in heat transfer is reduced; hence, the duty or enthalpy rise decreases, thereby avoiding steaming.

V. EXCESS COMBUSTION AIR

A. Excess Air

The total combustion airflow to a boiler is generally controlled by adjusting forced and induced draft fan dampers in relation to the fuel flow. *Excess air* is the amount of additional combustion air over that theoretically required to burn a given amount of fuel. The benefits of increasing excess air include increased combustion intensity, reduced carbon loss or CO formation, or both and reduced slagging conditions. Disadvantages include increased fan power consumption, increased heat loss up the stack, increased tube erosion, and possibly increased NO_x formation.

For most coal ashes, particularly those from eastern U. S. bituminous coals, the solid-to-liquid phase changes occur at lower temperatures if free oxygen is not present (reducing conditions) around the ash particles. As a result, more slagging occurs in a boiler operating with insufficient excess air during which localized reducing conditions can occur. For some fuels, including western U.S. subbituminous coals, the oxidizing–reducing temperature differential is much less.

Localized tube metal wastage may also occur in furnace walls under low excess air conditions, but the impact is less clearly defined. The absence of free oxygen (a reducing atmosphere) and the presence of sulfur (from the fuel) are known causes of tube metal wastage. The sulfur combines with hydrogen from the fuel to form hydrogen sulfide (H_2S). The H_2S reacts with the iron in the tube metal and forms iron sulfide, which is subsequently swept away with the flue gas. Chlorine also promotes tube wastage. Although most conventional fuels contain very little chlorine, it is a problem in refuse-derived fuels.

B. Excess Air for Combustion

Perfect or stoichiometric combustion is the complete oxidation of all the combustible constituents of a fuel, consuming exactly 100% of the oxygen contained in the combustion air. Excess air is any amount above that theoretical quantity.

Commercial fuels can be burned satisfactorily only when the amount of air supplied to them exceeds that which is theoretically calculated as required from equations showing the chemical reactions involved. The quantity of excess air provided in any particular case depends on the following:

1. The physical state of the fuel in the combustion chamber
2. Fuel particle size, or oil viscosity
3. The proportion of inert matter present
4. The design of furnace and fuel-burning equipment

For complete combustion, solid fuels require the greatest, and gaseous fuels the least, quantity of excess air. Fuels that are finely subdivided on entering the furnace burn more easily and require less excess air than those induced in large lumps or masses. Burners, stokers, and furnaces having design features producing a high degree of turbulence and mixing of the fuel with the combustion air require less excess air.

TABLE 9.3 Typical Excess Air at Fuel-Burning Equipment

Fuels	Type of furnace or burners	% excess air
Pulverized coal	Completely water-cooled furnace—wet or dry ash removal	15–20
	Partially water-cooled furnace	15–40
Crushed coal	Cyclone furnace: pressure or suction	13–20
	Fluidized-bed combustion	15–20
Coal	Spreader stoker	25–35
	Water-cooled vibrating grate stoker	25–35
	Chain grate and traveling grate	25–35
	Underfeed stoker	25–40
Fuel oil	Register-type burners	3–15
Natural gas, coke oven, and refinery gas	Register-type burners	3–15
Blast furnace gas	Register-type burners	15–30
Wood/bark	Traveling grate, watercooled vibrating grate	20–25
Bagasse	All furnaces	25–35
Refuse-derived fuels	Completely water-cooled furnace traveling grate	40–60
Municipal solid waste	Water-cooled, refractory-covered furnace with reciprocating grate	80–100
	Rotary kiln furnace	60–100
Black liquor	Recovery furnaces for kraft and soda pulping processes	15–20

Source: Refs. 1 and 13.

Table 9.3 indicates the range in values for the excess-air percentage commonly employed by the designer. These are expressed in percentage of theoretical air, and are understood to be at the design load condition of the boiler. (At lower loads, both in design and in operation, higher percentages are sometimes used.)

VI. CONTROLLING EXCESS COMBUSTION AIR [5]

A. O_2 Trim: Oxygen Analysis

The percentage of oxygen, by volume, in the combustion effluent can be used as a guide in improving boiler fuel/air ratios. Typically, automatic oxygen analysis systems have been used only in large boiler installations that could justify the expense for these controls. However, the rising prices and limited availability of fuel have prompted a reassessment of the economic facts, and oxygen analysis, combined with fuel/air ratio adjustment, is now used on many packaged boilers.

B. Benefits of Oxygen Analysis

Packaged boilers are usually equipped with single-actuator combustion control systems in which a jackshaft is mechanically linked to the fuel valve and air damper. Because the actuator normally cannot monitor or adjust fuel/air ratios, these ratios are determined by a series of combustion tests before the control system is installed. The tests dictate where the air damper must be set for each fuel valve position. The fixed fuel/air ratio will be correct as long as the many variables that can affect combustion are properly controlled to remain at the values established by the tests. Tables 9.4 and 9.5 list the most critical variables for natural gas and oil. Deviation of the variables from the established points can decrease boiler efficiency, resulting in significant increase in fuel consumption.

Deviation in a typical boiler would be about 50% of the maximum shown in the tables, and load factor would be about 75%. An example of the annual fuel loss, caused by variations in heating value, air temperature, and the rest in a "normal" situation for natural gas would be

$$\$12,400 \times 0.50 \times 0.75 = \$4,650.00 \text{ per } 10,000 \text{ lb/hr streaming capacity.}$$

Fluctuations of combustion variables are reflected by increases in the excess oxygen level of the stack gases. Fuel losses can be controlled by installing a system to maintain excess oxygen at its optimum value. Tables or curves indicating the optimum oxygen reading for each boiler load are available from the boiler or burner manufacturer. The operator must not adjust the fuel/air ratio without reference to these data. If such data are not available, new combustion tests must be conducted to determine optimum oxygen levels at various loads.

Automatic control systems use an electronic function generator to establish a setpoint, based on boiler load and desired oxygen level for the fuel being fired, for the fuel/air ratio controller. Measured oxygen serves as the feedback signal to the controller. The function generator electronically controls combustion to follow the oxygen and load curve throughout the boiler's firing range. The system is continually trimmed to produce the optimum conditions for maximum combustion efficiency.

See Tables 9.4 and 9.5 for natural gas and fuel oil loss.

TABLE 9.4 Natural Gas Loss Associated with Deviation of Typical Combustion Variables from Design Conditions[a]

Variable	Normal range of deviation	Maximum increase in excess oxygen (%)	Maximum decrease in boiler efficiency (%)	Loss/yr/ 10M#/hr ($)
Gas temperature	90°–50°F	1.00	0.67	3,400.00
Heating value	1150–950 (Btu/ft³)	0.50	0.33	1,100.00
Gas specific gravity	0.60–0.70	1.67	1.10	4,000.00
Combustion air temperature	50°–90°F	1.00	0.67	2,200.00
Combustion air, RH	0–100% at 70°F	0.50	0.33	1,700.00

[a] Based on 75% boiler efficiency, 100% load, 500°F flue gas temp., and fuel at $3.00/1000scf

TABLE 9.5 Fuel Oil Loss Associated with Deviation of Typical Combustion Variables from Design Conditions[a]

Variable	Normal range of deviation	Maximum increase in excess oxygen (%)	Maximum decrease in boiler efficiency (%)	Loss/yr/ 10M#/hr ($)
Heating value	20,000– 17,500 Btu/lb	0.50	0.33	2,428.00
Combustion air temperature	50°–90°F	1.00	0.67	4,857.00
Combustion air, RH	0–100% at 70°F	0.50	0.33	2,428.00

[a] Based on 75% boiler efficiency, 100% load, 500°F flue gas temp., and fuel cost of $1.00/gal

C. Variables Affecting Combustion [5]

1. Natural Gas

Fuel pressure	Barometric pressure
Fuel temperature	Air temperature
Fuel heating value	Air relative humidity
Fuel specific gravity	Combustion air fan cleanliness
Burner linkage wear	

2. Fuel Oil

Fuel pressure	Barometric pressure
Fuel temperature	Air temperature
Fuel heating value	Air relative humidity
Fuel specific gravity	Combustion air fan cleanliness
Fuel viscosity	Burner linkage wear

D. Combustion Air and Excess Air [1,17]

Combustion air is the amount of air required for complete combustion in time, temperature, and turbulence for a given fuel. Excess air is the amount of air above that required for perfect combustion (stoichiometric conditions). Because of imperfect mixing of air and fuel, some excess air is required in all combustion situations. See Figure 9.3 for further information.

1. Reduction of Excess Air

The reduction of excess air is a major step in improving efficiency. The lower limit of excess air is reached whenever there is incomplete combustion or flame impingement on the tubes. The main causes of excess air follow:

- Air leaks
- Improper draft control
- Faulty burner operation

a. Analysis and Solutions. Before excess air can be reduced, its sources must be identified. This can be done by analysis of the flue gas for O_2 or CO_2. Analysis for O_2 is preferred because O_2 readings are more sensitive to the exact amount of excess air present. Gas samples should be taken from the firebox as well as the stack.

- Low O_2 in the firebox and high O_2 at the stack indicates leaks in the furnace casing or ductwork.
- High O_2 in both firebox and stack indicates an excessive amount of air entering the firebox.

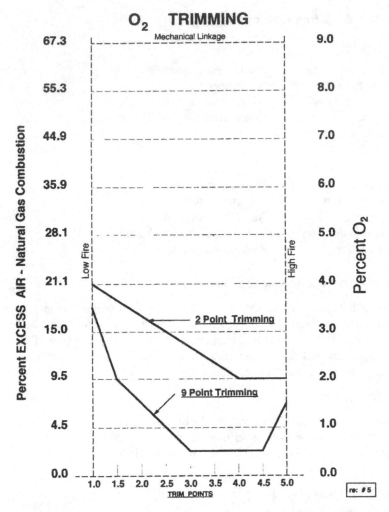

FIGURE 9.3 The effect of O_2 trimming in controlling excess combustion air. (From Ref. 5.)

The CO_2 readings would be opposite the O_2 readings in the foregoing analysis.

Once the sources of excess air are determined, they should be eliminated. Leaks can be sealed by replacing gaskets, using aluminum tape or sealing cements to cover cracks, and by replacing badly warped doors. Excessive air entering the firebox can be reduced by adjusting the draft. Furnace draft is properly controlled when the damper is adjusted so that the pressure underneath the convection tubes is 2- to 3-mm (0.1-in.) water column pressure below atmospheric pressure. When a strong or gusty wind makes the draft fluctuate, the damper should be opened slightly so that the vacuum is not less than 2 mm during the fluctuation.

2. Burner Design and Excess Air

Sometimes faulty burners or insufficient maintenance prevent a furnace from achieving efficient low-excess–air operation. For example, a faulty burner can have a good flame pattern, but still smoke. Many times this is compensated by the addition of excess air, which reduces efficiency. To trace the cause of faulty burners, operate the unit so that all burners have their individual fuel shutoff valves wide open and their air registers opened the same amount. In the burners that have a poor flame pattern, the problem can usually be traced to plugging, enlarged fuel orifices, or incorrect gun position. The burner gun should be far enough into the burner so that the flame barely touches the muffler block. Poor atomization can be suspected if there are a large number of oil drips under the burner.

Closing the air doors of burners that are out of service, such as for cleaning, is an important practice for good burner operation. If the air registers are rusty or difficult to move, they should be replaced or covered to stop air leakage.

In particular, for oil-firing excess air is directly related to combustion air control and oil atomization. These two variables must work together if the burner–boiler combination is to perform satisfactorily. Modern fuel-burning systems offer many years experience in design and operating capabilities. Through research and development several burner manufacturers have established that steam is the best medium for atomization. Also, the steam consumption for atomizing is a function of fuel input in pounds per hour, rather than a percentage of the steam capacity produced. Experimentation to determine the proper oil gun tip spray angle has developed a flame pattern for firing large heat inputs in a relatively confined space.

These and other design parameters have confirmed that 15% excess air is the best design condition for burning fuel oil. The flame pattern within the furnace is important to achieve complete combustion. Improper flame pattern can result in carbon buildup on the tube surfaces which will restrict thermal transfer and result in high exit gas temperatures and loss in efficiency. The purpose of a boiler is to produce steam at the lowest operating cost. To operate a boiler at the most economical level and greatest efficiency, the fuel must be completely burned with a minimum amount of excess air. An increase in excess air of only 10% reduces the efficiency 0.5% and the fuel consumption becomes 40 lb/hr greater. For a small investment, compared to the boiler cost, an oxygen analyzer and recorder can be installed and not only monitor, but also control, the excess air.

3. Combustion Controls

Combustion controls, mainly the fuel/air ratio, are also an important factor in excess air and energy conservation. On small package boilers, a simple jackshaft–positioning-type system with adjustable fuel control valves is more than adequate to sense steam pressure, position control valves, and fan damper linkage as steam demand increases or decreases. Systems with separate drive units for the fuel valves and air dampers can be supplied with fuel/air ratio adjustment on a control panel for use on larger package boiler units. The large units can benefit most from complete automatic control systems that monitor steam flow, steam pressure, fuel pressure, fuel flow, stack O_2 and CO, and other parameters. Combustion controls become even more important in controlling fuel and air for special conditions, such as simultaneously firing of two different fuels. Regardless of the boiler size involved, controls should be selected to meet the plant needs with specific emphasis on low excess air operation.

4. Keep It Clean

Even though well-designed equipment is purchased and installed, good maintenance and cleaning practices must be developed and followed. For instance, a small component, such as a fuel oil burner tip, costing less than $200.00, can become worn from oil abrasion and cause the entire system to operate uneconomically. A clean and well-insulated boiler can also be an energy conserver. On oil firing, carbon buildup on the gas side of tubes can impede thermal transfer and cause high exit gas temperatures. For this reason, the units are equipped with sootblowers. When firing fuel oil, these sootblowers should be operated at least once a shift (three times a day). Failure to blow the soot off the tube surfaces on a schedule makes the cleaning operation that much harder when blowing is eventually done. Always remember that soot buildup on tube surfaces directly affects the efficiency of the steam generator.

- Ash, soot, and mineral deposits are poor conductors; in fact, most are insulators. If deposits build up on or in the tubes, heat transfer from the hot gas flow to the steam, water, or airflow is retarded. The heat that should have gone into the steam ends up being lost out the stack.
- Blocking the passages in the convection section with deposits inhibits draft.
- Because a full volume of air is not driven into the firebox, a normal volume of fuel cannot be burned. This incomplete combustion will result in decreased steam production, increased smoke production, or both.

The degree of cleanliness of the convection section can be estimated by measuring the draft loss between the firebox and the stack. A high draft loss indicates restriction in the passageway.

The water side of the tubes should also be checked for scale buildup. If a scale deposit is formed on the inside tube surfaces, heat cannot pass effectively into the water. This restricts heat absorption and tube failures could result. Preventing scale requires close cooperation with a reputable water treatment laboratory, and close surveillance of the feedwater system.

VII. PREHEATING FEEDWATER [12]

Although a feedwater heater acts, to some extent, as a purifier, its primary function is that of heating the water. As the heat content of live steam ranges from 1100 to 1300 Btu/lb above 32°F, 1% less heat is required to evaporate the feedwater into steam for every 11°–13°F the water is heated. The decrease in fuel consumption, or saving in fuel, owing to heating the feedwater will vary with the overall efficiency of the boiler unit. Ordinarily, the temperature of the feedwater does not appreciably affect the overall efficiency, but with some types of boilers, changes in temperature reduce or increase the rate of heat transfer and, hence, the efficiency.

If H represents the heat content of the boiler steam above 32°F, t_0 and t the initial and final temperature of the feedwater, respectively, e the overall efficiency of the boiler unit, then S, the percentage saving in fuel due to preheating, may be expressed as

$$S = 100 \frac{(t - t_0) e}{H - (t_0 - 32)}$$

Example. 400 psig steam at 1205 Btu, makeup water temperature of 80°F, deaerator outlet water temperature of 250°F, and 75% boiler efficiency.

$$S = 100 \frac{(250 - 80) \times 0.75}{1205 - 80 - 32)} = 11\% \text{ fuel savings}$$

VIII. CONDENSATE AND BLOWDOWN [1,19,52]

A. Condensate

Condensate refers to steam that had been condensed by heat transfer. This condensate is too expensive to just dump to sewer. The value of this condensate consists of its remaining heat and the cost of treating water to replace the condensate. The condensate is collected by a piping system and returned to a condensate receiver. Most condensate does not require treatment before reuse and is pumped directly to the deaerator. Makeup water is added directly to the condensate to form boiler feedwater. In some cases, however, especially where steam is used in industrial processes, the steam condensate is contaminated by corrosion products or by the in-leakage of cooling water or substances in the process.

Demineralizer systems installed to purify condensate are known as condensate-polishing systems. A condensate-polishing system is a requisite to maintain the purity required for satisfactory operation of large once-through boilers.

B. Blowdown

Blowdown is the removal of a small fraction of the recirculating water in steam drum boilers. The primary function of the blowdown system is to control the dissolved solids of the recirculating water. This, in turn, controls carryover and corrosive attack of the boiler. Blowdown also reduces the concentration of solid corrosion products (metal oxides), but solid particle removal efficiency is very limited. Complete blowdown recovery is not economical unless the flow is very large. However heat recovery from blowdown is very economical, and the heat is usually used to preheat makeup water to the deaerator.

If continuous blowdown is not provided and, particularly, if the feedwater is of poor quality, the boiler should be blown down at frequent intervals to prevent serious scale accumulations. If it is necessary to blowdown the water wall headers, it should be done only when the boiler is under very light load, or no load, or only in accordance with specific instructions from the boiler manufacturer. When blowing down a boiler in which a valve and cock are used in the blowoff line, first open the cock and then open the valve *slowly*. After blowing down one gauge of water, the valve should be closed *slowly* before the cock is closed, to avoid water hammer. *Do not leave the boiler during the blowdown operation.*

C. Blowdown Heat Recovery

This involves the recovery of waste heat energy contained in drum water blowdown and expelled condensate. This heat energy can be effectively used to preheat boiler feedwater instead of just being discarded. Approximately a 10°F increase in feedwater temperature will result in a 1% improvement in efficiency. Nominal costs for automatic blowdown systems and condensate recovery units are 1000 and 8000 dollars, respectively.

IX. FLASH STEAM HEAT RECOVERY [20]

Flash steam is a form of heat recovery from a hot water source. In an efficient steam system, this source is primarily the condensate-return system. It is possible to use the boiler blowdown also. Let us consider both.

A. Condensate Return

High pressure condensate forms at the same temperature as the high-pressure steam from which it condenses as the latent heat (enthalpy of evaporation) is removed. When this condensate is discharged to a lower pressure, the energy it contains is greater than it can hold while remaining as liquid water. The excess energy reevaporates some of the water as steam at the lower pressure. Conventionally, this steam is referred to as *flash steam*, although, in fact, it is perfectly good steam even if at a lower pressure. The quantity of flash steam available from each pound of condensate (or boiler blowdown) can be calculated as follows;

Example. 10,000 lb/hr of condensate
Condensate pressure of 300 psig
Flash tank pressure of 17.5 psig

Solution.

Sensible heat at 300 psig	398.9 Btu/lb
Sensible heat at 17.5 psig	223.1 Btu/lb
Heat available for flashing	175.8 Btu/lb
Latent heat at 17.5 psig	942.9 Btu/lb

Proportion evaporated to steam $\dfrac{175.8}{942.9} = 0.18645$ or 18.65%

Flash steam available:
$= 0.18645 \times 10{,}000$ lb/hr
$= 1864$ lb/hr steam at 17.5 psig—saturated

This means that condensate from high-pressure sources usually should be collected and led to a flash tank that operates at a lower pressure. Remember that all of the flashing off does not normally take place in the flash vessel. It begins within the seat of the steam trap and continues in the condensate line. Only when the high-pressure traps are very close to the flash vessel does any flashing at all take place within the flash tank. Instead, the flash vessel is primarily a flash steam separator. Its shape and dimensions are chosen to encourage separation of the considerable volume of low-pressure steam from the small volume of liquid.

Some uses of flash steam from condensate include

1. Deaerator
2. Feedwater heater
3. Heating coils
4. Absorption chiller

B. Boiler Blowdown

This source of high-temperature water for flash steam is often overlooked. The possibility of carryover of undesirable elements to the deaerator must be carefully studied. This flash steam can be used in heat exchangers for feedwater heating or for heating other liquid or gaseous products.

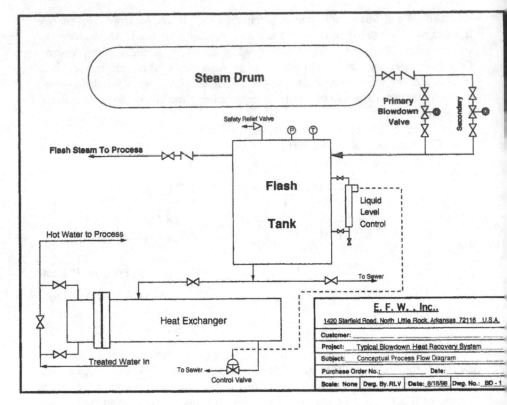

FIGURE 9.4 Schematic of a typical blowdown heat recovery system. (Courtesy of E.F.W., Inc.)

The use of a heat exchanger for heat recovery from boiler blowdown is the alternative to flashing it to steam. Either way, this valuable source of heat should not be wasted.

For more information see Table 9.4 on flash steam and Figure 9.4a heat recovery system.

X. STEAM GENERATOR OVERALL EFFICIENCY

1. Steam Generator Overall Efficiency $= \dfrac{\text{Output, Btu/hr}}{\text{Input, Btu/hr}}$

2. Output, Btu/hr $= S(h_g - h_{f1} + B(h_{f3} - h_{f1})$

   S = steam flow (lb/hr)

B = blowdown (lb/hr)

Input (Btu/hr) = F × H

F = fuel input (lb/hr) (as fired)

H = fuel higher heating value (Btu/lb) (as fired)

Example.

Steam production per hour (S)	= 56,000 lb
Steam conditions	= 600 psig at 750°F
(Btu/lb) (h_g)	= 1379.6
Continuous blowdown (B)	= 5% = 2800 lb/hr
Sensible (Btu/lb) (h_{f3})	= 474.8
Feedwater entering economizer	= 300°F
Sensible (Btu/lb) (h_{f1})	= 269.6
Fuel (lb/hr) (F)	= 12,000
Fuel (Btu/lb) (H)	= 6,500

$$\text{Efficiency} = \frac{[56,000 \times (1379.6 - 269.6)] + [2,800 \times (474.8 - 269.6)]}{12,000 \times 6.500}$$

= 0.8043 or 80.43%

For more information see Table 9.6 on boiler steam usage.

TABLE 9.6

Annual Energy Cost of boiler steam useage w/ various boiler efficiencies — Ralph L. Vandagriff
Date: October 7, 1997 — Run Date 5/200 12:56

Enter Data:

Lb-Hr Steam	100,000.00
Steam Pressure (Psig)	270
Steam Temperature (Deg F)	sat
Btu-Lb Steam	1203.60
Feedwater Temp (Deg F)	230
Feedwater (BTU)	198.00
Operating Hours (per year)	8600
Customer:	
Location:	
N.Gas: BTU/Cu.Ft.	1000.00
Total Fuel BTU input per pound of steam	1005.60

Boiler Operating Efficiency / Annual Energy Cost of steam (Dollars)

Cost of Fuel (Dollars per Million Btu)	75.00%	76.00%	77.00%	78.00%	79.00%	80.00%	81.00%	82.00%	83.00%	84.00%	85.00%
Energy BTU/1000 lb steam	1,340,800.00	1,323,157.89	1,305,974.03	1,289,230.77	1,272,911.39	1,257,000.00	1,241,481.48	1,226,341.46	1,211,566.27	1,197,142.86	1,183,058.82
$1.90	2,190,867.20	2,162,040.00	2,133,961.56	2,106,603.08	2,079,937.22	2,053,938.00	2,028,560.74	2,003,841.95	1,979,699.28	1,956,131.43	1,933,118.12
$2.00	2,306,176.00	2,275,831.58	2,246,275.32	2,217,476.92	2,189,407.59	2,162,040.00	2,135,348.15	2,109,307.32	2,083,893.98	2,059,085.71	2,034,861.18
$2.10	2,421,484.80	2,389,623.16	2,358,589.09	2,328,350.77	2,298,877.97	2,270,142.00	2,242,115.56	2,214,772.68	2,188,088.67	2,162,040.00	2,136,604.24
$2.20	2,536,793.60	2,503,414.74	2,470,902.86	2,439,224.62	2,408,348.35	2,378,244.00	2,348,882.96	2,320,238.05	2,292,283.37	2,264,994.29	2,238,347.29
$2.30	2,652,102.40	2,617,206.32	2,583,216.62	2,550,098.46	2,517,818.73	2,486,346.00	2,455,650.37	2,425,703.41	2,396,478.07	2,367,948.57	2,340,090.35
$2.40	2,767,411.20	2,730,997.89	2,695,530.39	2,660,972.31	2,627,289.11	2,594,448.00	2,562,417.78	2,531,168.78	2,500,672.77	2,470,902.86	2,441,833.41
$2.50	2,882,720.00	2,844,789.47	2,807,844.16	2,771,846.15	2,736,759.49	2,702,550.00	2,669,185.19	2,636,634.15	2,604,867.47	2,573,857.14	2,543,576.47
$2.60	2,998,028.80	2,958,581.05	2,920,157.92	2,882,720.00	2,846,229.87	2,810,652.00	2,775,952.59	2,742,099.51	2,709,062.17	2,676,811.43	2,645,319.53
$2.70	3,113,337.60	3,072,372.63	3,032,471.69	2,993,593.85	2,955,700.25	2,918,754.00	2,882,720.00	2,847,564.88	2,813,256.87	2,779,765.71	2,747,062.59
$2.80	3,228,646.40	3,186,164.21	3,144,785.45	3,104,467.69	3,065,170.63	3,026,856.00	2,989,487.41	2,953,030.24	2,917,451.57	2,882,720.00	2,848,805.65
$2.90	3,343,955.20	3,299,955.79	3,257,099.22	3,215,341.54	3,174,641.01	3,134,958.00	3,096,254.81	3,058,495.61	3,021,646.27	2,985,674.29	2,950,548.71
$3.00	3,459,264.00	3,413,747.37	3,369,412.99	3,326,215.38	3,284,111.39	3,243,060.00	3,203,022.22	3,163,960.98	3,125,840.96	3,088,628.57	3,052,291.76
$3.10	3,574,572.80	3,527,538.95	3,481,726.75	3,437,089.23	3,393,581.77	3,351,162.00	3,309,789.63	3,269,426.34	3,230,035.66	3,191,582.86	3,154,034.82
$3.20	3,689,881.60	3,641,330.53	3,594,040.52	3,547,963.08	3,503,052.15	3,459,264.00	3,416,557.04	3,374,891.71	3,334,230.36	3,294,537.14	3,255,777.88
$3.30	3,805,190.40	3,755,122.11	3,706,354.29	3,658,836.92	3,612,522.53	3,567,366.00	3,523,324.44	3,480,357.07	3,438,425.06	3,397,491.43	3,357,520.94
$3.40	3,920,499.20	3,868,913.68	3,818,668.05	3,769,710.77	3,721,992.91	3,675,468.00	3,630,091.85	3,585,822.44	3,542,619.76	3,500,445.71	3,459,264.00
$3.50	4,035,808.00	3,982,705.26	3,930,981.82	3,880,584.62	3,831,463.29	3,783,570.00	3,736,859.26	3,691,287.80	3,646,814.46	3,603,400.00	3,561,007.06
$3.60	4,151,116.80	4,096,496.84	4,043,295.58	3,991,458.46	3,940,933.67	3,891,672.00	3,843,626.67	3,796,753.17	3,751,009.16	3,706,354.29	3,662,750.12
$3.70	4,266,425.60	4,210,288.42	4,155,609.35	4,102,332.31	4,050,404.05	3,999,774.00	3,950,394.07	3,902,218.54	3,855,203.86	3,809,308.57	3,764,493.18
$3.80	4,381,734.40	4,324,080.00	4,267,923.12	4,213,206.15	4,159,874.43	4,107,876.00	4,057,161.48	4,007,683.90	3,959,398.55	3,912,262.86	3,866,236.24
$3.90	4,497,043.20	4,437,871.58	4,380,236.88	4,324,080.00	4,269,344.81	4,215,978.00	4,163,928.89	4,113,149.27	4,063,593.25	4,015,217.14	3,967,979.29
$4.00	4,612,352.00	4,551,663.16	4,492,550.65	4,434,953.85	4,378,815.19	4,324,080.00	4,270,696.30	4,218,614.63	4,167,787.95	4,118,171.43	4,069,722.35
$4.10	4,727,660.80	4,665,454.74	4,604,864.42	4,545,827.69	4,488,285.57	4,432,182.00	4,377,463.70	4,324,080.00	4,271,982.65	4,221,125.71	4,171,465.41
$4.20	4,842,969.60	4,779,246.32	4,717,178.18	4,656,701.54	4,597,755.95	4,540,284.00	4,484,231.11	4,429,545.37	4,376,177.35	4,324,080.00	4,273,208.47
$4.30	4,958,278.40	4,893,037.89	4,829,491.95	4,767,575.38	4,707,226.33	4,648,386.00	4,590,998.52	4,535,010.73	4,480,372.05	4,427,034.29	4,374,951.53
$4.40	5,073,587.20	5,006,829.47	4,941,805.71	4,878,449.23	4,816,696.71	4,756,488.00	4,697,765.93	4,640,476.10	4,584,566.75	4,529,988.57	4,476,694.59
$4.50	5,188,896.00	5,120,621.05	5,054,119.48	4,989,323.08	4,926,167.09	4,864,590.00	4,804,533.33	4,745,941.46	4,688,761.45	4,632,942.86	4,578,437.65
$4.60	5,304,204.80	5,234,412.63	5,166,433.25	5,100,196.92	5,035,637.47	4,972,692.00	4,911,300.74	4,851,406.83	4,792,956.14	4,735,897.14	4,680,180.71
$4.70	5,419,513.60	5,348,204.21	5,278,747.01	5,211,070.77	5,145,107.85	5,080,794.00	5,018,068.15	4,956,872.20	4,897,150.84	4,838,851.43	4,781,923.76
$4.80	5,534,822.40	5,461,995.79	5,391,060.78	5,321,944.62	5,254,578.23	5,188,896.00	5,124,835.56	5,062,337.56	5,001,345.54	4,941,805.71	4,883,666.82
$4.90	5,650,131.20	5,575,787.37	5,503,374.55	5,432,818.46	5,364,048.61	5,296,998.00	5,231,602.96	5,167,802.93	5,105,540.24	5,044,760.00	4,985,409.88
$5.00	5,765,440.00	5,689,578.95	5,615,688.31	5,543,692.31	5,473,518.99	5,405,100.00	5,338,370.37	5,273,268.29	5,209,734.94	5,147,714.29	5,087,152.94
$5.10	5,880,748.80	5,803,370.53	5,728,002.08	5,654,566.15	5,582,989.37	5,513,202.00	5,445,137.78	5,378,733.66	5,313,929.64	5,250,668.57	5,188,896.00
$5.20	5,996,057.60	5,917,162.11	5,840,315.84	5,765,440.00	5,692,459.75	5,621,304.00	5,551,905.19	5,484,199.02	5,418,124.34	5,353,622.86	5,290,639.06
$5.30	6,111,366.40	6,030,953.68	5,952,629.61	5,876,313.85	5,801,930.13	5,729,406.00	5,658,672.59	5,589,664.39	5,522,319.04	5,456,577.14	5,392,382.12
$5.40	6,226,675.20	6,144,745.26	6,064,943.38	5,987,187.69	5,911,400.51	5,837,508.00	5,765,440.00	5,695,129.76	5,626,513.73	5,559,531.43	5,494,125.18
$5.50	6,341,984.00	6,258,536.84	6,177,257.14	6,098,061.54	6,020,870.89	5,945,610.00	5,872,207.41	5,800,595.12	5,730,208.43	5,662,485.71	5,595,868.24
$5.60	6,457,292.80	6,372,328.42	6,289,570.91	6,208,935.38	6,130,341.27	6,053,712.00	5,978,974.81	5,906,060.49	5,834,903.13	5,765,440.00	5,697,611.29
$5.70	6,572,601.60	6,486,120.00	6,401,884.68	6,319,809.23	6,239,811.65	6,161,814.00	6,085,742.22	6,011,525.85	5,939,597.83	5,868,394.29	5,799,354.35
$5.80	6,687,910.40	6,599,911.58	6,514,198.44	6,430,683.08	6,349,282.03	6,269,916.00	6,192,509.63	6,116,991.22	6,043,292.53	5,971,348.57	5,901,097.41
$5.90	6,803,219.20	6,713,703.16	6,626,512.21	6,541,556.92	6,458,752.41	6,378,018.00	6,299,277.04	6,222,456.59	6,147,467.23	6,074,302.86	6,002,840.47

10

Electricity Generation and Cogeneration

Cogeneration; Cogeneration Using Biomass; Cogeneration Report: Survey of Cogeneration in Texas: Gulf Coast Cogeneration Association, Houston, Texas; Conclusions.

I. COGENERATION

What is cogeneration? Cogeneration is the use of energy in a sequential fashion to produce simultaneously thermal energy and power: specifically, steam and electricity. An example is a gas engine driving an electrical power generation system with the exhaust from the engine passing through a waste heat recovery boiler to generate process steam. Another example is the burning of a fuel in a boiler to produce steam, running the steam through a backpressure steam engine or steam turbine, and then taking the steam to process. The amount of fuel required to produce the power and the heat in either cogeneration system is less than the amount of fuel that would have been required to produce the same amounts of electricity and heat separately.

The two common forms of cogeneration are called "bottoming cycle" and "topping cycle." Bottoming cycle refers to the system in which fuel is burned to produce steam for process requirements and waste heat is recovered for the production of power. Topping cycle is when fuel is burned for steam to produce power first and the waste heat is used for process requirements.

Efficient usage of bottoming cycles requires a high-temperature waste heat stream and this somewhat limits the opportunities. Topping cycles can be used almost any place that process heat is required.

Cogeneration will improve the overall system efficiency in a steam system. Some of the more common cogeneration systems include the following:

1. Drop higher-pressure boiler steam through a backpressure turbine–generator to a lower steam pressure to process.
2. Drop higher-pressure boiler steam through a backpressure turbine–generator to a lower steam pressure to an absorption chiller for refrigeration.
3. Recover waste heat from the exhaust of a gas turbine–generator, generate steam, and drop the steam through a backpressure turbine–generator to process.
4. Recover waste heat from the exhaust and cooling water of a gas engine generator, generate steam, and drop the steam through a backpressure turbine–generator to process.
5. Recover waste heat from a source, generate steam with a heat recovery boiler, drop the steam through a backpressure turbine–generator to process.
6. Use a condensing turbine–generator in the foregoing examples when process steam demand does not exist.
7. Replace steam pressure-reducing valves with backpressure turbine–generator sets.

A. Examples of Cogeneration

Figures 10.1 to 10.15 illustrate various types of cogeneration and their parameters.

II. COGENERATION USING BIOMASS

A. Published Reports

1. Sugar Cane Bagasse: Jamaica

By: Office of Energy, Bureau for Science and Technology, United States Agency for International Development.

Report number: 89–10
Title: Jamaica Cane/Energy Project Feasibility Study
Date published: September 1986

Project limiting factor: Quantity of sugar cane produced annually, and the derived supply of bagasse and barbojo biofuel

Operating schedule: 206 days sugarcane harvesting season
                         87 days of fuel from storage facility
                         42 days scheduled maintenance
                         30 days forced outage

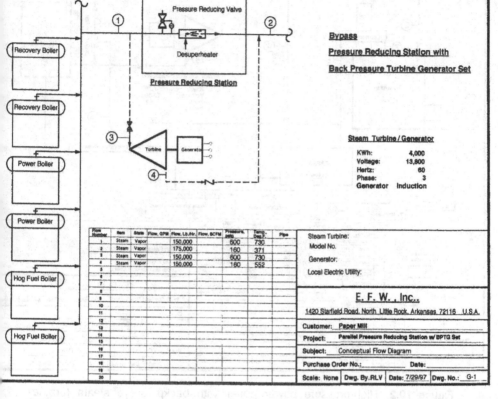

FIGURE 10.1 Parallel a pressure-reducing station with backpressure steam turbine–generator.

Steam flow	150,000 lb/hr
Steam pressure/temp	600 psig/730°F
Utility $/kW	$0.075
Plant load, kWh	25,000
Generation	4,000 kWh, 13,800 V, 60 Hz, 3 ph

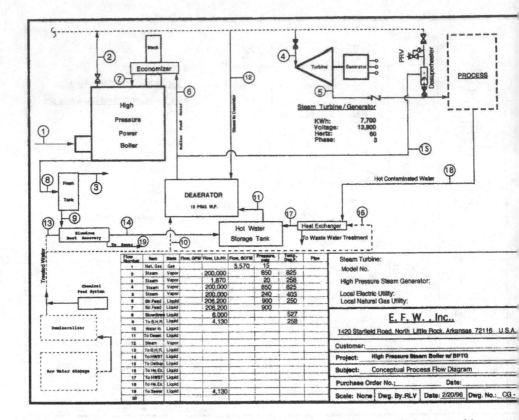

FIGURE 10.2 High-pressure power boiler with backpressure steam turbine–generator.

Fuel	Natural gas
Process steam	Required
Steam flow	200,000 lb/hr
Steam pressure/temp	850 psig/850°F
Deaerator	15 psig
BFW temp	250°F
Natural gas; SCFM	5,570
$/MCF	$2.00
Utility $/kW	$0.075
Plant load, kWh	9,000
Generation	7,700 kWh, 13,800 V, 60 Hz, 3 ph

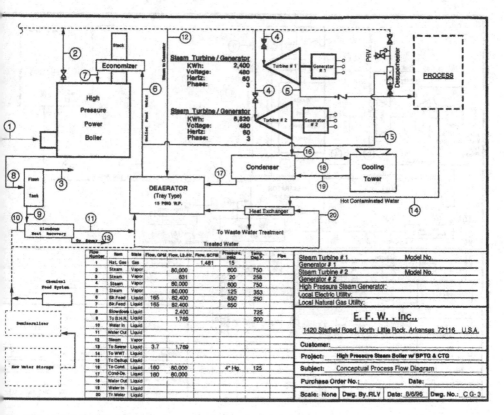

FIGURE 10.3 High-pressure power boiler with BP turbine–generator and condensing turbine–generator.

Fuel	Natural gas
Process steam	Intermittent
Steam flow	80,000 lb/hr
Steam pressure/temp	600 psig/750°F
Deaerator	15 psig
BFW temp	250°F
Natural gas; SCFM	1,481
$/MCF	$2.20
Utility $/kW	$0.065
Plant load, kWh	3,000
Generation: backpressure turbine–generator	2,400 kWh, 480 V, 3 ph, 60 Hz
Generation: condensing turbine–generator	6,820 kWh, 480 V, 3 ph, 60 Hz

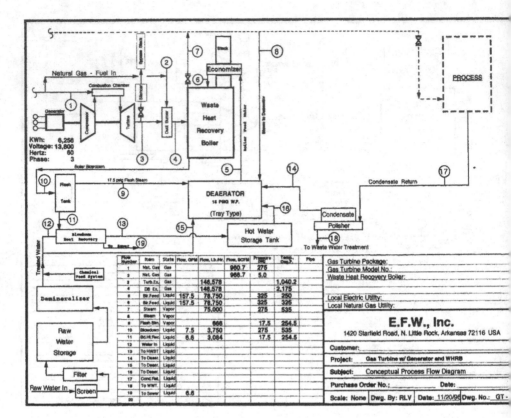

FIGURE 10.4 Gas turbine with duct burner and waste heat recovery boiler.

Fuel	Natural gas
Process steam	Continuous
Steam flow	75,000 lb/hr
Steam pressure/temp:	275 psig/535°F
Deaerator	15 psig
BFW temp	250°F
Natural gas; SCFM	1,928
$/MCF	$2.20
Utility $/kW	$0.07
Plant load, kWh	23,000
Generation	6,256 kWh, 13,800 V, 3 ph, 60 Hz

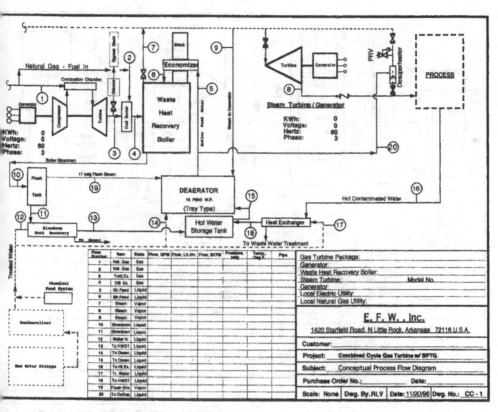

FIGURE 10.5 Gas turbine with duct burner, WHRB, and backpressure steam turbine–generator.

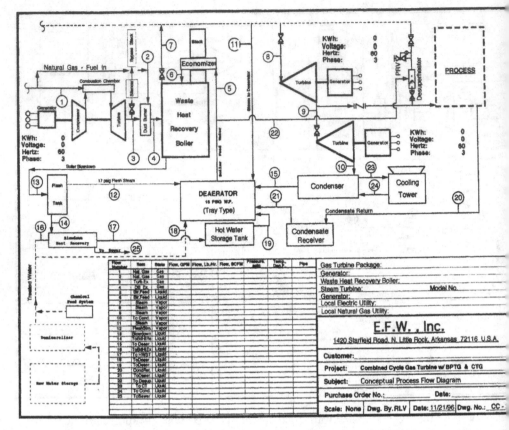

FIGURE 10.6 Gas turbine with duct burner, WHRB, BPTG, and CTG in series.

Fuel: Bagasse (51%), barbojo (24%), no. 6 fuel oil (25%)
[Bagasse is the waste left from processed sugar cane; barbojo is sun-dried cane tops and leafs.]
Fuel moisture content: 35%

Annual fuel production: 750,000 tons of bagasse at average 50% moisture content
625,000 tons (approximate) of barbojo at average of 50% moisture content

MW of electricity generated: Approximately 35 MW

Steam equipment: Two steam boilers, spreader stoker fired with oil burners

Steam: 165,000 lb/hr(each)   Conditions: 900 psig at 900°F

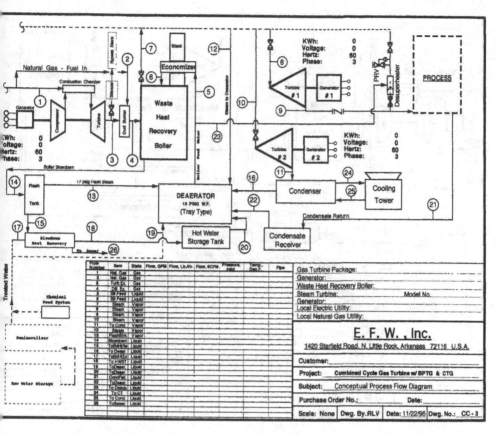

FIGURE 10.7 Gas turbine with duct burner, WHRB, BPTG, and CTG in parallel.

Generating equipment: One tandem compound steam turbine generator (autoextraction)

Steam supply: 329,500 lb/hr   Conditions: 850 psig at 900°F

Speed: 3,000 rpm   Exhaust: 3.5 in. Hga

Generator rating: 13,800 V, 40,000 kVA, 0.85 power factor

Auxiliary equipment: Condensate storage; deaerator with three boiler feed pumps (50% capacity each)

Feedwater preheater; demineralizer–softener chemical feed system

Condenser with pumps; cooling tower with pumps

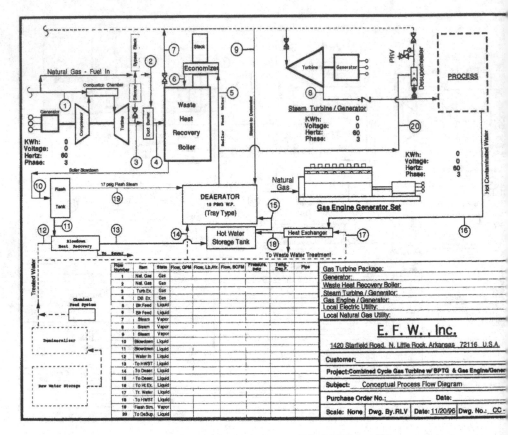

FIGURE 10.8 Gas turbine with duct burner, WHRB, BPTG, and gas engine generator set.

Operating conditions: Cogeneration mode

                      200 psig process steam   146,000 lb/hr

                      20 psig process steam   44,000 lb/hr

                      Power output   21,067 kW

          Electricity production mode only

                      Process steam   0

                      Power output   34,250 kW (gross)

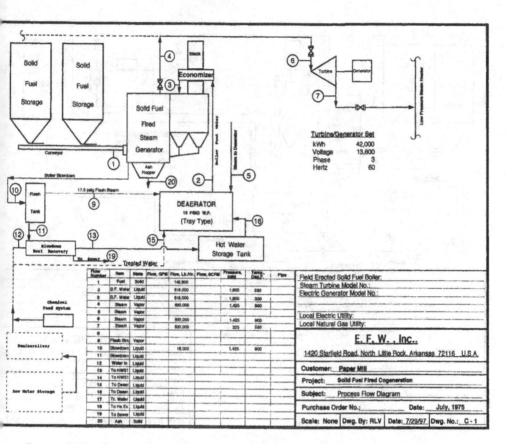

FIGURE 10.9 Solid fuel-fired field-erected high-pressure boiler with BPTG.

Fuel	Hogged fuel
Process steam	Continuous
Steam flow	600,000 lb/hr
Steam pressure/temp	1425 psig/900°F
Deaerator	15 psig
BFW Temp	250°F
Fuel, lb/hr	148,800
$/ton, delivered	$10.00
Utility, $/kW	$0.07
Plant load, kWh	83,000
Generation	42,000 kWh, 13,800 V, 3 ph, 60 Hz

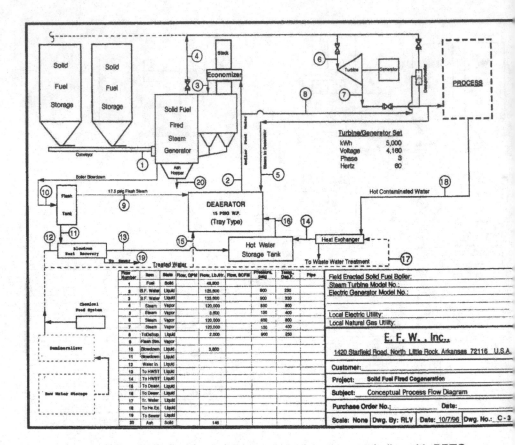

FIGURE 10.10 Solid fuel-fired, field-erected, high-pressure boiler with BPTG.

Fuel	Hogged fuel
Process steam	Continuous
Steam flow	120,000 lb/hr
Steam pressure/temp:	850 psig/800°F
Deaerator	15 psig
BFW temp	250°F
Fuel, lb/hr	48,800
$/ton, delivered	$10.00
Utility, $/kW	$0.065
Plant load, kWh	8,000
Generation	5,000 kWh, 4,160 V, 3 ph, 60 Hz

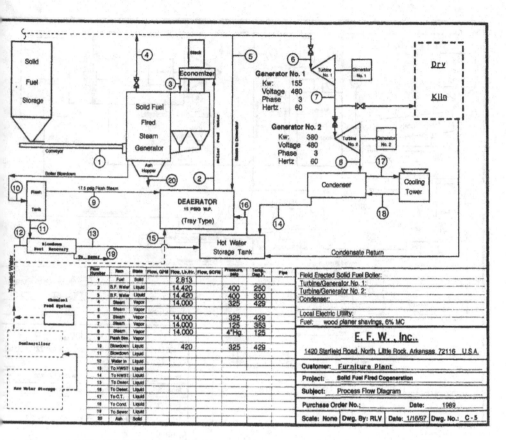

FIGURE 10.11 Solid fuel-fired field-erected, high-pressure boiler with BPTG and CTG.

Fuel	Dry wood shavings
Process steam	Intermittent
Steam flow	14,400 lb/hr
Steam pressure/temp	325 psig/429°F
Deaerator	15 psig
BFW temp	250°F
Fuel, lb/hr	2,813
$/ton, delivered	$0.00
Utility, $/kW	$0.075
Plant load, kWh	3,000
Generation	535 kWh, 480 V, 3 ph, 60Hz

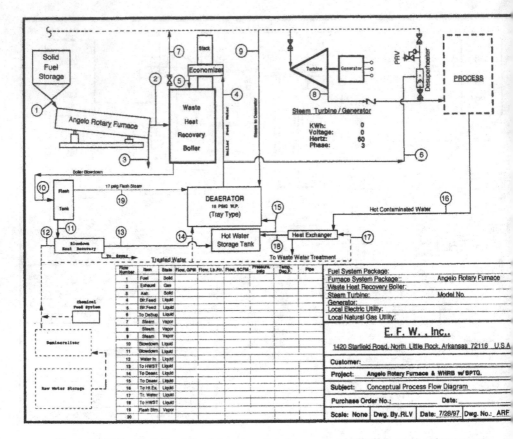

The table within the diagram:

Flow Number	Item	State	Flow, GPM	Flow, Lb/Hr.	Flow, SCFM	Pressure, psig	Temp. Deg.F.	Pipe
1	Fuel	Solid						
2	Exhaust	Gas						
3	Ash.	Solid						
4	Blr.Feed	Liquid						
5	Blr.Feed	Liquid						
6	To DeSup.	Liquid						
7	Steam	Vapor						
8	Steam	Vapor						
9	Steam	Vapor						
10	Blowdown	Liquid						
11	Blowdown	Liquid						
12	Water In	Liquid						
13	To HWST	Liquid						
14	To Deaer.	Liquid						
15	To Deaer.	Liquid						
16	To Ht.Ex.	Liquid						
17	Tr. Water	Liquid						
18	To HWST	Liquid						
19	Flash Stm.	Vapor						
20								

Fuel System Package:
Furnace System Package::   Angelo Rotary Furnace
Waste Heat Recovery Boiler:
Steam Turbine: Model No.
Generator:
Local Electric Utility:
Local Natural Gas Utility:

E. F. W. , Inc..

1420 Starfield Road, North Little Rock, Arkansas 72116 U.S.A

Customer:
Project: **Angelo Rotary Furnace & WHRB w/ BPTG.**
Subject: Conceptual Process Flow Diagram
Purchase Order No.:   Date:
Scale: None | Dwg. By.RLV | Date: 7/28/97 | Dwg. No.: ARF

FIGURE 10.12   Solid fuel-fired angelo rotary furnace with WHRB and back pressure turbine–generator.

Fuel	
Process steam	
Steam flow	lb/hr
Steam pressure/temp	
Deaerator	15 psig
BFW temp	250°F
Fuel, lb/hr	
$/ton, delivered	$0.00
Utility, $/kW	$0
Plant load, kWh	
Generation:	kWh, 480 V, 3 ph, 60 Hz

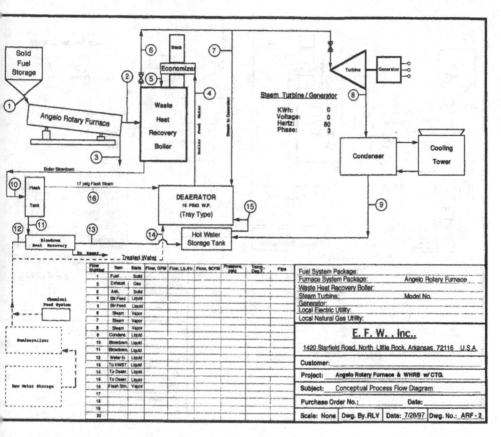

FIGURE 10.13 Solid fuel-fired angelo rotary furnace with WHRB and condensing turbine–generator.

Fuel	Municipal solid waste
No process steam	
Steam flow	13,000 lb/hr
Steam pressure/temp	250 psig/406°F
Deaerator:	15 psig
BFW temp	250°F
Fuel, lb/hr	10,000
$/ton, delivered	−$2.00
Utility, $/kW	$0.095
Plant load, kWh	140
Generation	1,100 kWh, 240 V, 3 ph, 50 Hz

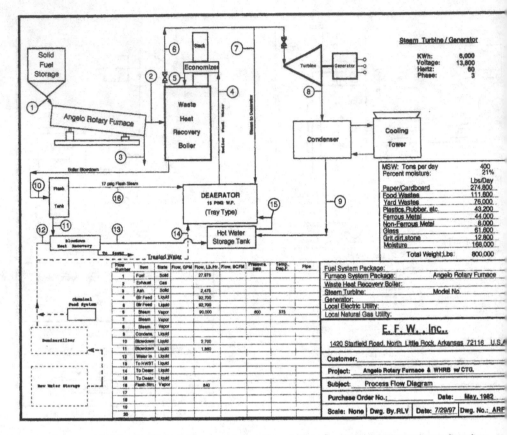

FIGURE 10.14 Solid fuel-fired angelo rotary furnace with WHRB and condensing turbine—generator.

Fuel	Municipal solid waste
No process steam	
Steam flow	90,000 lb/hr
Steam pressure/temp	600 psig/575°F
Deaerator	15 psig
BFW temp	250°F
Fuel, lb/hr	27,975
$/ton; delivered	−$3.00
Utility, $/kW	$0.075
Plant load, kWh	800
Generation	6,000 kWh, 13,800 V, 3 ph, 60 Hz

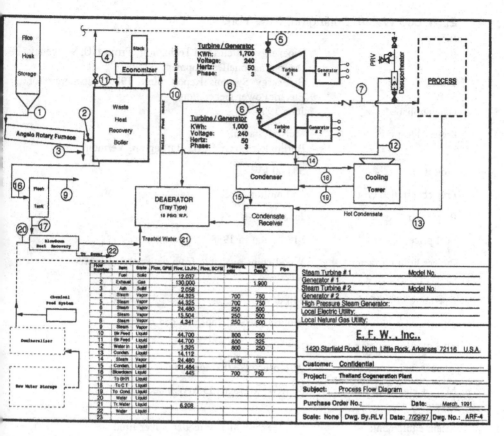

Flow Number	Item	State	Flow, GPM	Flow, Lb./Hr.	Flow, SCFM	Pressure, psia	Temp. Deg.F.	Pipe
1	Fuel	Solid		12,037				
2	Exhaust	Gas		130,000			1,900	
3	Ash	Solid		2,058				
4	Steam	Vapor		44,325		700	750	
5	Steam	Vapor		44,325		700	750	
6	Steam	Vapor		24,480		250	500	
7	Steam	Vapor		15,504		250	500	
8	Steam	Vapor		4,341		250	500	
9	Steam	Vapor						
10	Blr Feed	Liquid		44,700		800	250	
11	Blr Feed	Liquid		44,700		800	325	
12	Water In	Liquid		1,325		800	250	
13	Conden.	Liquid		14,112				
14	Steam	Vapor		24,480		4"Hg	125	
15	Conden.	Liquid		21,464				
16	Blowdown	Liquid		445		700	750	
17	To BHR	Liquid						
18	To CT	Liquid						
19	To Cond.	Liquid						
20	Water	Liquid						
21	Tr. Water	Liquid		6,208				
22	Water	Liquid						
23								

Steam Turbine # 1   Model No.
Generator # 1
Steam Turbine # 2   Model No.
Generator # 2
High Pressure Steam Generator:
Local Electric Utility:
Local Natural Gas Utility:

E. F. W., Inc.

1420 Starfield Road, North Little Rock, Arkansas 72116   U.S.A.

Customer: Confidential

Project: Thailand Cogeneration Plant

Subject: Process Flow Diagram

Purchase Order No.:   Date: March, 1991

Scale: None | Dwg. By.RLV | Date: 7/29/97 | Dwg. No.: ARF-4

FIGURE 10.15   Solid fuel-fired angelo rotary furnace with WHRB, BPTG, and CTG.

Fuel	Rice husks
Process steam	Continuous
Steam flow	44,325 lb/hr
Steam pressure/temp	700 psig/750°F
Deaerator	15 psig
BFW temp	250°F
Fuel, lb/hr	12,037
$/ton, delivered	$0.00
Utility, $/kW	$0.11
Plant load, kWh	Unknown
Generation	2,700 kWh, 240 V, 3 ph, 50 Hz

2. Wood Waste: Burlington, New York

By:	Bioenergy Systems and Technology Project, U. S. Agency for InternationalDevelopment
Title:	Bioenergy Systems Report, June 1983: Bioenergy for ElectricPower Generation
Date published:	June 1983
Project location:	McNeil Station Burlington Electric Co., Burlington, Vermont
Project size:	50,000 kW
Project installed cost:	$60 million U. S.
Project on line:	June 1984
Fuel price:	$18.00/ton in 1988
Project limiting factor:	Wood chips available from national forest preserve
Operating schedule:	348 days/yr
Fuel:	Biomass, wood chips, hogged bark, etc.
Fuel moisture content:	35–50%
Annual fuel production:	92 ton/hr; 773,000 tons/yr
MW of electricity generated:	50 MW, average; 53 MW, maximum
Steam equipment:	One steam boiler, spreader stoker fired
	Steam: 480,000 lb/hr; Conditions: 1275 psig at 950°F
	Manufacturer: Erie City Energy Div., Zurn Industries
Generating equipment:	Unknown
	Speed: unknown; Exhaust: unknown
	Generator rating: unknown
Availability:	95% in year 1988
Auxiliary equipment:	Unknown
Operating conditions:	Major problems—none
	Minor problems—NO_x at low load under some conditions

3. Wood Waste—TVA

By: TVA Generating Group—Southeastern Regional Biomass Energy Program
Title: Biomass Design Manual: Industrial Size Systems
Date published: Reprint, 1991

Section No. 2   Components of Wood Energy Systems

a. **Wood Fuel Receiving, Handling, and Storage Facilities (Condensed)***

Wood fuel may also contain tramp iron and other metals, large rocks, sand, and other detrimental materials. To facilitate control of the combustion process and protect the combustion and fuel-handling equipment, the fuel entering the combustion chamber should be as uniform in particle size as is practical and should be free of materials that could cause damage. The fuel screening and sizing system consists of a metal detector that shuts down the system when metal is detected, a magnet separator to remove tramp metal, a screening device to separate oversized wood pieces, and equipment to reduce oversized pieces to a size that can be handled by the wood fuel combustion equipment-stoking system.

Dry wood fuel storage (15% moisture or less) must be stored in a covered area to prevent moisture absorption and fuel degradation.

Large quantities of sawmill residues and other wet or green wood can be stored outdoors in an uncovered area. The natural angle of repose of the wood will tend to cause the piles to shed water, leaving the interior of the pile relatively dry.

Care should be taken not to allow wood to stand in a silo for extended periods, such as over the summer.

Open storage is the least expensive method of storing large quantities of wet or green fuel. A concrete pad, sloped to drain away from the fuel pile, is recommended.

b. **Fuel-Handling Equipment (Condensed)***

Front-end loaders are used in many wood energy systems to place wood in storage, retrieve it, and feed it into the system. The front-end loaders should be equipped with oversized buckets.

Pneumatic conveyors using high-velocity air are well suited to conveying small particles of wood fuel, such as sawdust, sander dust, finely hogged fuel, etc. This is the most economical conveying system for these fuels over long distances.

---

* *Note*: See publication for complete text.

Screw conveyors have two distinct advantages over other mechanical conveyors—they can convey up steep inclines and they can meter the
amount of fuel being conveyed. The major disadvantage is cost. Screw
conveyors also have difficulty in conveying stringy wood or wood fuel
particles larger than 2 in.[2]

Belt conveyors are the least expensive and have lowest energy requirements. The major drawback is that they should not be inclined more than
15 degrees owing to spillback of the wood fuel.

Drag or flight chain conveyors can handle a steeper incline than belt conveyors. They are versatile and rugged and their operational energy requirement is relatively low. The cost of chain conveyors falls between
that for belt and screw conveyors.

4. Wood Waste: Crossett, Arkansas

By:	Forest Products Research Society, Madison, Wisconsin Proceedings No. P-79-22
Title:	Hardware for Energy Generation in the Forest Products Industry
Paper title:	Description and Operation of the Wood Waste Storage Distribution System and 9A Wood Waste Boiler, T. O. Rytter, Mgr. Utilities and Engineering, Georgia–Pacific Corp., Crossett Devision, Crossett, Arkansas

Date Published: 1979

Project location: Georgia–Pacific Paper Mill, Crossett, Arkansas

Project size: 42,000 kW

Project installed cost: $12 million U. S.

Project on line: July 1975

Operating schedule: 355 days/yr

Fuel: Biomass, wood chips, hogged bark, sander dust, etc.

Fuel moisture content: 35–50%

Tons per day of hogged   1560–1680
   fuel and sander dust:

MW of electricity generated: 41 MW, average; 43 MW, maximum.

Boiler design data:

Capacity: 600,000 lb/hr on wood and no. 6 fuel oil; 400,000 lb/hr on wood only;
             200,000 lb/hr no. 6 fuel oil only

Design pressure: 1425 psig; operating pressure: 900 psig

Final steam temperature: 900°F; feedwater temperature: 350°F

Wood and no. 6 oil firing: (600,000 lb/hr)
    Gas temperature leaving boiler: 729°F; gas temperature leaving economizer: 534°F; gas temperature leaving air preheater: 307°F; efficiency: 76.5%

Wood firing: (400,000 lb/hr)
    Gas temperature leaving boiler: 663°F; gas temperature leaving economizer: 498°F; gas temperature leaving air preheater: 298°F; efficiency: 71.15%

Boiler manufacturer: Erie City Energy Division, Zurn Industries
Traveling grate manufacturer: Detroit Stoker
F.D. fan and I.D. fan manufacturer: Buffalo Forge
Generating equipment manufacturer: Unknown

5. Wood Waste: Wright City, Oklahoma

By: Forest Products Research Society, Madison, Wisconsin
                  Proceedings No. P-80-26
Title: Energy Generation and Cogeneration from Wood
Paper title: CoGeneration in a Wood Products Mill, A. L. Vraspir, Project Mgr.
                  Weyerhaeuser Company, Tacoma, Washington

Date Published: 1980

Project location: Weyerhaeuser Forest Products Facility, Wright City Plant, Wright
                      City, Oklahoma

Project size: 5000 kW

Project installed cost: $8.5 million, U. S.

Project on line: 1979

Operating schedule: 355 days/yr

Fuel: Biomass, wood chips, hogged bark, sander dust, etc.

Fuel moisture content: 35–50%

Ton/day of hogged fuel,   380 (approximately 211 bone dry tons)
  chips, etc:

MW of electricity generated: 4.5 MW, average; 5.2 MW, maximum

Boiler design data:
  Capacity: 120,000 lb/hr on wood only

  Design pressure: 900 psig; operating pressure: 850 psig

  Final steam temperature: 825°F; Feedwater temperature: 250°F

Wood firing:
  Gas temperature leaving boiler: unknown
  Gas temperature leaving air preheater: 350°F; Efficiency: 71% (approximate)

Boiler manufacturer:	Erie City Energy Division, Zurn Industries
Traveling grate manufacturer:	Unknown
F.D. fan and I.D. fan manufacturer:	Unknown
Generating equipment manufacturer:	Generator: Electric Machinery
	Turbine: Turbodyne 150 psig back-pressure
	exhaust

Fuel savings per year: Approximately $800,000.00 (minimum) (at $1/mcf-nat.gas)

Electricity savings per year: Approximately $420,000.00 (minimum) (at $0.02/kWh)

Note: Operator training time per operator: 300 hr, classroom and field
    three operators per shift
    Unscheduled downtime since start-up is less than 1%

For more information see Tables 10.1–10.6 and Figure 10-16.

III. COGENERATION REPORT: SURVEY OF COGENERATION IN TEXAS: GULF COAST COGENERATION ASSOCIATION, HOUSTON, TEXAS

A. Summary

The results of a survey conducted by the Gulf Coast Cogeneration Association (GCCA) show that, for the 30 installations in Texas currently included in the survey, the availability and capacity factors have averaged 96 and 84%, respectively, over their operating lifetimes. This performance meets or exceeds that of central station utility power generation. The survey covers 30 systems that were placed in service between 1929 and 1986.

These results also demonstrate that the cogeneration systems continued to operate through cycles of business conditions and fuel-pricing changes during the 1970s and 1980s and are operating today, sometimes under much different conditions than originally envisioned by the project initiators.

The majority of the projects in the survey were designed to reduce purchased power cost, although some large projects were built primarily for firm power sales to utilities.

Overall, cogeneration appears to be both reliable and cost effective under a variety of economic conditions.

TABLE 10.1

Boiler House Notes:					Page No.:		
By: Ralph L. Vandagriff				Estimated Steam Turbine/Generator Output			
Date: 6/25/97							
File: EstTGOutputr1				Customer:			
				Location:			
	Typical Efficiencies for Steam Turbines						
	Single Stage	5 Stage	7 Stage	9 Stage			
Efficiency:	30.0%	55.0%	65.0%	75.0%			
Turbine No.				One	Two	Three	Four
	INPUT VARIABLES						
Steam Flow: Lb./Hour:				200,000	200,000	75,000	75,000
	Turbine Inlet						
Steam Pressure in:Psig				850	850	600	600
Initial Steam Temp:deg. F.				825	825	750	750
Enthalpy @ Inlet:			h1	1410.14	1410.14	1378.92	1378.92
Entropy @ Inlet:				1.599	1.599	1.61	1.61
	Turbine Outlet						
Steam Pressure Out: Psig				240	240	215	215
Estimated Outlet Steam Temp:deg.F.				402.7	496	393.6	470
Estimated: Enthalpy @ Outlet:			h2	1201.3	1260.00	1200	1249
(w/ inlet entropy)				(Saturated)	(Superheat)	(Saturated)	(Superheat)
Estimated Generator Efficiency:			ng	90.0%	90.0%	90.0%	90.0%
Estimated Turbine Efficiency:			nt	70.0%	70.0%	70.0%	70.0%
	RESULTS						
Theoretical Steam Rate:				16.3414576	22.7304516	19.0741672	26.2680881
3412.75/(h1-h2)							
Actual Steam Rate: lb/Kwh				25.9388215	36.0800819	30.2764559	41.6953779
3412.75/(nt x (h1-h2))/ng							
Estimated Kw/Hour Generated:				7,710	5,543	2,477	1,799
	Estimated Cost of Generated Electricity						
		Operating Hours per Year		8,700			
Assumptions:		Natural Gas Cost per MCF		$2.50			
Lb. per hour or steam:				200,000	200,000	75,000	75,000
Estimated Cost of High Pressure/Superheat Boiler				$1,350,000	$1,350,000	$550,000	$550,000
Estimated Cost of Standard Pressure/Saturated Boiler				$1,100,000	$1,100,000	$450,000	$450,000
Estimated Cost of Back pressure turbine generator set				$2,340,000	$2,340,000	$500,000	$500,000
Estimated Kw/hour generated				7,710	6,543	2,477	1,799
Plant saturated steam conditions				240/sat	240/sat	215/sat	215/sat
Steam enthalpy:				1201.3	1201.3	1200	1200
Proposed high pressure/superheated steam conditons:				850/825F	850/825F	600/750F	600/750F
Steam enthalpy:				1410.14	1410.14	1378.92	1378.92
Boiler feedwater temperature, F:				250	250	250	250
Estimated Boiler combustion efficiency:				83.00%	83.00%	84.00%	84.00%
Annual cost of fuel; plant saturated steam conditions:				$5,153,440	$5,153,440	$1,907,009	$1,907,009
Annual cost of fuel; High pressure/superheated conditions:				$6,247,963	$6,247,963	$2,254,465	$2,254,465
Difference in cost of fuel:				$1,094,523	$1,094,523	$347,456	$347,456
Annual KwH generated:				67,080,919	48,226,055	21,551,400	15,649,217
Annual Additional Costs:		Turbine Generator Set(10Yr.)		$234,000	$234,000	$50,000	$50,000
		Difference:HPBlr.vsStd.Blr.(10yr)		$25,000	$25,000	$10,000	$10,000
		Annual extra fuel cost		$1,094,523	$1,094,523	$347,456	$347,456
		Total:		$1,353,523	$1,353,523	$407,456	$407,456
KW Cost = Total Extra Annual Cost divided by KwH				$0.020177	$0.028066	$0.018906	$0.026037

TABLE 10.2

BOILER HOUSE NOTES:
Steam Turbine: Theoretical Pounds of Steam per kWh

Ralph L. Vandagriff   5/22/98

									Page No. re:	No. 8
Initial Pressure, psig	150	250	400	600	600	850	850	900	900
Initial Temperature, deg. F.	365.9	500.0	650.0	750.0	825.0	825.0	900.0	825.0	900.0
Initial Superheat, deg. F	0.0	94.0	201.9	261.2	336.2	297.8	372.8	291.1	366.1
Initial Enthalpy, Btu/lb.	1195.5	1261.8	1334.9	1379.6	1421.4	1410.6	1453.5	1408.4	1451.6
THEORETICAL STEAM RATES - LB/KWH - CONDENSING										
Turbine Exhaust Pressure Inchs Hg absolute										
2.0	10.520	9.070	7.831	7.083	6.761	6.580	6.282	6.555	6.256
2.5	10.860	9.343	8.037	7.251	6.916	6.723	6.415	6.696	6.388
3.0	11.200	9.582	8.217	7.396	7.052	6.847	6.590	6.819	6.502
4.0	11.760	9.996	8.524	7.644	7.282	7.058	6.726	7.026	6.694
THEORETICAL STEAM RATES - LB/KWH - BACKPRESSURE										
Turbine Exhaust Pressure Lbs./Sq.Inch - Gage										
5	21.69	16.57	13.01	11.05	10.42	9.838	9.288	9.755	9.209
10	23.97	17.90	13.83	11.84	10.95	10.300	9.705	10.202	9.617
20	28.63	20.44	15.33	12.68	11.90	11.100	10.430	10.982	10.327
30	33.69	22.95	16.73	13.63	12.75	11.800	11.080	11.670	10.952
40	39.39	25.52	18.08	14.51	13.54	12.460	11.660	12.304	11.520
50	46.00	28.21	19.42	15.36	14.30	13.070	12.220	12.900	12.060
60	53.90	31.07	20.76	16.18	15.05	13.660	12.740	13.470	12.570
75	69.40	35.77	22.81	17.40	16.16	14.50	13.51	14.28	13.30
80	75.90	37.47	23.51	17.80	16.54	14.78	13.77	14.55	13.55
100		45.21	26.46	19.43	18.05	15.86	14.77	15.59	14.50
125		57.88	30.59	21.56	20.03	17.22	16.04	16.87	15.70
150		76.50	35.40	23.83	22.14	18.61	17.33	18.18	16.91
160		86.80	37.57	24.79	23.03	19.17	17.85	18.71	17.41
175			41.16	26.29	24.43	20.04	18.66	19.52	18.16
200			48.24	29.00	26.95	21.53	20.05	20.91	19.45
250			69.10	35.40	32.89	24.78	23.08	23.90	22.24
300				43.72	40.62	28.50	26.53	27.27	25.37
400				72.20	67.00	38.05	35.43	35.71	33.22
425				84.20	78.30	41.08	38.26	38.33	35.65
600						78.50	73.10	68.11	63.40

TABLE 10.3

File:	TG				Turbine - Generator Sets						Boiler House Notes:		Page:
Run Date:	Feb. 10, 1999				(Projects done since 1981)								
By:	R.L. Vandagriff												
Date:	Supplier	SteamFlow Lb./Hour	Inlet Pr. Psig.	Inlet Temp. Deg. F.	Extraction Lb./Hour	Extr.Pr. Psig	Exhaust Pressure	Ex.Temp Deg.F.	Kw/Hour Generated	Voltage	Cost: $ T/G Set	$/Kw	Turbine WR lb/kw/hour
7/28/81	General Electric	500,000	1,800	1,000	500,000	125	3.5Hga	120	63,400	13,800	$8,610,000.00	$135.80	7.8864
8/26/81	General Electric	500,000	1,800	1,000			125		34,000	13,800	$4,790,000.00	$140.88	14.7059
4/4/90	Dresser-Rand	300,000	900	750			230	410	9,150	13,800	$2,400,000.00	$262.30	32.7869
3/4/90	Dresser-Rand	200,000	900	950			300	460	6,200	4,160	$1,700,000.00	$274.19	32.2581
7/13/90	Dresser-Rand	150,000	850	825			4"HgA	125	15,450	13,800	$3,634,000.00	$235.21	9.7087
4/4/86	Murray	150,000	875	825			380	710	2,800	4,160	$725,000.00	$258.93	53.5714
1/26/93	Murray	150,000	600	730			160	552	4,000	2,400	$980,000.00	$245.00	37.5000
10/26/83	Turbodyne	130,000	850	825			4" HgA	125	13,444	13,800	$2,400,000.00	$178.52	9.6697
10/26/83	Turbodyne	130,000	850	825	120,000	25	4" HgA	125	8,677	13,800	$2,400,000.00	$276.59	14.9821
10/26/83	Turbodyne	130,000	850	825	90,000	25	4" HgA	125	9,869	13,800	$2,400,000.00	$243.19	13.1726
10/26/83	Turbodyne	130,000	850	825	60,000	25	4" HgA	125	11,061	13,800	$2,400,000.00	$215.98	11.7530
3/4/90	Dresser-Rand	120,000	900	950			125	362	5,800	480	$1,500,000.00	$258.62	20.6897
8/11/95	Murray	120,000	450	460			170	376	1,260	480	$235,000.00	$186.51	95.2381
8/11/95	Murray	110,000	600	750			235	401	1,550	480	$325,000.00	$209.68	70.9677
6/1/82	General Electric	100,000	875	825	50,000	300	2.5" HgA	109	5,100	4,160	$2,270,000.00	$445.10	19.5078
12/4/81	General Electric	100,000	850	825			125	478	4,090	4,160	$1,620,000.00	$396.09	24.4499
12/4/81	General Electric	100,000	850	825	100,000	125	2.5" HgA	109	11,100	13,800	$2,530,000.00	$227.93	9.0090
0/26/92	Elliott	100,000	600	730			160		2,100	4,160	$312,000.00	$148.57	47.8190
2/26/92	Coppus	100,000	600	730			160	552	1,825	4,160	$350,000.00	$191.78	54.7945
7/13/90	Dresser-Rand	90,000	850	825			4" HgA	125	8,680	13,800	$2,277,000.00	$282.33	10.3687
5/16/83	General Electric	90,000	850	825			2.5" HgA	109	9,850	13,800	$2,930,000.00	$297.46	9.1371
5/16/83	General Electric	90,000	600	750			2.5" HgA	109	9,200	13,800	$2,620,000.00	$306.52	9.7826
7/28/83	Turbodyne	90,000	600	750	90,000	125			2,922	13,800	$1,500,000.00	$513.35	30.8008
7/28/83	Turbodyne	90,000	600	750			4" HgA	125	7,673	13,800	$1,500,000.00	$195.49	11.7294
4/4/86	Murray	72,000	380	710			4" HgA	125	6,000	13,800	$1,100,000.00	$183.33	12.0000
0/22/85	Terry	70,150	600	750	20,000	70	3" HgA	115	5,750	13,800	$1,175,000.00	$204.35	12.2000
0/22/85	Terry	70,070	600	750			3" HgA	115	7,150	13,800	$1,115,000.00	$155.94	9.5000
4/29/83	Turbodyne	65,000	825	850			4" HgA	125	6,813	4,160	$1,600,000.00	$234.85	9.5406
4/29/83	Turbodyne	65,000	825	850	60,000	160	4" HgA	125	2,833	4,160	$1,600,000.00	$564.77	22.9439
3/2/84	Turbodyne	61,000	850	825			4" HGA	125	6,310	4,160	$1,480,000.00	$234.55	9.6672
4/25/84	Terry	60,000	825	850			4" HgA	125	6,510	4,160	$1,130,000.00	$173.58	9.2166
5/3/68	Dresser-Rand	60,000	600	700			70	449	2,339	480	$411,832.00	$176.07	25.6520
5/3/68	Dresser-Rand	60,000	400	449			4" HgA	125	3,491	480	$557,877.00	$159.75	17.1871
3/2/84	Turbodyne	50,000	850	825			4" HgA	125	5,150	4,160	$1,480,000.00	$287.38	9.7087
4/29/91	Turbodyne	50,000	600	750			15	302	2,750	480	$600,000.00	$218.18	18.1818
4/29/91	Turbodyne	50,000	600	750	\		125	490	1,500	480	$325,000.00	$216.67	33.3333
4/31/84	Skinner	50,000	600	650			140	445	1,040	480	$105,100.00	$101.06	48.0769
1/14/95	Murray	50,000	450	460			125	353	670	480	$205,000.00	$305.97	74.6269
4/22/85	Terry	49,960	600	750			3" HgA	115	4,900	13,800	$895,000.00	$182.65	10.2000
1/26/92	Coppus	49,000	380	710			180	615	510	2,400	$215,000.00	$421.57	96.0784
4/29/84	Terry	48,011	270	413			15	250	1,480	480	$290,700.00	$196.42	32.4399
7/21/86	Coppus	36,000	165	373			50	298	365	480	$105,000.00	$272.73	93.5065
90/89	Murray	35,000	230	399			125	353	237	480	$80,000.00	$337.55	147.6793
90/89	Murray	35,000	230	399			4" HgA	125	2,150	480	$450,000.00	$209.30	16.2791
90/89	Murray	35,000	230	399			4" HgA	125	1,940	480	$420,000.00	$216.49	18.0412
90/89	Dresser-Rand	35,000	230	399			125	353	232	480	$81,923.00	$353.12	150.8621
90/89	Dresser-Rand	35,000	230	399			4" HgA	125	1,977	480	$498,591.00	$252.20	17.7036
90/89	Dresser-Rand	35,000	230	399			4" HgA	125	1,829	480	$505,516.00	$276.39	19.1361
1/4/82	Turbodyne	32,000	600	750	32,000	120			893	480	$199,000.00	$222.84	35.8343
4/86	Coppus	31,000	380	710			180	400	350	480	$108,500.00	$310.00	88.5714
1/83	Terry	30,000	700	750			140	361	641	480	$68,000.00	$106.08	46.8019
15/83	Turbodyne	30,000	600	750			3.5" HgA		2,500	480	$537,000.00	$214.80	12.0000
15/83	Turbodyne	30,000	600	750	30,000	140			575	480	$537,000.00	$933.91	52.1739
23/83	Skinner	30,000	600	750			120	350	1,046	480	$305,000.00	$291.59	28.6807
10/83	Coppus	27,000	600	486			15	250	750	480	$89,500.00	$92.67	36.0000
14/95	Murray	20,700	450	460			125	353	265	480	$133,500.00	$503.77	78.1132
14/84	Skinner	16,822	285	417			15	250	400	480	$64,500.00	$161.25	47.0550
90/89	Murray	17,500	230	399			4" HgA	125	960	480	$320,000.00	$333.33	18.2292
90/89	Murray	17,500	230	399			125	353	90	480	$70,000.00	$777.78	194.4444
90/89	Dresser-Rand	17,500	230	399			125	353	110	480	$78,600.00	$714.55	159.0909
90/89	Dresser-Rand	17,500	230	399			4"HgA	125	936	480	$367,307.00	$392.42	18.6966
90/89	Dresser-Rand	17,500	125	353			4"HgA	125	854	480	$353,118.00	$413.49	20.4918
14/89	Dresser-Rand	16,320	20	259			4" HgA	125	528	480	$295,000.00	$558.71	30.9091
14/89	Dresser-Rand	16,320	20	259			4" HgA	125	337	480	$178,000.00	$528.19	48.4273
15/63	Skinner	16,000	600	750			0	212	850	480	$60,000.00	$94.12	18.6235
6/84	Skinner	14,426	600	750			0	212	791	480	$89,400.00	$113.02	18.2377
2/84	Skinner	12,992	200	387			10	239	282	480	$41,200.00	$146.10	46.0709
4/83	Skinner	12,700	600	750			4" HgA	125	800	480	$66,100.00	$82.63	15.8750
1/85	Murray	12,000	300	421			10"HgA	192	375	480			32.0000
1/85	Murray	14,000	300	421			100 psig	338	155	480			90.3226
4/84	Skinner	4,190	200	388			10	239	96	480	$36,600.00	$381.25	43.6458
3/84	Skinner	3,306	250	406			6	230	80	480	$40,100.00	$501.25	41.3250

GAS TURBINES - Currently Available - 1998 - (Partial List)

By: Ralph L. Vandagriff   Date: 6/16/98   File: GTGT1

Capacity (kw)	Manufacturer	Model No.	Compression Ratio	Required Gas Pressure	Rating Condition	Turbine Speed (RPM)	Heat Rate (KJ/Kw-Hr)	Heat Rate (Btu/Kw-Hr)	Efficiency (Approximate) (3413/HR)	Turbine Inlet Temp (F)	Exhaust Temp (C)	Exhaust Temp (F)	Exhaust Flow (Kg/Sec.)	Exhaust Flow (lb/sec.)	Fuel	Fuel Cost/Kw w/ $2.00 per MCF Net Gas	Unfired Waste Approximate Steam #/Hr. 125psig/sat Feedwater
1,140	Solar	Saturn 20		180	ISO	15000/1800	14,400	14,075	24.25%	60-100	530	907	10.90	24.03	Gas	$0.028150	7,994
2,000	General Electric	PGT 2	2.8:1	400	ISO			13,648	25.01%	60-100		1,022		28.40	Gas	$0.027297	15,319
2,790	U.S. Turbine	UST 2800	8.5:1		ISO	12,830		13,476	25.33%	60-100		1,053		33.20	Gas	$0.029952	19,212
3,424	U.S. Turbine	UST 3500	9.3:1		ISO	14,589		12,854	26.55%	60-100		1,002		41.03	Gas	$0.025708	22,037
3,515	Solar	Centaur 40		200	ISO			12,240	27.89%	60-100		819		35.30	Gas	$0.024480	18,916
3,952	U.S. Turbine	UST 4000	10.2:1		ISO	14,589		12,256	27.45%	60-100		1,069		44.09	Gas	$0.024512	24,438
4,400	Dresser-Rand	DR-990	12.4:1		ISO	17,800	11,813	11,196	30.48%	60-100	491	916	20.00	47.11	Gas	$0.022393	23,925
5,000	Solar	Taurus 60			ISO			11,230	30.34%	60-100		895		44.40	Gas	$0.022500	28,062
5,025	U.S. Turbine	UST 5000	12.7:1	400	ISO	14,589		11,223	30.41%	60-100	523	997	24.40	53.79	Gas	$0.022446	27,580
5,220	General Electric	PGT 5	9.2:1	400	ISO	15000/1800	13,380	12,682	26.91%	60-100		973		52.00	Gas	$0.023363	32,182
6,412	US Turbine	UST 6500	10.2:1		ISO	11,500		10,583	32.25%	60-100		985		40.70	Gas	$0.021166	31,697
6,464	US Turbine	UST 6600	10.2:1		ISO	14,600		8,481	40.24%	60-100		1,045		56.23	Gas	$0.018962	27,211
6,800	Solar	Taurus 70			ISO			10,900	31.31%	60-100		910		31.80	Gas	$0.021600	31,119
7,814	U.S. Turbine	UST 7700	13.5:1		ISO	14,800		8,619	39.80%	60-100		995		66.40	Gas	$0.017238	32,077
7,887	U.S. Turbine	UST 8000	19.4:1		ISO	11,500		10,527	32.42%	60-100		940		86.35	Gas	$0.021054	37,577
9,290	Solar	Mars 90			ISO			10,765	31.70%	60-100		868		92.81	Gas	$0.021530	43,762
10,140	General Electric	PGT 10	14.0:1	400	ISO	7,800	11,860	11,051	30.88%	60-100	464	903	42.10	91.81	Gas	$0.022103	49,194
10,695	Solar	Mars 100			ISO			10,505	32.49%	60-100		910		100.31	Gas	$0.021010	50,509
13,390	General Electric	PGT 16	21.5:1	450	ISO	7,800	10,230	9,696	35.20%	60-100	493	919	43.50	97.48	Gas	$0.019392	54,726
13,090	Dresser-Rand (GE) (LM-1600)	DR-606	22.3:1	460	ISO	12,827	9,886	9,180	27.18%	60-100	487	909	44.40	149.91	Dual	$0.018361	32,452
21,810	General Electric	PGT 25	17.8:1	400	ISO	6,500	10,155	9,625	35.46%	60-100		975		149.91	Gas	$0.019250	89,927
22,350	General Electric	LM 2500	18.0:1	400	ISO	3,600	9,970	9,450	36.12%	60-100		973		148.59	Gas	$0.018699	89,927
25,100	Dresser-Rand (GE) (LM 2500)	DR-61	18.1:1	400	ISO	9,340	9,608	9,106	37.48%	60-100	536	997		148.59	Gas	$0.018213	92,305
29,260	Dresser-Rand (GE) (LM 2500)	DR-61G	18.1:1	400	ISO	9,340	9,388	9,088	37.56%	60-100	523	973			Gas	$0.018175	88,848
29,300	General Electric	MS 5001	10.5:1	375	ISO	5,094	12,650	11,990	28.47%	60-100	487	909	122.50	270.06	Gas	$0.023979	144,699
27,500	Dresser-Rand (GE) (LM2500+)	DR-61G Plus	22.1:1	430	ISO	9,340	9,646	9,144	37.32%	60-100		927	83.10	183.20	Gas	$0.018289	101,371
36,340	General Electric	MS 6001	11.8:1	375	ISO	5,094	11,460	10,862	31.42%	60-100	539	1,002	137.00	302.03	Gas	$0.021724	189,070
	General Electric	LM 6000	30.0:1	500	ISO	9,800	9,230	8,747	38.93%	60-100	519	946	122.80	270.95	Gas	$0.017534	160,168

TABLE 10.4

TABLE 10.5a

Date: 1/26/98R3									Boiler House Notes:		Page:		
GAS TURBINES													
By: Ralph L. Vandagriff			File: GT										
Capacity (kw)	Manufacturer	Model No.	Compression Ratio	Required Gas Pressure	Rating Condition	Turbine Speed (RPM)	Heat Rate (Btu/kwh)	Efficiency (Approximate) (3413/HR)	Exhaust Heat (Btu/kwh)	Turbine Inlet Temp.(F)	Exhaust Temp. (F)	Exhaust Flow (lb/sec.)	Fuel
80	Klock./Hum./Deutz	TA 216	2.6:1		ISO	3,600	30,925	11.04%			1,202	2.00	Gas
101	Turbomeca	OREDON IV	3.6:1		ISO	6,000	22,260	15.33%			1,087	2.00	Gas
205	Kawasaki H.I.	S1A-02	9.0:1		ISO	1,800	21,000	16.26%		1,670	968	3.90	Gas
205	Kawasaki H.I.	S1B-02	9.0:1		ISO	3,600	21,000	16.25%		1,670	968	3.90	Gas
210	TurboSystems Intl.	TG-5	9.0:1		ISO	53,000	19,666	17.36%			968	4.00	Gas
220	Klock./Hum./Deutz	KA 215	9.0:1		ISO	1,600	18,672	18.20%			964	4.00	Gas
294	Noel Penny Turbines	NPT 403	5.8:1		ISO	3,600	20,000	17.07%		1,712	1,080	5.00	Gas
330	Turbomeca	Astazou IV	5.6:1		ISO	43,500	19,090	17.96%			914	6.00	Dual
480	Norwalk	TG-7	7.0:1	135	ISO		18,400	18.56%			1,050	6.00	Gas
510	Onan	560GTU	11.0:1	200	ISO		17,100	19.96%			923	8.00	Gas
545	Garrett	IM831-800	10.6:1	210	ISO	41,730	17,100	19.96%			1,080	8.00	Gas
565	Centrix Gas Turbine	CS 600-2	5.25:1		ISO	22,000	15,890	21.48%			977	11.00	Gas
750	Turbomeca	Turmo III	4.9:1		ISO	31,100	19,750	17.28%			950	12.00	Dual
1,000	Turbomeca	Turmo XII	8.2:1		ISO	31,240	16,350	20.87%			842	16.00	Dual
1,120	Avco Lycoming	TF 16	13.3:1		ISO	3,000	11,545	29.56%			895	11.50	Gas
1,140	Solar	Saturn 20		180	ISO		14,075	24.25%			980	14.29	Gas
1,295	UST/Kawasaki	UST 1200	9.3:1		ISO	22,000	14,617	23.86%			858	18.10	Gas
1,277	Ruston	TA 1750	4.5:1		ISO	6,000	20,410	16.72%		1,517	950	25.00	Gas
1,390	Dresser-Rand	KG2-3R	3.9:1		ISO	18,000	13,190	25.88%		1,517	572	27.80	Gas
1,400	Konigs./Dresser	KG-3	9.0:1		ISO	35,000	13,100	26.06%		2,012	1,091	14.30	Gas
1,440	TurboSystems Intl.	TG-9BT	8.2:1		ISO	26,500	14,805	23.06%		1,340	1,055	16.00	Gas
1,472	UST/Kawasaki	UST 1500	9.4:1		ISO	22,000	14,100	24.21%			995	17.90	Gas
1,475	Dresser-Rand	KG2-3C	3.9:1		ISO	18,000	21,620	15.79%		1,517	1,049	26.40	Gas
1,840	Kongsberg	KG 3	9.0:1		ISO	35,000	14,620	23.34%			1,034	15.00	Dual
1,840	Dresser-Rand	KG2-3E	9.0:1		ISO	16,000	19,965	17.08%			1,020	33.00	Gas
1,995	Ruston	TA2500	5.1:1		ISO	11,800	17,570	19.10%			940	29.00	Gas
2,000	General Electric	PGT 2	2.8:1	400	ISO	1500/1800	13,648	25.01%	60 - 100		1,022	24.03	Gas
2,043	UST/Kawasaki	UST 2100	11.2:1		ISO	22,000	13,878	24.96%			1,087	20.50	Gas
2,260	Klock./Hum./Deutz	KT 134	8.0:1		ISO		16,187	21.06%			941	34.60	Gas
2,365	UST/Kawasaki	UST250000	9.0:1		ISO	22,000	10,450	32.66%			1,058	18.90	Gas
2,695	UST/Allison	501-KB3	6.5:1		ISO	12,850	13,522	25.24%			1,076	26.10	Gas
2,795	U.S. Turbine	UST 2600	8.5:1		ISO	12,850	13,476	25.33%			1,053	26.40	Gas
3,000	Dresser-Rand	KG 5			ISO		16,100	21.20%			932	46.50	Gas
3,110	Kongsberg	KG 5	6.5:1		ISO	17,400	17,880	19.09%			898	48.00	Dual
3,424	U.S. Turbine	UST 3500	9.3:1		ISO	14,589	12,854	26.55%			1,002	35.20	Gas
3,460	UST/Allison	501 KB4	9.3:1	250	ISO	14,400	12,768	26.73%			987	34.50	Gas
3,515	Solar	Centaur 40/40S		200	ISO		12,240	27.96%			819	41.03	Gas
3,568	Ruston	TB5000	6.8:1		ISO	10,400	14,860	22.97%			918	46.00	Gas
3,944	UST/Allison	501 KB5 S	10.2:1	250	ISO	14,400	12,236	27.89%			1,072	34.90	Gas
3,982	U.S. Turbine	UST 4000	10.2:1		ISO	14,589	12,256	27.86%			1,060	35.30	Gas
4,064	General Electric	LM 500	14.5:1		ISO	16,000	12,010	28.42%			955	35.00	Gas
4,220	Dresser-Rand	DR-990	12.0:1		ISO	17,800	11,780	28.97%			898	45.00	Gas
4,350	Solar	Centaur 50/50S		200	ISO		11,866	28.76%			927	41.93	Gas
4,400	Dresser - Rand	DR - 990	12.4:1		ISO	17,900	11,196	30.48%	60-100		916	44.09	Gas
4,610	Allison	501-KB7			ISO	14,600	12,120	28.16%			996	44.90	Gas
4,877	Allison	570 KA	12.7:1	275	ISO	9,000	12,810	26.64%			1,049	42.00	Gas
4,907	EuropeanGasTurbines	Typhoon			ISO		11,142	30.63%					
5,000	Solar	Taurus 60/60S			ISO		11,250	30.34%			896	47.11	Gas
5,025	U.S. Turbine	UST 5000	12.7:1		ISO	14,589	11,223	30.41%			997	44.40	Gas
5,220	General Electric	PGT 5	9.2:1	400	ISO	1500/1600	12,682	26.91%	60-100		973	53.79	Gas
5,365	UST/Allison	571 KA	12.7:1	275	ISO	11,500	11,287	30.37%			1,065	49.20	Gas
5,797	UST/Allison	501-KH	9.3:1		ISO	14,600	9,249	36.90%			963	40.70	Gas
6,230	Ruston	Tornado	12.0:1	155	ISO	11,085	12,120	28.16%			877	60.00	Dual

B. Introduction

There has been considerable testimony at the Public Utility Commission of Texas and other public statements (especially in newspaper articles) that picture cogeneration as being unreliable and likely to shutdown in the future, causing a negative impact on ratepayers. These statements are speculative and, therefore, difficult to prove or disprove. Accordingly, the Gulf Coast Cogeneration Association felt that the public would be better served if factual information about the historical

TABLE 10.5b

GAS TURBINES, continued;													
By: Ralph L. Vandagriff			File: GT										
Capacity (kw)	Manufacturer	Model No.	Compression Ratio	Required Gas Pressure	Rating Condition	Turbine Speed (RPM)	Heat Rate (Btu/kwh)	Efficiency (Approximate) (3413/HR)	Exhaust Heat (Btu/kwh)	Turbine Inlet Temp.(F)	Exhaust Temp. (F)	Exhaust Flow (lb/sec.)	Fuel
6,249	EuropeanGasTurbines	Tornado			ISO		11,285	30.80%					Gas
6,300	Solar	Taurus 70/70S			ISO		10,900	31.31%			910	56.23	Gas
6,392	UST/Allison	501-KH 5	10.2:1		ISO	14,600	9,245	36.92%			1,090	41.10	Gas
6,412	U.S. Turbine	UST 6500	10.2:1		ISO	11,500	10,583	32.28%		60-100	965	52.00	Gas
6,496	U.S. Turbine	UST 8800	10.2:1		ISO	14,600	8,461	40.34%			1,045	40.70	Gas
6,800	Solar	Taurus 70			ISO		10,900	31.31%		60-100	910	56.23	Gas
7,490	EuropeanGasTurbines	Tempest			ISO		10,876	31.38%					
7,614	U.S. Turbine	UST 7700	13.5:1		ISO	14,600	8,619	39.60%			995	51.80	Gas
7,697	U.S. Turbine	UST 8000	19.4:1		ISO	11,500	10,527	32.42%			940	66.40	Gas
8,050	Westinghouse	W101PG				6,000	15,340	22.26%	9,550				
8,630	Fiat TTG	TG 7	8.0:1		ISO	6,000	15,680	21.77%			800	134.00	Dual
8,670	UST/	THM 1304D	10.0:1		ISO	6,000	12,695	26.89%			959	100.80	Gas
9,225	Mitsubishi	MW-101	8.0:1		ISO	6,000	15,770	21.64%			772	151.00	Gas
9,290	Solar	Mars 90/90S			ISO		10,785	31.70%			888	86.35	Gas
9,760	General Electric	G3142 RECUP				6,500	10,520	32.44%	5,040				
9,760	Hispano-Suiza	THM 1304	10.0:1		ISO	12,000	13,070	26.11%			959	101.00	Dual
10,140	General Electric	PGT 10	14.0:1	400	ISO	7,900	11,051	30.88%		60-100	903	92.81	Gas
10,200	General Electric	G3142				6,500	13,540	25.21%	8,020				
10,480	General Electric	MS3002J	7.1:1		ISO	6,500	14,800	23.08%			979	115.00	Gas
10,895	Solar	Mars 100/100S		490	ISO		10,505	32.49%			910	91.81	Gas
11,568	Rolls Royce	SPEY				5,900	10,560	32.32%	4,210				
11,802	Rolls Royce	AVON78G				5,000	12,720	26.83%	5,330				
12,066	Coberra	2348			ISO	5,200	13,720	24.89%			766	166.00	Gas
12,529	UST/	MF-111A	12.6:1		ISO	9,880	11,285	30.24%			1,042	107.70	Gas
12,700	Ishikawajima	IM2000			ISO	3,700	11,240	30.36%			773	126.00	Gas
12,990	General Electric	LM 2500-2D	7.0:1	218			11,580	29.47%			766	125.00	Gas
13,002	Rolls Royce	AVON101G				5,200	12,910	26.44%	5,650				
13,070	Mitsui Engineering	5860	12.4:1		ISO	6,760	12,360	27.61%			671	129.00	Gas
13,360	General Electric	PGT 16	21.5:1	450	ISO	7,900	9,896	36.20%		60-100	919	100.31	Gas
13,440	General Electric	LM 1600			ISO	7,000	9,545	36.76%			909	100.00	Gas
13,390	Dresser-Rand	DR-60G	22.3:1	480	ISO	12,827	9,180	37.18%		60-100	909	97.66	Gas
14,476	UST/	MF-111B	14.7:1		ISO	9,880	11,053	30.88%			1,011	125.20	Gas
14,718	Rolls Royce	AVON121G				4,760	12,250	27.86%	5,490				
15,428	Dresser Clark	DF-200R			ISO	5,400	12,930	26.40%			861	170.00	Gas
15,554	Coberra	2556			ISO	4,950	12,880	26.50%			875	169.00	Gas
15,810	Mitsui Engineering	5890	6.9:1		ISO	5,475	14,260	23.95%			857	189.00	Gas
16,890	UST/	MF-111AB	14.7:1		ISO	9,880	11,000	31.03%			1,075	114.00	Gas
17,700	Westinghouse	W191PG	7.5:1		ISO	4,912	14,870	22.95%			780	270.00	Gas
18,565	Mitsubishi	MJ-191	7.0:1		ISO	4,912	14,390	23.72%			748	270.00	Gas
18,860	Fiat TTG	TG 16	7.0:1		ISO	4,650	14,120	24.17%			766	262.00	Gas
19,191	Rolls Royce	RB211-22				4,760	10,310	33.10%	4,010				
20,000	Curtis Wright	P0020	18.0:1	400	ISO		10,890	31.34%			688	144.00	Gas
20,580	Dresser Clark	DJ-125R			ISO	5330	13,180	25.90%			845	180.00	Gas
20,920	Rolls Royce	RB211-24A				5000	10,220	33.40%	4,100				
21,100	Ishikawajima	IM2500			ISO	3700	10,520	32.44%			868	147.00	Gas
21,587	Rolls Royce	RB211-24B				5000	10,160	33.58%	4,100				
21,793	Dresser Clark	DJ-270G			ISO	5400	10,480	32.57%			965	149.00	Gas
21,910	General Electric	PGT 25	17.8:1	400	ISO	6500	9,625	35.46%		60-100	975	149.91	Gas
22,190	Mitsubishi	MW-151	11.0:1		ISO	6546	12,980	26.29%			959	212.00	Gas
22,560	Dresser-Rand	DR-61			ISO		9,450	36.12%			966	152.00	Gas
22,900	General Electric	LM 2500			ISO	3600	9,280	36.76%			975	152.00	Gas
24,797	Coberra	6456			ISO	4950	10,680	31.96%			887	198.00	Gas
25,364	Coberra	6482			ISO	4800	10,430	32.72%			887	198.00	Gas
25,564	Dresser Clark	DJ-290G			ISO	5000	10,470	32.60%			870	199.00	Gas
25,900	General Electric	MS5001P	10.2:1		ISO	5105	13,450	25.38%			916	267.00	Gas

operating performance of existing cogeneration systems was collected and presented.

It should be emphasized that the results of the GCCA cogeneration survey presented in this report represent a first pass at defining the operating history of cogeneration in Texas. The GCCA intends to continue to add to and update the basic survey results as more companies participate in the survey, new projects come on-line, and the old projects continue their excellent operations.

TABLE 10.6

Gas Engines						
Ralph L. Vandagriff			**Boiler House Notes**			Page:
date: 5/8/96						
Natural Gas @ 1000.00 Btu/cu.ft.						
KwH	Manufacturer	Model No.	Natural Gas Cu.Ft./Hour	Btu/KwH	Exhaust Lb./Hour	Exhaust Temp.,deg.F.
74	Cummins	G-495	942	12,729.7		
85	Caterpillar	3306-LCR	1,188	13,976.5		
100	Caterpillar	3306-HCR	1,252	12,520.0		
117	Cummins	G-743	1,405	12,008.5		
135	Caterpillar	3306-TALC	1,744	12,918.5		
135	Caterpillar	G342-LCR	1,762	13,051.9		
141	Cummins	G-855	1,646	11,673.8		
150	Caterpillar	3306-TAHC	1,783	11,886.7		
150	Caterpillar	G342-HCR	1,801	12,006.7		
171	Cummins	GTA-743	1,997	11,678.4		
179	Cummins	GTA-855-A	2,188	12,223.5		
185	Caterpillar	G342-TALC	2,347	12,686.5		
200	Caterpillar	G342-TAHC	2,378	11,890.0		
205	Caterpillar	G379-LCR	2,836	13,834.1		
219	Cummins	GTA-855B	2,625	11,988.3		
230	Caterpillar	G379-HCR	2,776	12,069.6		
255	White Superior	6G-510	3,132	12,282.4	2580	1200
278	Cummins	GTA-1150	3,279	11,795.0		
290	Waukesha	2900G	3,100	10,689.7		
300	Caterpillar	G379-TALC	3,797	12,656.7		
315	Caterpillar	G398-LCR	4,075	12,936.5		
325	Caterpillar	G379-TAHC	3,831	11,787.7		
350	Caterpillar	G398-HCR	4,079	11,654.3		
360	Waukesha	3600G	3,800	10,555.6	2490	1080
415	Caterpillar	G399-LCR	5,709	13,756.6		
425	White Superior	6G-825	4,650	10,941.2	5220	1250
450	Caterpillar	G398-TALC	5,600	12,444.4		
450	Waukesha	2900GSI	5,000	11,111.1	3110	1100
460	Caterpillar	G399-HCR	5,560	12,087.0		
492	Cummins	GTA-1710	5,251	10,672.8		
500	Caterpillar	G398-TAHC	6,073	12,146.0		
500	Waukesha	5200G	5,500	11,000.0	3440	1020
503	White Superior	40-X-6	4,906	9,756.5	8400	760
503	White Superior	40-GDX-6	4,629	9,203.2	7200	740
550	Waukesha	3600GSI	6,200	11,272.7	3920	1130
565	Caterpillar	G399-TALC	7,233	12,801.8		
566	White Superior	8G-825	6,200	10,954.1	6300	1330
584	White Superior	8GTLA/B	5,940	10,171.2	9060	1058
595	Waukesha	5900G	6,300	10,588.2	4050	1040
615	Superior-Cooper	6GTLB	5,940	9,658.5	8940	1058
650	Caterpillar	G399-TAHC	7,700	11,846.2		
669	White Superior	40-GDX-8	6,143	9,181.6	9600	740
725	Waukesha	7100G	7,500	10,344.8	4960	1060
779	White Superior	8GTLA/B	7,920	10,166.9	11760	955
800	Waukesha	5200GSI	9,000	11,250.0	5340	1070
820	Superior-Cooper	8GTLB	7,920	9,658.5	12120	955
850	White Superior	12G-825	9,300	10,941.2	9960	1240
900	Waukesha	5900GSI	9,900	11,000.0	6170	1100
900	Waukesha	VHP 5900	10,380	11,533.3		
975	Waukesha	9500G	10,900	11,179.5	7340	1180
1,000	White Superior	8SGTB	9,653	9,652.5	13080	1015
1,100	Waukesha	7100GSI	12,200	11,090.9	7730	1130
1,133	White Superior	16G-825	12,400	10,944.4	12660	1330
1,168	White Superior	40-X-12	11,253	9,634.4	18000	725
1,416	White Superior	12SGT	14,400	10,169.5	17100	880
1,475	Waukesha	9500GSI	15,900	10,779.7	11100	1220
1,556	White Superior	40-X-16	15,004	9,630.3	24000	725
1,877	White Superior	16SGTA/B	18,815	10,024.0	28080	765
2,635	DeLaval	R-46	22,500	8,538.9	37200	885
3,100	Cooper Bessemer	KSV12-GDT	27,488	8,867.1	45900	865
3,339	DeLaval	R5-L6	28,768	8,615.8	52313	870
3,515	DeLaval	R-46	30,000	8,534.9	49600	885
3,625	Cooper Bessemer	LSVB12SGC	30,351	8,372.7	48000	900
4,125	Cooper Bessemer	KSV16-SGC	35,020	8,489.7	54400	910
4,454	DeLaval	R5-L8	38,378	8,616.5	69789	870
4,850	Cooper Bessemer	LSVB18SGC	40,468	8,343.9	64000	900
5,175	Cooper Bessemer	KSV20-SGC	43,826	8,468.8	68000	910
6,050	Cooper Bessemer	LSVB20SGC	50,585	8,361.2	80000	900

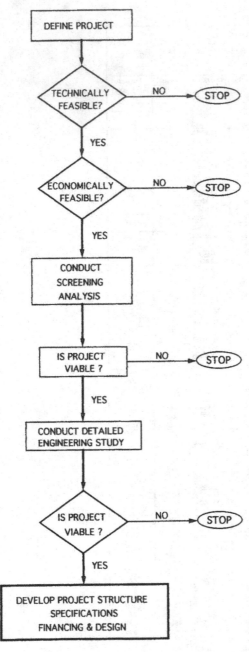

FIGURE 10.16 Recommended cogeneration decision making process.

The survey covers cogeneration systems that have been in service since 1929. Fuels used include bark wood waste, rice hulls, diesel, no. 2 fuel oil, kraft pulping liquor, natural gas, and by-product gas. Natural gas is the dominant fuel.

The GCCA attempted to locate and contact as many pre-PURPA cogenerators as possible, but there is a lot of this "old" cogeneration that is yet to be included in the survey.

C. Discussion

1. General

The purpose of the GCCA survey was to determine some of the historic operating and reliability characteristics of cogeneration systems in Texas. The survey results were broken down into three types of systems: (1) engine–heat recovery, (2) boiler–steam turbine, and (3) gas turbine–heat recovery (with and without combined cycle). The engine–heat recovery systems were primarily used in institutional applications. The steam turbine and gas turbine categories were primarily industrial; however, some smaller gas turbines were used in institutional applications. A gas turbine–combined cycle "system" composed of more than one gas turbine–heat recovery boiler set and operated as a group or unit were sometimes reported as one installation.

2. Survey Results

The results of the survey are summarized in Table 10.7. There were a total of 30 projects reported: 5 engines, 10 boiler–steam turbines (STG), and 15 gas turbines–combined cycle (GTG/GTGCC). Projects with less than 1 year of operation were excluded from the survey. year of start-up varies from 1929 to 1986.

The *availability factor* is defined simply as the percentage of hours that the system was available for operation since start-up. The average availability factor for engine systems is 88%, for GTG/GTGCC is 96%, and for STG systems is 97%. The *capacity factor* is defined as the average annual energy produced in kilowatt-hours divided by the product of the rated capacity of the system times 8760 hr/yr. The average capacity factor for engine systems is 41%, for STG 75%, and for GTG/GTGCC systems is 85%. Overall, the availability of cogenerated power, as indicated by the survey is 96%, and the capacity factor is 84%. The engine system availability is much higher than the capacity factor because the projects surveyed in this category were operated primarily to meet thermal load. For two projects, the thermal load requirement existed less than 15% of the time. Note that the capacity factor for engine systems without these two projects is 63%, which is felt to be more representative of this category of cogeneration systems.

Of the 30 projects, 87% have operated continuously since start-up and 97% are in operation today. *Continuous* was defined as operation with routine outages

TABLE 10.7 Aggregate Results–GCCA Survey of Cogeneration in Texas

	Engine/heat boiler recovery	Gas turbine/total/steam heat recovery		
		Turbine[a]		
Number of projects	5	10	15	30
Capacity (mw)	18.5	205.6	2902.3	3126.4
Year of initial operation	1978–1985	1929–1985	1954–1986	1929–1986
Availability factor (%)[b]	87.8	96.8	95.7	95.7
Capacity factor (%)[c]	40.5(63.2)[d]	74.5	85.0	84.1
Projects currently in-service or replacement in-service (%)	100.0	100.0	93.0	97.0
Projects in-service since start-up (%)	60.0	100.0	87.0	87.0
Projects for self-generation (%)	100.0	60.0	40.0	53.0
Projects for self-generation–as-available power sales (%)	0.0	30.0	27.0	27.0
Projects for firm–as-available power sales (%)	0.0	10.0	33.0	20.0
Capacity installed (%) by decade				
1920s		1.5		0.1
1930s		9.7		0.6
1940s		31.6		2.1
1950s		27.0	9.2	10.2
1960s		12.2	8.7	8.9
1970s	62.0	0.0	2.3	2.5
1980s	38.0	18.0	79.8	75.6

[a] Includes combined cycle systems.
[b] Defined as percentage of time system was available since start-up.
[c] Defined as average annual energy produced (MWHRS) divided by the product of capacity (rated) and 8760 hr.
[d] Capacity factor excluding two projects with less than 20% capacity factors.

only. Of the individual system categories, 60% of the engine systems have operated continuously, and 100% are currently in service. The STG systems have the best-operating record with 100% of those surveyed continuously operating since start-up and 100% currently in operation. The GTG/GTGCC category is almost as impressive, with 87% operating continuously since start-up and 93% currently in service.

The majority of the projects surveyed were installed to meet internal electrical requirements—some 53% of the projects. An additional 27% of projects surveyed were basically self-generation projects, with some as-available power sales. Only 20% of the projects involved firm power sales, with or without as-available sales. Although this might surprise some, it really reinforces the diversity of approach to cogeneration by different companies (i.e., some want to merely reduce cost of production [self-generators]), whereas others want to compete with utilities to supply the lowest-cost power (firm sales). By category, 100% of the engine systems were self-generation projects, versus 90% for STG systems and 67% for GTG/GTGCC systems.

IV. CONCLUSIONS

The survey addresses several reliability issues raised by critics of cogeneration:

1. One such issue is that the variation in fuel prices over time may reduce availability of cogenerated power. Many of the projects surveyed have operated through the low–gas-cost era, the embargoes of the 1970s, the rapid escalation of the early 1980s, and the ultimate collapse of energy prices in 1985 and 1986. Other projects, especially some of the large-firm–sales projects, were put in service in the early 1980s and have seen energy prices fall from all-time highs. These projects have, no doubt, been helped economically by the decrease in their main operating cost, but were based on much higher fuel cost projections. The surveyed systems seem to be quite flexible in varying fuel price scenarios, as demonstrated by the span of start-up years and the availability and capacity factors over time.

2. Another reliability issue, mentioned by critics of cogeneration, is that the thermal output users (e.g., chemical process) may reduce requirements or go out of business, causing the cogeneration system to be uneconomical to operate. Out of the 30 projects in the survey, only 1 was shut down permanently. Overall, the survey indicates that 97% of the projects are currently in service, and 87% have been in service continuously since start-up.

3. Are cogeneration system availability–capacity factors competitive with utilities? Certainly, the survey results speak for themselves in this area—availability of 96% and a capacity factor of 84% exceed utility company performance records, even for comparable combined-cycle systems.

4.   One final issue, frequently mentioned, is that power generation frequently is not the mainline business of the cogenerator and, accordingly, will not receive proper operating and maintenance attention. For firm sales projects, power generation *is* the mainline business of that particular installation. This statement is generally aimed at self-generators. Once again, the survey data indicate that cogenerators have a good operating record in terms of longevity, availability, and capacity factor. One must remember that the "other" product of cogeneration, a thermal output, often places a higher availability--capacity factor requirement on the cogeneration system than the electrical demand. Self-generators install cogeneration as the most economical way to provide energy (both electrical and thermal) to their processes. Without reliable, economical energy supply, the processes cannot maintain competitiveness required in todays international marketplace. Hence, self-generators require their systems to maintain a very high operating factor.

Appendix

I. CONVERSION FACTORS [32]

A. Heat, Work, and Power

1 British thermal unit (Btu)	= 778.26 ft-lb
	= 107.6 kg-m
	= 0.2520 cal
1 Btu/lb	= 0.556 cal/kg
1 Btu/ft^3	= 8.90 cal/m^3
1 Btu/ft^2	= 2.712 cal/m^2
1 Btu/ft^2, °F	= 4.88 cal/m^2, °C
1 Btu/hr, ft^2, °F/ft	= 1.488 cal/hr, m^2, °C/m
1 Btu/hr, ft^2, °F/in.	= 0.1240 cal/hr, m^2, °C/m
1 therm	= 100,000 Btu
1 calorie	= 3,088 ft-lb
	= 427 kg-m
	= 3.968 Btu
1 cal/kg	= 1.8 Btu/lb
1 cal/m^3	= 0.1124 Btu/ft^3
1 cal/m^2	= 0.3687 Btu/ft^2
1 cal/m^2, C°	= 0.2048 Btu/ft^2, °F
1 cal/hr, m^2, C°/m	= 0.672 Btu/hr, ft^2, °F/ft
	= 8.06 Btu/hr, ft^2, °F/in.
1 ft-lb	= 0.1383 kg-m
1 kilogram-meter (kg-m)	= 7.23 ft-lb

1 kilowatt (kW)	= 738 ft-lb/sec
	= 102 kg-m/sec
	= 1.341 hp
	= 1.360 metric hp
1 horsepower (hp)	= 33,000 ft-lb/min
	= 550 ft-lb/sec
	= 76.04 kg-m/sec
	= 0.746 kW
	= 1.014 metric hp
1 metric hp	= 32,550 ft-p/min
	= 542 ft-lb/sec
	= 75 kg-m/sec
	= 0.735 kW
	= 0.986 hp
1 kilowatt hour (kWh)	= 3,412.75 Btu (3413)
	= 860 cal
1 hp-hr	= 2545.1 Btu (2545)
1 metric hp-hr	= 632 cal
1 lb/hp-hr	= 0.447 kg/metric hp-hr
1 kg/metric hp-hr	= 2.235 lb/ hp-hr
1 boiler hp	= 10 ft^2 of boiler heating surface
	= 34.5 lb/hr evaporated from and at 212°F
100% boiler rating	= 3,348 Btu (i.e., 3.45 lb evaporation from and at 212°F)/ft^2 heating surface/hr

B. Weight [32]

1 US long ton	= 2240 lb
	= 1,016 kg
1 US short ton	= 2000 lb
	= 907 kg
1 pound (lb)	= 16 oz
	= 7,000 grains
	= 0.454 kg
1 ounce (oz)	= 0.0625 lb
	= 28.35 g
1 grain	= 64.8 mg
	= 0.0023 oz
1 lb/ft	= 1.488 kg/m
1 metric ton	= 1000 kg
	= 0.984 long ton
	= 1.102 U.S. short tons
	= 2205 lb

1 kg	= 1,000 g
	= 2.205 lb
1 g	= 1000 mg
	= 0.03527 oz
	= 15.43 grains
1 kg/m	= 0.672 lb/ft

C. Density

1 ft³/lb	= 0.0624 m³/kg
1 lb/ft³	= 16.02 kg/m³
1 grain/ft³	= 2.288 g/m³
1 grain/U. S. gal	= 17.11 g/m³
	= 17.11 mg/Liter
1 m³/kg	= 16.02 ft³/lb
1 kg/m³	= 0.0624 lb/ft³
1 g/m³	= 0.437 grain/ft³
	= 0.0584 grain/U. S. gal
1 g/L	= 58.4 grain/U. S. gal
1 kg/m³	= 1 g/L
	= 1 part/thousand
1 g/m³	= 1 mg/L
	= 1 part per million

D. Length and Area [32]

1 statute mile	= 1,760 yards (yd)
	= 5,280 ft
	= 1.609 km
1 yard (yd)	= 3 ft
	= 0.914 m
1 foot (ft)	= 12 in.
	= 30.48 cm
1 inch (in.)	= 25.40 mm
100 ft/min	= 0.508 m/sec
1 mile²	= 640 acres
	= 259 hectares (ha)
1 acre	= 4,840 yd
	= 0.4047 ha
1 yd²	= 9 ft²
	= 0.836 m²
1 ft²	= 144 in.²
	= 0.0929 m²

1 in.^{2}	$= 6.45$ cm^2
1 nautical mile	$= 6,080$ ft
	$= 1.853$ km
1 nautical mile/hr	$= 1$ knot
1 kilometer (km)	$= 1000$ m
	$= 0.621$ statute miles
1 meter (m)	$= 100$ cm
	$= 1,000$ mm
	$= 1.094$ yd
	$= 3.281$ ft
	$= 39.37$ in.
1 micron (μm)	$= 0.001$ mm
	$= 0.000039$ in.
1 m/sec	$= 196.9$ ft/min
1 km^2	$= 100$ ha
	$= 0.3861$ mile^2
1 hectare (ha)	$= 10,000$ m^2
	$= 2.471$ acres
1 m^2	$= 10,000$ cm^2
	$= 1.196$ yd^2
	$= 10.76$ ft^2
1 cm^2	$= 100$ mm^2
	$= 0.1550$ in.^{2}

E. U. S. Customary System

1. Units of Volume

Unit	Multiply by	To obtain	Unit	Multiply by	To obtain
inch^3	0.00433	gallon	feet^3	0.0370	yard^3
inch^3	0.000579	feet^3	feet^3	0.0000230	acre:feet
inch^3	0.0000214	yard^3	yard^3	46,656	inch^3
gallon	231	inch^3	yard^3	202	gallon
gallon	0.1337	feet^3	yard^3	27	feet^3
gallon	0.00495	yard^3			
gallon	0.00000307	acre:feet	acre:feet	325,800	gallon
gallon	0.0238	barrel (oil)	barrel (oil)	42	gallon
gallon	0.8327	imperial gallon	imperial gallon	1.2	gallon
foot^3	1728	inch^3	acre:feet	43,560	feet^3
foot^3	7.48	gallon			

2. Volume: Flow Rates

Unit	Multiply by	To obtain	Unit	Multiply by	To obtain
feet³/second	60	feet³/minute	gallon/second	8.022	feet³/minute
feet³/second	3600	feet³/hour	gallon/second	481.3	feet³/hour
feet³/second	7.48	gallon/second	gallon/second	60	gallon/minute
feet³/second	448.8	gallon/minute	gallon/second	3600	gallon/hour
feet³/second	26,930	gallon/hour	gallon/minute	0.00223	feet³/second
feet³/second	646,317	gallon/day	gallon/minute	0.1337	feet³/minute
feet³/second	1,983	acre:feet/day	gallon/minute	8.022	feet³/hour
feet³/minute	0.01667	feet³/second	gallon/minute	0.01667	gallon/second
feet³/minute	60	feet³/hour	gallon/minute	60	gallon/hour
feet³/minute	0.1247	gallon/second	gallon/minute	499.925	pound/hour*
feet³/minute	7.48	gallon/minute	gallon/hour	0.0000371	feet³/second
feet³/minute	448.8	gallon/hour	gallon/hour	0.00223	feet³/minute
feet³/hour	0.0002778	feet³/second	gallon/hour	0.1337	feet³/hour
feet³/hour	0.01667	feet³/minute	gallon/hour	0.0002778	gallon/second
feet³/hour	0.002078	gallon/second	gallon/hour	0.01667	gallon/minute
feet³/hour	0.1247	gallon/minute	barrel/minute (oil)	42	gallon/minute
feet³/hour	7.48	gallon/hour	barrel/day (oil)	0.0292	gallon/minute
gallon/second	0.1337	feet³/second	acre:feet/day	0.5042	feet³/second

(*Water at 68°F)

3. Pressure Units

Unit	Multiply by	To Obtain
inches of water	0.0833	feet of water
inches of water	0.0736	inches of mercury
inches of water	0.5776	ounces/sq.inch
inches of water	82.98	ounces/ft²
inches of water	0.03602	pound/in.² (psi)
inches of water	5.1869	pound/ft²
inches of water	0.0025	bar
inches of water	2.4910	millibar
inches of water	0.2491	kilopascals
inches of water	0.0025	kilograms/cm²
feet of water	12	inches of water
feet of water	0.8832	inches of mercury
feet of water	995.8	ounces/ft²
feet of water	6.936	ounces/in.²
feet of water	0.4322	pound/in.² (psi)
feet of water	62.24	pound/ft²
feet of water	0.02989	bar
feet of water	29.89	millibar
feet of water	2.989	kilopascals
feet of water	0.0305	kilograms/cm²
feet (any liquid)	0.4322 × specific gravity	pound/in.² (psi)
inches of mercury	13.57	inches of water
inches of mercury	1.131	feet of water

Unit	Multiply by	To obtain
ounces/in²	0.0625	pounds/in.² (psi)
ounces/in.²	1.73	inches of water
ounces/in.²	0.144	feet of water
ounces/in.²	0.127	inches of mercury
ounces/in.²	0.00431	bar
ounces/in.²	4.309	millibar
ounces/in.²	0.4309	kilopascals
ounces/in.²	0.0044	kilograms/cm²
ounces/ft²	0.01205	inches of water
ounces/ft²	0.001004	feet of water
ounces/ft²	0.000887	inches of mercury
ounces/ft²	0.000434	pound/in.² (psi)
ounces/ft²	0.0625	pound/ft²
pound/in.² (psi)	27.762	inches of water
pound/in.² (psi)	2.314	feet of water
pound/in.² (psi)	2.314/sp. gravity	feet (any liquid)
pound/in.² (psi)	2.043	inches of mercury
pound/in.² (psi)	16	ounces/in.²
pound/in.² (psi)	230.4	ounces/ft²
pound/in.² (psi)	144	pound/ft²
pound/in.² (psi)	0.06802	atmosphere
pound/in.² (psi)	0.06895	bar

Unit	Multiply by	To Obtain	Unit	Multiply by	To obtain
inches of mercury	7.858	ounces/in.²	pound/in.² (psi)	68.95	millibar
inches of mercury	1128	ounces/ft²	pound/in.² (psi)	6.895	kilopascals
inches of mercury	0.4894	pound/in.² (psi)	pound/in.² (psi)	0.0703	kilograms/cm²
inches of mercury	70.47	pound/ft²	pound/ft²	0.1928	inches of water
inches of mercury	0.03342	atmosphere	pound/ft²	0.01607	feet of water
inches of mercury	0.03386	bar	pound/ft²	0.01419	inches of mercury
inches of mercury	33.86	millibar	pound/ft²	16	ounces/ft²
inches of mercury	3.386	kilopascals	pound/ft²	0.00694	pound/in.² (psi)
inches of mercury	0.03453	kilograms/cm²	atmosphere	14.7	pound/in.² (psi)
atmosphere	29.92	inches of mercury			

Note: U. S. units use water and mercury at 68°F.

Unit	Multiply by	To Obtain	Unit	Multiply by	To obtain
bar	14.50	pounds/in.²	kilopascals	0.1450	pounds/in.²
bar	401.15	inches of water	kilopascals	4.015	inches of water
bar	33.45	feet of water	kilopascals	0.3345	feet of water
bar	29.53	inches of mercury	kilopascals	0.2953	inches of mercury
bar	232	ounces/in.²	kilopascals	2.32	ounces/in.²
bar	1000	millibar	kilopascals	0.01	bar
bar	100	kilopascals	kilopascals	10	millibar
bar	1.020	kilograms/cm²	kilopascals	0.0102	kilograms/cm²
millibar	0.0145	pounds/in²	kilograms/cm²	14.22	pounds/in.²
millibar	0.4015	inches of water	kilograms/cm²	393.7	inches of water
millibar	0.03345	feet of water	kilograms/cm²	32.81	feet of water
millibar	0.02953	inches of mercury	kilograms/cm²	28.96	inches of mercury
millibar	0.232	ounces/in.²	kilograms/cm²	227.5	ounces/in.²
millibar	0.001	bar	kilograms/cm²	0.9807	bar
millibar	0.100	kilopascals	kilograms/cm²	980.7	millibar
millibar	0.00102	kilograms/cm²	kilograms/cm²	98.07	kilopascals

II. SELECTED ADDITIONAL CONVERSION FACTORS [32]

A. Temperature, Measured

degree Fahrenheit (°F) = (temperature Celsius × 1.8) + 32
$\qquad$ = temperature Rankin − 459.67
$\qquad$ = 1.8 × (temperature Kelvin − 273.16) + 32
degree Celsius (°C) $\qquad$ = (temperature Fahrenheit − 32) × 5/9
$\qquad$ = temperature Kelvin − 273.16
$\qquad$ = 5/9 (temperature Rankin) − 273.16
degree Rankin (°R) $\qquad$ = 459.67 + temperature Fahrenheit
$\qquad$ = (1.8 × temperature Celsius) + 491.69
$\qquad$ = 1.8 × temperature Kelvin
degree Kelvin (K) $\qquad$ = (temperature Fahrenheit − 32) × 5/9 + 273.16
$\qquad$ = 273.16 + temperature Celsius
$\qquad$ = 5/9 × temperature Rankin

B. Water at Maximum Density, 39.2°F (4.0°C)

1 cubic foot (ft^3)	= 62.4 pounds (lbs)
1 cubic meter (m^3)	= 1,000 kilogram
1 pound (lb)	= 0.01602 cubic feet (ft^3)
1 liter (L)	= 1.0 kilogram (kg)
1 kilogram/cubic meter (kg/m^3)	= 1 gram/liter (g/L)
	= 1 part per thousand
1 gram/cubic meter (g/m^3)	= 1 part per million (ppm)

C. Thermal Conductivity

1 Btu ft/ft^2 hr °F	= 1.730 W/m K
	= 1.488 kcal/m hr K
1 Btu in./ft^2 hr °F	= 0.1442 W/m K
1 Btu/ft^2 hr °F	= 0.004139 cal cm/cm^2 sec °C
1 Btu in./ft^2 hr °F	= 0.0003445 cal cm/cm^2 sec °C
1 W/m K	= 0.5778 Btu ft/ft^2 hr °F
	= 6.934 Btu in./ft^2 hr °F
1 cal cm/cm^2 sec °C	= 241.9 Btu ft/ft^2 hr °F
	= 2903 Btu in./ft^2 hr °F
	= 418.7 W/m K

III. MISCELLANEOUS CONVERSION FACTORS

A. Weights and Measures in 1947 [15]

1. Troy Weight

1 carat	= 3.086 grains (200 mg)
24 grains	= 1 pennyweight (pwt)
20 pennyweight (pwt)	= 1 ounce (oz)
12 ounces (oz.)	= 1 pound (373.24 grams)

(used for weighing gold, silver, and jewels)

2. Apothecaries' Weight

20 grains	= 1 scruple
3 scruples	= 1 dram
8 drams	= 1 ounce
12 ounces	= 1 pound (373.24 grams)

(the ounce and pound in this are the same as in Troy weight)

3. Avoirdupois Weight

27 11/32 grains	= 1 dram
16 drams	= 1 ounce
16 ounces	= 1 pound
25 pounds	= 1 quarter
4 quarters	= 1 hundredweight (cwt)

4. Dry Measure

2 pints	= 1 quart
8 quarts	= 1 peck
4 pecks	= 1 bushel
36 bushels	= 1 chaldron

5. Liquid Measure

4 gills	= 1 pint
2 pints	= 1 quart
4 quarts	= 1 gallon
31.5 gallon	= 1 barrel
2 barrels	= 1 hogshead
42 gallons	= 1 barrel (petroleum)

6. Cloth Measure

2.25 inches = 1 nail
4 nails   = 1 quarter
4 quarters = 1 yard

7. Mariner's Measure

6 feet   = 1 fathom
120 fathoms   = 1 cable length
7.5 cable lengths = 1 mile
6,080.27 feet   = 1 nautical mile

8. Miscellaneous

3 inches   = 1 palm
4 inches   = 1 hand
9 inches   = 1 span
18 inches   = 1 cubit
1 rick (wood) = 64 ft^3
2 ricks (wood) = 1 cord (wood) (128 ft^3)

IV. Properties of Air

Degrees F	Volume, in ft³, of 1 lb dry air at atm. pressure of 14.7 psi	Volume, in ft³, of 1 lb of dry air + vapor to saturate it	Weight, in lb, of 1 ft³, of dry air at atm. pressure of 14.7 psi	Weight, in lb of saturated vapor per ft³	Weight of saturated vapor, lb per lb of dry air
0	11.58	11.59	0.086331	0.000067	0.000781
32	12.39	12.47	0.080728	0.000303	0.003782
40	12.59	12.70	0.079439	0.000410	0.005202
50	12.84	13.00	0.077884	0.000588	0.007640
62	13.15	13.40	0.076097	0.000887	0.011880
70	13.35	13.69	0.074950	0.001153	0.015780
80	13.60	14.09	0.073565	0.001580	0.022260
90	13.86	14.55	0.072230	0.002137	0.031090
100	14.11	15.08	0.070942	0.002855	0.043050
120	14.62	16.52	0.068500	0.004920	0.081300
140	15.13	18.84	0.066221	0.008130	0.153200
160	15.64	23.09	0.064088	0.012940	0.298700
180	16.16	33.04	0.062090	0.019910	0.657700
200	16.67	77.24	0.060210	0.029720	2.295300
210	16.86		0.059313		
212	16.91		0.059131		
220	17.11		0.058442		
240	17.61		0.056774		
260	18.12		0.055200		
280	18.62		0.053710		

300	19.12	0.052297
320	19.62	0.050959
340	20.13	0.049686
360	20.63	0.048476
380	21.13	0.047323
400	21.63	0.046223
425	22.26	0.044920
450	22.89	0.043686
475	23.52	0.042520
500	24.15	0.041414
525	24.77	0.040364
550	25.40	0.039365
575	26.03	0.038415
600	26.66	0.037500
650	27.91	0.035822
700	29.17	0.034280
750	30.43	0.032810
800	31.68	0.031561
850	32.94	0.030358
900	34.20	0.029242
950	35.45	0.028206
1000	36.81	0.027180
1500	49.37	0.020295
2000	61.94	0.016172
2500	74.56	0.013441
3000	87.13	0.011499

Source: Ref. 15.

V. Quick Reference Tables

A. Weights of Carbon Steel Plates in Fractional Thicknesses

(For thickness over 1 in., multiply inches times bottom line and then add fractional inch.)

Decimals of an inch	Thickness — Fractions of an Inch						Weight	
	32 Thirtyseconds	16 Sixteenths	10 Tenths	8 Eighths	4 Fourths	2 Halves	lb/ft²	lb/in.²
0.03125	1						1.275	0.008854
0.06250	2	1					2.550	0.017708
0.09375	3						3.825	0.026563
0.10000			1				4.080	0.028333
0.12500	4	2		1			5.100	0.035417
0.15625	5						6.375	0.044271
0.18750	6	3					7.650	0.053125
0.20000			2				8.160	0.056667
0.21875	7						8.925	0.061979
0.25000	8	4		2	1		10.200	0.070833
0.28125	9						11.475	0.079688
0.30000			3				12.240	0.085000
0.31250	10	5					12.750	0.088542
0.34375	11						14.025	0.097396
0.37500	12	6		3			15.300	0.106250

0.40000	16.320	0.113333
0.40625	16.575	0.115104
0.43750	17.850	0.123958
0.46875	19.125	0.132813
0.50000	20.400	0.141667
0.53125	21.675	0.150521
0.56250	22.950	0.159375
0.59375	24.225	0.168229
0.60000	24.480	0.170000
0.62500	25.500	0.177083
0.65625	26.775	0.185938
0.68750	28.050	0.194792
0.70000	28.560	0.198333
0.71875	29.325	0.203646
0.75000	30.600	0.212500
0.78125	31.875	0.221354
0.80000	32.640	0.226667
0.81250	33.150	0.230208
0.84375	34.425	0.239063
0.87500	35.700	0.247917
0.90000	36.720	0.255000
0.90625	36.975	0.256771
0.93750	38.250	0.265625
0.96875	39.525	0.274479
1.00000	40.800	0.283333

Additional index columns:

13 14 15 16 17 18 19 20 21 22 23 24 25 26 27 28 29 30 31 32

7 8 9 10 11 12 13 14 15 16

4 5 6 7 8 9 10

4 5 6 7 8

2 3 4

1 2

B. Weight of Carbon Steel Bars per Foot of Length

Size (in.)	Round	Square	Hex	Size (in.)	Round	Square	Hex	Size (in.)	Round	Square	Hex
1/16	0.010			1	2.670	3.400	2.940	4	42.73	54.40	47.11
5/64	0.016			1 1/16	3.010	3.840	3.320	4 1/8	45.44	57.85	50.09
3/32	0.023			1 1/8	3.380	4.300	3.730	4 1/4	48.23	61.41	53.18
7/64	0.032			1 3/16	3.770	4.800	4.150	4 3/8	51.11	65.08	56.35
1/8	0.042	0.053	0.050	1 1/4	4.170	5.310	4.600	4 1/2	54.08	68.85	59.63
9/64	0.053			1 5/16	4.600	5.860	5.070	4 5/8	57.12	72.73	62.97
5/32	0.065	0.083	0.070	1 3/8	5.050	6.430	5.570	4 3/4	60.25	76.71	66.42
11/64	0.079			1 7/16	5.520	7.030	6.090	4 7/8	63.46	80.80	69.97
3/16	0.094	0.120	0.104	1 1/2	6.010	7.650	6.630	5	66.76	85.00	73.60
13/64	0.109			1 9/16	6.520	8.300	7.180	5 1/8	70.14	89.30	77.33
7/32	0.128	0.163	0.141	1 5/8	7.050	8.980	7.780	5 1/4	73.60	93.71	81.14
15/64	0.147			1 11/16	7.600	9.680	8.390	5 3/8	77.15	98.23	85.05
1/4	0.167	0.213	0.184	1 3/4	8.180	10.410	9.010	5 1/2	80.78	102.85	89.06
17/64	0.189			1 13/18	8.770	11.170	9.670	5 5/8	84.49	107.58	
9/32	0.211	0.269	0.233	1 7/8	9.390	11.950	10.350	5 3/4	88.29	112.41	97.34
19/64	0.235			1 15/16	10.020	12.760	11.050	5 7/8	92.17	117.35	101.61
5/16	0.261	0.332	0.288	2	10.68	13.60	11.78	6	96.13	122.40	105.98
11/32	0.316	0.402	0.348	2 1/16	11.36	14.46	12.52	6 1/4	104.31	132.81	
23/64	0.345			2 1/8	12.06	15.35	13.30	6 1/2	112.82	143.65	
3/8	0.376	0.478	0.414	2 3/16	12.78	16.27	14.09	6 3/4	121.67	154.91	
25/64	0.405			2 1/4	13.52	17.21	14.91	7	130.85	166.60	
13/32	0.441			2 5/16	14.28	18.18	15.74				
27/64	0.476			2 3/8	15.06	19.18	16.61				
7/16	0.511	0.651	0.564								

15/32	0.587	0.850	0.740	2 7/16	15.87	20.20	17.49	7 1/2	150.21	191.25
31/64	0.627	1.080	0.930	2 1/2	16.69	21.25	18.40	8	170.90	217.60
1/2	0.668	1.200	1.150	2 9/16	17.53	22.33	19.33	8 1/2	192.90	245.65
33/64	0.710	1.330	1.390	2 5/8	18.40	23.43	20.29	9	216.30	275.40
17/32	0.754	1.610	1.660	2 11/16	19.29	24.56	21.26	9 1/2	241.00	306.85
9/16	0.845	1.910	1.940	2 3/4	20.19	25.71	22.27	10	267.04	340.00
37/64	0.893	2.240	2.250	2 13/16	21.13	26.90	23.29	11	323.11	411.40
19/32	0.942	2.600	2.590	2 7/8	22.07	28.10	24.34	12	384.53	489.60
39/64	0.992	2.990		2 15/16	23.04	29.34	25.40	13	451.29	574.60
5/8	1.040			3	24.03	30.60	26.50	14	523.39	666.40
41/64	1.100			3 1/16	25.05			15	600.83	765.00
21/32	1.150			3 1/8	26.08	33.20	28.75	16	683.61	870.40
11/16	1.260			3 3/16	27.13			17	771.73	982.60
23/32	1.380			3 1/4	28.21	35.91	31.10	18	865.20	1101.60
47/64	1.440			3 5/16	29.30			19	964.00	1227.40
3/4	1.500			3 3/8	30.42	38.73	33.54	20	1068.10	1360.00
49/64	1.570			3 7/16	31.55			21	1177.60	1499.40
25/32	1.630			3 1/2	32.71	41.65	36.07	22	1292.50	1645.60
51/64	1.700			3 9/16	33.89			23	1412.60	1798.60
13/16	1.760			3 5/8	35.09	44.68	38.69	24	1538.10	1958.40
27/32	1.900			3 11/16	36.31			25	1669.00	2125.00
7/8	2.040			3 3/4	37.55	47.81	41.41			
29/32	2.190			3 13/16	38.81					
15/16	2.350			3 7/8	40.10	51.05	44.21			
31/32	2.510			3 15/16	41.40					

Source: Steel Plates. Bethlehem Steel Co., 1948.

VI. Volume of Cylinders

Inside diameter (in.)	Volume (in.³/in. of cylinder length)	Volume (gal/in. of cylinder length)	Weight of water/ in. of cylinder length at 60°F. (8.334 lb/gal) (lb)	Inside diameter (in.)	Volume (in.³/in. cylinder length)	Volume in (gal/in. cylinder length)	Weight of water/ in. of cylinder length at 60°F. (8.334 lb/gal) (lb)	Inside diameter (in.)	Volume (in.³/in. cylinder length)	Volume (gal/in. cylinder length)	Weight of water/ in. of cylinder length at 60°F. (8.334 lb/gal) (lb)
1.00	0.7854	0.0034	0.0283	11.00	95.0334	0.4114	3.4286	22.00	380.1336	1.6456	13.7144
1.25	1.2272	0.0053	0.0443	11.25	99.4022	0.4303	3.5862	22.50	397.6088	1.7213	14.3449
1.60	1.7672	0.0077	0.0638	11.50	103.8692	0.4497	3.7474	23.00	415.4766	1.7986	14.9895
1.75	2.4053	0.0104	0.0868	11.75	108.4343	0.4694	3.9121	23.50	433.7372	1.8777	15.6483
2.00	3.1416	0.0136	0.1133	12.00	113.0976	0.4896	4.0803	24.00	452.3904	1.9584	16.3213
2.25	3.9761	0.0172	0.1434	12.25	117.8591	0.5102	4.2521	24.50	471.4364	2.0409	17.0084
2.50	4.9088	0.0213	0.1771	12.50	122.7188	0.5313	4.4274	25.00	490.8750	2.1250	17.7098
2.75	5.9396	0.0257	0.2143	12.75	127.6766	0.5527	4.6063	25.50	510.7064	2.2109	18.4252
3.00	7.0686	0.0306	0.2550	13.00	132.7326	0.5746	4.7887	26.00	530.9304	2.2984	19.1549
3.25	8.2958	0.0359	0.2993	13.25	137.8868	0.5969	4.9747	26.50	551.5472	2.3877	19.8987
3.50	9.6212	0.0417	0.3471	13.50	143.1392	0.6197	5.1642	27.00	572.5566	2.4786	20.6567
3.75	11.0447	0.0478	0.3985	13.75	148.4897	0.6428	5.3572	27.50	593.9588	2.5713	21.4288
4.00	12.5664	0.0544	0.4534	14.00	153.9384	0.6664	5.5538	28.00	615.7536	2.6656	22.2151
4.25	14.1863	0.0614	0.5118	14.25	159.4853	0.6904	5.7539	28.50	637.9412	2.7617	23.0156
4.50	15.9044	0.0689	0.5738	14.50	165.1304	0.7149	5.9576	29.00	660.5214	2.8594	23.8302
4.75	17.7206	0.0767	0.6393	14.75	170.8736	0.7397	6.1648	29.50	683.4944	2.9589	24.6591
5.00	19.6350	0.0850	0.7084	15.00	176.7150	0.7650	6.3755	30.00	706.8600	3.0600	25.5020
5.25	21.6476	0.0937	0.7810	15.25	182.6546	0.7907	6.5898	30.50	730.6184	3.1629	26.3592
5.50	23.7584	0.1029	0.8572	15.50	188.6924	0.8169	6.8076	31.00	754.7694	3.2674	27.2305
5.75	25.9673	0.1124	0.9368	15.75	194.8283	0.8434	7.0290	31.50	779.3132	3.3737	28.1160
6.00	28.2744	0.1224	1.0201	16.00	201.0624	0.8704	7.2539	32.00	804.2496	3.4816	29.0157
6.25	30.6797	0.1328	1.1069	16.25	207.3947	0.8978	7.4824	32.50	829.5788	3.5913	29.9295
6.50	33.1832	0.1437	1.1972	16.50	213.8252	0.9257	7.7144	33.00	855.3006	3.7026	30.8575
6.75	35.7848	0.1549	1.2910	16.75	220.3538	0.9539	7.9499	33.50	881.4152	3.8157	31.7996
7.00	38.4846	0.1666	1.3884	17.00	226.9806	0.9826	8.1890	34.00	907.9224	3.9304	32.7560
7.25	41.2826	0.1787	1.4894	17.25	233.7056	1.0117	8.4316	34.50	934.8224	4.0469	33.7264
7.50	44.1788	0.1913	1.5939	17.50	240.5288	1.0413	8.6778	35.00	962.1150	4.1650	34.7111
7.75	47.1731	0.2042	1.7019	17.75	247.4501	1.0712	8.9275	35.50	989.8004	4.2849	35.7099
8.00	50.2656	0.2176	1.8135	18.00	254.4696	1.1016	9.1807	36.00	1,017.88	4.4064	36.7229
8.25	53.4563	0.2314	1.9286	18.25	261.5873	1.1324	9.4375	36.50	1,046.35	4.5297	37.7501
8.50	56.7452	0.2457	2.0472	18.50	268.8032	1.1637	9.6979	37.00	1,075.21	4.6546	38.7914
8.75	60.1322	0.2603	2.1694	18.75	276.1172	1.1953	9.9617	37.50	1,104.47	4.7813	39.8469
9.00	63.6174	0.2754	2.2952	19.00	283.5294	1.2274	10.2292	38.00	1,134.12	4.9096	40.9166
9.25	67.2008	0.2909	2.4245	19.25	291.0398	1.2599	10.5001	38.50	1,164.16	5.0397	42.004
9.50	70.8824	0.3069	2.5573	19.50	298.6484	1.2929	10.7746	39.00	1,194.59	5.1714	43.0984
9.75	74.6621	0.3232	2.6937	19.75	306.3551	1.3262	11.0527	39.50	1,225.42	5.3049	44.2106

10.00	78.5400	0.3400	2.8336	20.00	314.1600	1.3600	40.00	11.3342	1,256.64	5.4400	45.3370
10.25	82.5161	0.3572	2.9770	20.50	330.0644	1.4289	40.50	11.9080	1,288.25	5.5769	46.4775
10.50	86.5904	0.3749	3.1240	21.00	346.3614	1.4994	41.00	12.4960	1,320.26	5.7154	47.6321
10.75	90.7628	0.3929	3.2745	21.50	363.0512	1.5717	41.50	13.0981	1,352.66	5.8557	48.8010
42.50	1,418.63	6.1413	51.1812	61.00	2,922.47	12.6514	79.50	105.4368	4,963.92	21.4889	179.0881
43.00	1,452.20	6.2866	52.3925	61.50	2,970.58	12.8597	80.00	107.1723	5,026.56	21.7600	181.3478
43.50	1,486.17	6.4337	53.6180	62.00	3,019.08	13.0696	80.50	108.9220	5,089.59	22.0329	183.6218
44.00	1,520.53	6.5824	54.8577	62.50	3,067.97	13.2813	81.00	110.6859	5,153.01	22.3074	185.9099
44.50	1,555.29	6.7329	56.1116	63.00	3,117.25	13.4946	81.50	112.4640	5,216.82	22.5837	188.2121
45.00	1,590.44	6.8850	57.3796	63.50	3,166.93	13.7097	82.00	114.2562	5,281.03	22.8616	190.5286
45.50	1,625.97	7.0389	58.6618	64.00	3,217.00	13.9264	82.50	116.0626	5,345.63	23.1413	192.8592
46.00	1,661.91	7.1944	59.9581	64.50	3,267.46	14.1449	83.00	117.8832	5,410.62	23.4226	195.2039
46.50	1,698.23	7.3517	61.2687	65.00	3,318.32	14.3650	83.50	119.7179	5,476.01	23.7057	197.5629
47.00	1,734.95	7.5106	62.5933	65.50	3,369.56	14.5869	84.00	121.5668	5,541.78	23.9904	199.9360
47.50	1,772.06	7.6713	63.9322	66.00	3,421.20	14.8104	84.50	123.4299	5,607.95	24.2769	202.3233
48.00	1,809.56	7.8336	65.2852	66.50	3,473.24	15.0357	85.00	125.3071	5,674.52	24.5650	204.7247
48.50	1,847.46	7.9977	66.6524	67.00	3,525.66	15.2626	85.50	127.1985	5,741.47	24.8549	207.1403
49.00	1,885.75	8.1634	68.0338	67.50	3,578.48	15.4913	86.00	129.1041	5,808.82	25.1464	209.5701
49.50	1,924.43	8.3309	69.4293	68.00	3,631.69	15.7216	86.50	131.0238	5,876.56	25.4397	212.0140
50.00	1,963.50	8.5000	70.8390	68.50	3,685.29	15.9537	87.00	132.9577	5,944.69	25.7346	214.4722
50.50	2,002.97	8.6709	72.2629	69.00	3,739.29	16.1874	87.50	134.9058	6,013.22	26.0313	216.9444
51.00	2,042.83	8.8434	73.7009	69.50	3,793.68	16.4229	88.00	136.8680	6,082.14	26.3296	219.4309
51.50	2,083.08	9.0177	75.1531	70.00	3,848.46	16.6600	88.50	138.8444	6,151.45	26.6297	221.9315
52.00	2,123.72	9.1936	76.6195	70.50	3,903.63	16.8989	89.00	140.8350	6,221.15	26.9314	224.4463
52.50	2,164.76	9.3713	78.1000	71.00	3,959.20	17.1394	89.50	142.8398	6,291.25	27.2349	226.9752
53.00	2,206.19	9.5506	79.5947	71.50	4,015.16	17.3817	90.00	144.8587	6,361.74	27.5400	229.5184
53.50	2,248.01	9.7317	81.1036	72.00	4,071.51	17.6256	90.50	146.8918	6,432.62	27.8469	232.0756
54.00	2,290.23	9.9144	82.6266	72.50	4,128.26	17.8713	91.00	148.9390	6,503.90	28.1554	234.6471
54.50	2,332.83	10.0989	84.1638	73.00	4,185.40	18.1186	91.50	151.0004	6,575.57	28.4657	237.2327
55.00	2,375.84	10.2850	85.7152	73.50	4,242.93	18.3677	92.00	153.0760	6,647.63	28.7776	239.8325
55.50	2,419.23	10.4729	87.2807	74.00	4,300.85	18.6184	92.50	155.1657	6,720.08	29.0913	242.4465
56.00	2,463.01	10.6624	88.8604	74.50	4,359.17	18.8709	93.00	157.2697	6,792.92	29.4066	245.0746
56.50	2,507.19	10.8537	90.4543	75.00	4,417.88	19.1250	93.50	159.3878	6,866.16	29.7237	247.7169
57.00	2,551.76	11.0466	92.0624	75.50	4,476.98	19.3809	94.00	161.5200	6,939.79	30.0424	250.3734
57.50	2,596.73	11.2413	93.6846	76.00	4,536.47	19.6384	94.50	163.6664	7,013.82	30.3629	253.0440
58.00	2,642.09	11.4376	95.3210	76.50	4,596.36	19.8977	95.00	165.8270	7,088.24	30.6850	255.7288
58.50	2,687.84	11.6357	96.9715	77.00	4,656.64	20.1586	95.50	168.0018	7,163.04	31.0089	258.4278
59.00	2,733.98	11.8354	98.6362	77.50	4,717.31	20.4213	96.00	170.1907	7,238.25	31.3344	261.1409
59.50	2,780.51	12.0369	100.3151	78.00	4,778.37	20.6856	96.50	172.3938	7,313.84	31.6617	263.8682
60.00	2,827.44	12.2400	102.0082	78.50	4,839.83	20.9517	97.00	174.6111	7,389.83	31.9906	266.6097
60.50	2,874.76	12.4449	103.7154	79.00	4,901.68	21.2194		176.8425			

References

1. SC Stultz, JB Kitto, eds., Steam, Its Generation and Use, 40th Edition, Barberton, Ohio: The Babcock & Wilcox Co., 1992.
2. Steam, Its Generation and Use, 38th Edition, New York: The Babcock & Wilcox Co., 1975.
3. ML Smith and KW Stinson, Fuels and Combustion, New York: McGraw-Hill, 1952.
4. North American Combustion Handbook, Vol. 1, Third Edition, Cleveland: North American Mfg. Co., 1986.
5. CJ Vorndran, Controlling Excess Combustion Air with O_2 Trimming, Cleveland: Cleveland Controls Co., 1979.
6. Thomas C. Elliott and the editors of Power Magazine, Standard Handbook of Powerplant Engineering, New York: McGraw-Hill, 1989.
7. CD Shields, Boilers—Types, Characteristics, and Functions, New York: McGraw-Hill, 1961 (1982 Reissue).
8. RH Perry, late ed., DW Green, ed. JO Maloney, asst ed. Perry's Chemical Engineer's Handbook, 6th Edition, New York: McGraw-Hill, 1984.
9. N Irving Sax, RJ Lewis, Sr., Hawley's Condensed Chemical Dictionary, 11th Edition, revised, New York: VanNostrand Reinhold, 1987.
10. RW Green, ed., and the Staff of Chemical Engineering, The Chemical Engineering Guide to—Corrosion Control in the Process Industries, New York: McGraw-Hill, 1986.
11. CR Westaway & AW Loomis, eds., Cameron Hydraulic Data, Ingersoll-Rand, New Jersey: Woodcliff Lake, 1981.
12. GF Gebhardt, Steam Power Plant Engineering, 6th Edition, New York: John Wiley & Sons, Inc., 1928.
13. JG Singer, ed. Combustion: Fossil Power Systems, 3rd Edition, Combustion Engineering Inc., 1000 Prospect Hill Road, Windsor, Connecticut, 1981.
14. GR Fryling, M.E., editor, Combustion Engineering, Revised Edition, New York: Combustion Engineering Inc., 1966.

15. O de Lorenze, ME, Editor, Combustion Engineering, New York: Combustion Engineering Company Inc., 1948.
16. T Baumeister, ed.-in-chief, EA Avallone, assoc. ed., Theodore Baumelster III, assoc. ed., Marks' Standard Handbook for Mechanical Engineers, 8th Edition, New York: McGraw-Hill, 1979.
17. JJ Dishinger, The Operating Engineer's Guide to Energy Conservation, Erie City Energy Division, Erie, PA: Zurn Industries, 1973.
18. R Goldstick, A Thumann, CEM, PE, Principles of Waste Heat Recovery, Fairmont Press, Atlanta, 1986.
19. LH Yaverbaum, ed., Energy Saving by Increasing Boiler Efficiency, Park Ridge, New Jersey: Noyes Data Corp., 1979.
20. Steam Utilization, Allentown, PA: Spirax-Sarco Inc., 1985.
21. IJ Karassik, Engineer's Guide to Centrifugal Pumps, Milan, Italy: HOEPLI, 1973.
22. R Carter, IJ Karassik & EF Wright, Pump Questions and Answers, 1st Edition, New York: McGraw-Hill, 1949.
23. Goulds Pump Data Sheet 781.1, (1968) and 766.6, (1959), Seneca Falls, New York: Goulds Pump Co.
24. HM Spring, Jr., AL Kohan, Boiler Operator's Guide, 2nd Edition, New York: McGraw-Hill, 1981.
25. SG Dukelow, The Control of Boilers, 2nd ed., Research Triangle Park, North Carolina: Instrument Society of America, 1994.
26. EB Woodruff, HB Lammers & TF Lammers, Steam Plant Operation, 5th Edition, New York: McGraw-Hill, 1984.
27. Steam, Its Generation and Use, 39th Edition, New York: The Babcock & Wilcox Co., 1978.
28. Field & Rolfe, Breighton Conference, London, England, 1970.
29. RHP Miller, Engineer, Forest Products Laboratory, USDA, 1951.
30. AJ Baker, Seminar Proceedings, U.S. Forest Service, East Lansing, Michigan, 11/11/82.
31. Steam, Its Generation and Use, 36th Edition, New York: The Babcock & Wilcox Company, 1923.
32. Steam, Its Generation and Use, 37th Edition, New York: The Babcock & Wilcox Company, 1963.
33. LE Young & CW Porter, General Chemistry, 3rd Edition, New York: Prentice-Hall, Inc., 1951.
34. CR Brunner, P.E., Handbook of Hazardous Waste Incineration, 1st Edition, TAB Professional & Reference Books, 1989.
35. CF Hirshfeld, M.M.E., & WN Barnard, M.E., Elements of Heat-Power Engineering, 2nd Edition, New York: John Wiley & Sons, 1915.
36. JR Allen and JA Bursley, Heat Engines, Fifth Edition, New York: McGraw-Hill, 1941.
37. RC King, B.M.E., M.M.E., D.Sc., P.E., Piping Handbook, 5th Edition, New York: McGraw-Hill Book Company, 1973.
38. Compressed Air & Gas Data, 2nd Edition, Charles W. Gibbs, editor, Woodcliff Lake, NJ: Ingersoll-Rand Company, 1971.

39. Compressed Air & Gas Handbook, 3rd Edition, New York, NY: Compressed Air & Gas Institute, 1973.

40. RE Gackenbach, Materials Selection for Process Plants, New York: Reinhold Publishing Corp., 1960.

41. Standard of Tubular Exchanger Manufacturers Association, Fifth Edition, New York: Tubular Exchanger Manufacturers Association, Inc., 1968.

42. Boiler Tube Company of America, Lyman, S. Carolina.

43. Pipe Friction Manual, New York: Hydraulic Institute, 1961.

44. Fan Engineering, 8th Edition, Edited by: R Jorgensen, Buffalo: Buffalo Forge Co., 1983.

45. Power Plant Theory & Design, 2nd Edition, New York: The Ronald Press Co., 1959.

46. Flow of Fluids-through Valves, Fittings, and Pipe, Tech. Paper No. 410, New York: Crane Co., 1972.

47. RL Mott, Univ. of Dayton, Applied Fluid Mechanics, 4th Edition, New York: Merrill, an Imprint of Macmillan Publishing Company, 1994.

48. See Reference No. 25. (Sam G. Dukelow, The Control of Boilers).

49. NP Chopey, ed., TG Hicks, P.E. series Ed., Handbook of Chemical Engineering Calculations, New York: McGraw-Hill, 1984.

50. WC Turner, E.E., M.E., P.E. & JF Malloy, M.E., P.E., Thermal Insulation Handbook, Robert E. Krieger Publishing Co., New York: McGraw-Hill, 1981.

51. Standard Handbook of Engineering Calculations, 2nd Edition, TG Hicks, P.E., Editor, SD Hicks, Coordinating Editor, New York: McGraw-Hill, 1985.

52. Power Boiler, Notes on Care & Operation, Hartford, Conn.: The Hartford Steam Boiler Inspection & Insurance Co., 1992.

53. Fundamentals of Boiler Efficiency, Houston: Exxon Company, 1976.

54. Engineering Letter No. A-6, Willowbrook, Illinois: The New York Blower Company, 1969.

55. J Karchesy, P Koch, Energy Production from Hardwoods Growing on Southern Pine Sites, USDA, Forest Service, GTR #SO-24, 1979.

56. RE Ketten, Operation and maintenance of deaerators in industrial plants, National Engineer, September 1986.

57. CR Wilson, Modern Boiler Economizers—Development and Applications, United Kingdom: E. Green & Son, 1981.

58. V Ganapathy, Steam Plant Calculations Manual, 2nd Edition, New York: Marcel Dekker, Inc., 1994.

59. V Ganapathy, Waste Heat Boiler Deskbook, Liburn, GA: The Fairmount Press, 1991.

Index

Absorption chiller, 294
Air
  aspirating, 29
  atomizing, 82
  compressed (*see* Compressed Air)
  excess, 181, 182
  heating, 35, 182, 192
  overfire, ports, 183
    primary, 43
    secondary, 43
  preheater, 267
  primary, 82
  properties, 338, 339
  purge, 4, 33
  register, 83
  required for combustion, 42, 44
  secondary, 107
  tertiary, 43
  underfire, 107
  windbox, 83
Anthracite (*see* Coal)
Ash, 23, 286
  handling, 6
  wood, 7
Atmospheric pressure, 34
Atomization, air, 82
  fuel oil, 82, 92
  mechanical, 82
  steam, 82

Auto-ignition
  minimum temperatures, 95
Auxiliary Drive, 2
Auxiliary steam turbine drives
  advantages, 2

Bagasse, 108, 109, 110, 111, 112
  air heater, 112
  clinker, 111
  combustion, 110
  composition, 109
  firing, 112
  furnace
    Cooke, 112
    Ward, 112
  green bagasse, 110
  hearth, 111
  induced draft fan, 112
  Jamaica, 294
  moisture, 110
  stokers, 112
Bar
  steel, round, square and hex, 342, 343
Bark
  hardwood, 137
  softwood, 137
Basic power plant checklist, 26

Printed in the United States
by Baker & Taylor Publisher Services